Surface Science of Adsorbents and Nanoadsorbents

Interface Science and Technology

Surface Science of Adsorbents and Nanoadsorbents

Properties and Applications in Environmental Remediation

Volume 34

Tawfik A. Saleh
Faculty Member, Chemistry Department, and IRC for Advanced Materials, King Fahd University of Petroleum and Minerals, Dhahran, Saudi Arabia

ELSEVIER

ACADEMIC PRESS
An imprint of Elsevier

Academic Press is an imprint of Elsevier
125 London Wall, London EC2Y 5AS, United Kingdom
525 B Street, Suite 1650, San Diego, CA 92101, United States
50 Hampshire Street, 5th Floor, Cambridge, MA 02139, United States
The Boulevard, Langford Lane, Kidlington, Oxford OX5 1GB, United Kingdom

ISBN: 978-0-12-849876-7
ISSN: 1573-4285

For information on all Elsevier publications visit our website at
https://www.elsevier.com/books-and-journals

Publisher: Susan Dennis
Acquisition Editor: Charles Bath
Editorial Project Manager: Aera F. Gariguez
Production Project Manager: Bharatwaj Varatharajan
Cover Designer: Mark Rogers

Typeset by TNQ Technologies

Interface Science and Technology

Series Editor: Arthur T. Hubbard

In this Series:

Vol. 1: Clay Surfaces: Fundamentals and Applications
Edited by F. Wypych and K.G. Satyanarayana

Vol. 2: Electrokinetics in Microfluids
By Dongqing Li

Vol. 3: Radiotracer Studies of Interfaces
Edited By G. Horányi

Vol. 4: Emulsions: Structure Stability and Interactions
Edited By D.N. Petsev

Vol. 5: Inhaled Particles
By Chiu-sen Wang

Vol. 6: Heavy Metals in the Environment: Origin, Interaction and Remediation
Edited By H.B. Bradl

Vol. 7: Activated Carbon Surfaces in Environmental Remediation
Edited By Teresa J. Bandosz

Vol. 8: Tissue Engineering: Fundamentals and Applications
By Yoshito Ikada

Vol. 9: Particles at Interfaces: Interactions, Deposition, Structure
By Zbigniew Adamczyk

Vol. 10: Interface Science in Drinking Water Treatment: Theory and Applications
Edited By G. Newcombe and D. Dixon

Vol. 11: Surface Complexation Modelling
Edited By Johannes Lützenkirchen

Vol. 12: Theory of Colloid and Interfacial Electric Phenomena
By Hiroyuki Ohshima

Vol. 13: Sorbent Deformation
By A.V. Tvardovskiy

Vol. 14: Advanced Chemistry of Monolayers at Interfaces: Trends in Methodology and Technology
Edited By Toyoko Imae

Vol. 15: Macromolecules in Solution and Brownian Relativity
By Stefano A. Mezzasalma

Vol. 16: The Properties of Water and their Role in Colloidal and Biological Systems
By Carel Jan van Oss

Vol. 17: Chemistry on Modified Oxide and Phosphate Surfaces: Fundamentals and Applications
By Robson Fernandes de Farias

Vol. 18: Interface Science and Composites
By Soo-Jin Park and Min-Kang Seo

Vol. 19: Nanoparticle Technologies: From Lab to Market
By Farid Bensebaa

Vol. 20: Particles at Interfaces
By Zbigniew Adamczyk

Vol. 21: Self-Assembly Processes at Interfaces: Multiscale Phenomena
By Vincent Ball

Vol. 22: Rheology of Emulsions: Electrohydrodynamics Principles
By Aleksandar M. Spasic

Vol. 23: Tailored Thin Coatings for Corrosion Inhibition using a Molecular Approach
By Simo Olavi Pehkonen and Shaojun Yuan

Vol. 24: Charge and Energy Storage in Electrical Double Layers
By Silvia Ahualli and Ángel V. Delgado

Vol. 25: Stimuli Responsive Polymeric Membranes
By Mihir Kumar Purkait, Manish Kumar Sinha, Piyal Mondal, and Randeep Singh

Vol. 26: Theory of Electrophoresis and Diffusiophoresis of Highly Charged Colloidal Particles
By Eric Lee

Vol. 27: Graphene Surfaces: Particles and Catalysts
By Karim Kakaei, Mehdi Esrafili and Ali Ehsani

Vol. 28: An Introduction to Green Nanotechnology
By Mahmoud Nasrollahzadeh, Mohammad Sajadi, Monireh Atarod, Mohadeseh Sajjadi and Zahra Isaabadi

Vol. 29: Emerging Natural and Tailored Nanomaterials for Radioactive Waste Treatment and Environmental Remediation
Edited by Changlun Chen

Vol. 30: Advanced Low-Cost Separation Techniques in Interface Science
Edited by George Z. Kyzas and Athanasios C. Mitropoulos

Vol. 31: Surface Science of Photocatalysis
Edited by Jiaguo Yu, Mietek Jaroniec and Chuanjia Jiang

Vol. 32: Photocatalysis: Fundamental Processes and Applications
Edited by Mehrorang Ghaedi

Vol. 33: Adsorption: Fundamental Processes and Applications
Edited by Mehrorang Ghaedi

Contents

2. Adsorption technology and surface science

3. Kinetic models and thermodynamics of adsorption processes: classification

Preface

Surface Science of Adsorbents and Nanoadsorbents: Properties and Applications in Environmental Remediation presents a unique collection of timely information on the surface science of adsorbents and nanoadsorbents. The book offers a perfect source to document developments and innovations, ranging from materials development and characterization of properties to applications that encompass the enhancement of sorption, degradation processes, and their usage for the removal of different pollutants, including heavy metals, dyes, pesticides, etc. It is written for postgraduate students, scientists in academia and industry, chemical engineers, and water-quality monitoring agencies working in water treatment, efficient materials, nanomaterials development, and quality control.

The increase of chemical structures is the focal point of research and technology that is mostly related to chemistry, materials science, physics, applied sciences, petroleum, and engineering. Advanced research on adsorbents and nanoadsorbents has mainly focused on the aspects of preparation of nanomaterials and adsorbents that have unique chemical, physical, thermal, and mechanical properties applicable to a wide range of applications related to environmental remediation. A variety of properties and phenomena has been investigated, and many of the studies have been directed toward understanding the properties and applications of nanomaterials. This becomes more interesting when nanomaterials are combined with polymers to form composites.

Due to their enhanced chemical and mechanical properties, the composites play promising roles in several applications. Nanomaterials have properties that are useful for enhancing surface-to-volume ratio, reactivity, strength, and durability. In pursuit of the same goal, this book offers detailed, up-to-date chapters on the overview of surface and interface science, adsorption technology, adsorption kinetic isotherms models, synthesis, properties, and technological developments of nanomaterials and adsorbents, and their applications in environmental remediation.

The book has been developed as a consequence of significant advances in the materials science community. In the excitement surrounding these materials and technologies, however, their potential has been frequently overhyped. The book explores these kinds of materials and forward-looking potential applications. The book is organized in a good way and makes extensive use of illustrations.

This book covers many aspects of nanomaterials, which are of current interest. This book is written for a large readership, including university students and researchers from diverse backgrounds such as chemistry, petroleum, materials science, physics, and engineering. It can be used not only as a textbook for both undergraduate and graduate students but also as a review and reference book for researchers in these fields. We hope that the chapters of this book will provide the readers with valuable insight into state-of-the-art advanced nanomaterials and technologies.

However, it is possible that some topics have been left out owing to constraints on the size of the book and possible errors in judgment. We trust that the preface will be useful to students, teachers, and researchers.

I welcome suggestions from readers toward improvements that can be incorporated in future editions of this book. I hope that you enjoy the book!

Tawfik A. Saleh

Department of Chemistry, King Fahd University of Petroleum and Minerals, Saudi Arabia

Acknowledgment

I would like to acknowledge the support provided by King Fahd University of Petroleum & Minerals. I would like to express my gratitude to the many people who saw me through this book; to all those who provided support, read, wrote, offered comments, allowed me to quote their remarks, and assisted in the editing, proofreading, and design. Above all, I want to thank my parents, relatives, wife, children, and the rest of my family, who supported and encouraged me in spite of all the time it took me away from them. Last and not least: I beg forgiveness of all those who have been with me over the course of the months and whose names I have failed to mention.

Many thanks.

Tawfik Abdo Saleh

Department of Chemistry, and IRC for Advanced Materials, King Fahd University of Petroleum and Minerals, Saudi Arabia

Chapter 1

Overview of surface and interface science

1. Introduction

Everything on the Earth (food, clothes, buildings, cars, communication devices, nanoelectronics, nanodevices, etc.) is made up of atoms. The development of new materials or nanomaterials (NMs) requires an understanding of the nature of the required properties of the targeted materials to be used for a specific application. By understanding the required properties, the raw precursors to be used for preparation can be determined. For this, the following are required: scientists who understand the atomic and molecular levels of the materials and how to break, rearrange, and form bonds and engineers who understand how to utilize the science of materials to control material processing, manufacturing, production, and design of large-scale amounts of materials. All materials are formed initially from atoms. Understanding the nature of the atoms and their chemical bonding and knowing how to control the conditions of the reactions between the precursors helps one to predict and control the ultimate properties of materials and NMs. Generally, the properties of any prepared materials mostly depend, to some extent, on the following factors: the type of atoms used to build and form the materials; the type of precursors used in the formation of the materials; the conditions and controlled factors used while conducting the reaction or synthesis of the materials; the functionality (moieties) created on the surface of the materials; further treatment and modification of the prepared materials; and the nature of the surface of prepared materials. For a better understanding, it is important to review the basic science and chemical terminologies of materials and their components.

1.1 Atoms and molecules

Molecules are made up of atoms that are made of subatomic particles, i.e., protons, neutrons, and electrons, which are, in turn, made of quarks (also

Surface Science of Adsorbents and Nanoadsorbents. https://doi.org/10.1016/B978-0-12-849876-7.00001-4

particles). An atom, such as that of hydrogen or iron, is the base unit of an element. Atoms have one or more negatively charged electrons that "orbit" a positively charged nucleus, which, in turn, comprises protons and neutrons. Atoms are the fundamental units of matter that can only be broken down into simpler forms by a nuclear reaction; they are the elemental particles of chemistry. Orbit is left in quotes above, as a distinct orbit does not exist in quantum mechanics, but only probabilities of a given particle being in a certain place. However, for the time being, thinking of them in orbits will paint a clear enough picture. Due to the different energy levels among electrons, they can be transferred to other atoms. Thus, two atoms become electrically charged due to the imbalance in charge between their nucleus and electrons, and they attract each other and form a bond. A conglomeration of atoms sticks together to form a molecule. Aside from complete transfer of electrons, atoms can bond by sharing electrons as well; such bonds are called covalent bonds. Regardless, it is the transfer or sharing of electrons that bond atoms together, and their conglomeration is a compound. They can be quite complicated with many different bonds between their various atoms (Pullman, 1998; Stern, 2005).

Molecules are compounds that comprise covalently bonded atoms. Some molecules comprise the same atom bonded to itself (Lewis, 1916; Demtröder, 2002). For example, an oxygen molecule, i.e., the formed oxygen that exists in the air mostly, comprises an oxygen atom bound to another oxygen atom. On the other hand, some molecules comprise different atoms such as a sodium chloride molecule, commonly known as table salt, that comprises sodium and chloride joined together in a chemical bond. Generally, a molecule, such as water or carbon dioxide, comprises two or more atoms bonded together (Table 1.1).

1.2 Element, compound, substance, and mixture

Atoms are organized into elements with common attributes. Simply put, elements strictly comprise atoms of that particular element in its original form (Fig. 1.1). For example, oxygen is an element; all its atoms are oxygen atoms. Nevertheless, they do not exist separately. Instead, they pair up to form oxygen gas, O_2, which is not an element but an oxygen molecule. Similarly, hydrogen atoms combine together to form hydrogen gas, H_2, which is a hydrogen molecule.

A compound is formed when two or more atoms of the same or different elements are bonded together ionically or covalently. For example, water is a molecule made up of an oxygen atom and two hydrogen atoms chemically bond together to form a water molecule, with chemical formula H_2O (Fig. 1.2).

A substance refers to anything composed of a pure element or compound. Its chemical composition is fixed, has individual properties, and can occur in solid, liquid, gaseous, or plasma states. An element made of a pure substance cannot be physically separated without breaking its chemical chain. For

TABLE 1.1 Some differences between atoms and molecules.

Term	Atoms	Molecules
Definition	An atom is a fundamental piece of matter. Everything in the universe is made of atoms. An atom itself is made up of three tiny kinds of particles called subatomic particles: Protons, neutrons, and electrons. The protons and the neutrons make up the center of the atom called the nucleus and the electrons fly around above the nucleus.	The smallest unit into which a substance can be divided without chemical charge, usually a group of two or more atoms.
Shape	An atom is spherical.	A molecule has various symmetries, depending on the arrangements of the atoms.
Orbital	All the orbitals on an atom are atomic orbitals.	All the orbitals on a molecule are molecular orbitals, although we often make some assumptions that the core orbitals are pretty much unaffected and are still mostly atomic orbitals.
Arrangement	All atoms can be sorted and identified in a single table, called the periodic table.	There are an infinite number of possible molecules made by combining different atoms in different ways.
Stability	Atoms are not stable alone and make chemical bonds with other atoms to become stable.	Molecules are stable alone.
Separation	Atoms cannot be separated into subatomic particles by chemical reactions.	Molecules can be separated into atoms by chemical reactions.
Foundation	Atoms are the foundation of the molecule.	Molecules have an equal number of positive and negative charges to form a neutral molecule.
Bonding	Atoms have no bonding.	Molecules have intermolecular force and intramolecular force.

Continued

TABLE 1.1 Some differences between atoms and molecules.—cont'd

Term	Atoms	Molecules
Types	Atoms have no further kind.	Molecules have homonuclear and heteronuclear.
Existence	Atoms may or may not exist independently.	Molecules always exist independently.
Examples	Hydrogen Oxygen Nitrogen	Water, H_2O Nitrogen dioxide, NO_2 Benzene, C_6H_6

example, both a pure block of iron and a corroded sheet of iron oxide are substances. Other examples of pure substances include water, gold, salt, sugar, and diamonds.

A mixture is formed when two or more things, substances, or elements are mixed together without forming chemical bonds. For example, water and oil, or sand and gravel. This mixing is not restricted to substances of the same physical state, for instance, a vapor is a mixture too; it is a mixture of water droplets and gas. Air is also a mixture. A mixture is either homogenous, where its constituents are distributed uniformly, for example, salt in water, or heterogeneous, where its constituents are not distributed uniformly, for example, sand in water. Unlike chemical compounds, mixtures can be separated into substances using physical methods such as filtration, freezing, and distillation. When a mixture forms, the enthalpy of mixing is very low; in other words, there is no energy change. Mixtures have variable compositions, while compounds have a definite formula with chemical bonds between their atoms. The properties of a mixture depend on the properties of the individual substances. A mixture can be formed of solids, liquids, and gases, as well as of gas and solid, liquid and solid, and gas and liquid.

1.3 Matter and material

Matter is any substance that has a mass and takes up space, as it has volume. Anything that has physical existence and can be observed is matter; and physicists further hypothesize the existence of dark matter, which cannot be observed. Elements, such as sulfur, gold, and potassium, and compounds such as water, air, and coffee are all examples of matter. Everything is matter, and matter can exist in a number of states.

Material broadly includes substances, both in pure chemical form and mixtures. The physical and chemical properties of materials are investigated in materials science. The material usually refers to some definite kind, quality, or quantity of matter, especially as intended for use (e.g., cotton material,

Term	Definition
Matter	The stuff from which everything is made.
Material	Any type of matter. It is a broad term for pure chemical substances or mixture of substances.
Atoms	Atoms are made of subatomic particles (protons, electrons, and neutrons), which are, in turn, made of quarks, which are also particles. An atom is the basic unit of an element.
Element	An element is made up of only one kind of atom. It is a substance that cannot further be subdivided into simpler substances. It is a material that is the same throughout, such as gold or silver. It can be classified into metals, non metal, metalloid, and noble gases.
Molecules	Molecules are made up of atoms. A molecule involves two or more atoms bonded together (all compounds, as well as some elements are molecules). Thus, a molecule is a unit of matter composed of multiple atoms, and is also the basic unit of a covalent chemical compound.
Compound	A substance that is made up of molecules, which are, in turn, made up of two or more different types of atoms. Salt (sodium chloride) is a compound.
Substance	A substance is anything that is composed of a pure element or compound.
Mixture	A mixture is formed when two or more things, substances, or elements are mixed together without forming chemical bonds.

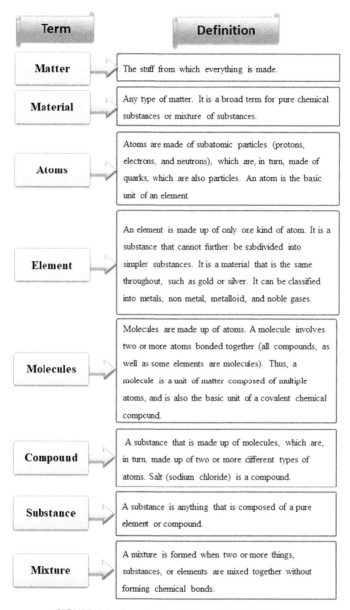

FIGURE 1.1 Some terminologies and their definitions.

explosive materials, a house built of poor materials). Materials are substances out of which a thing or a combination of things is made. Simply put, material

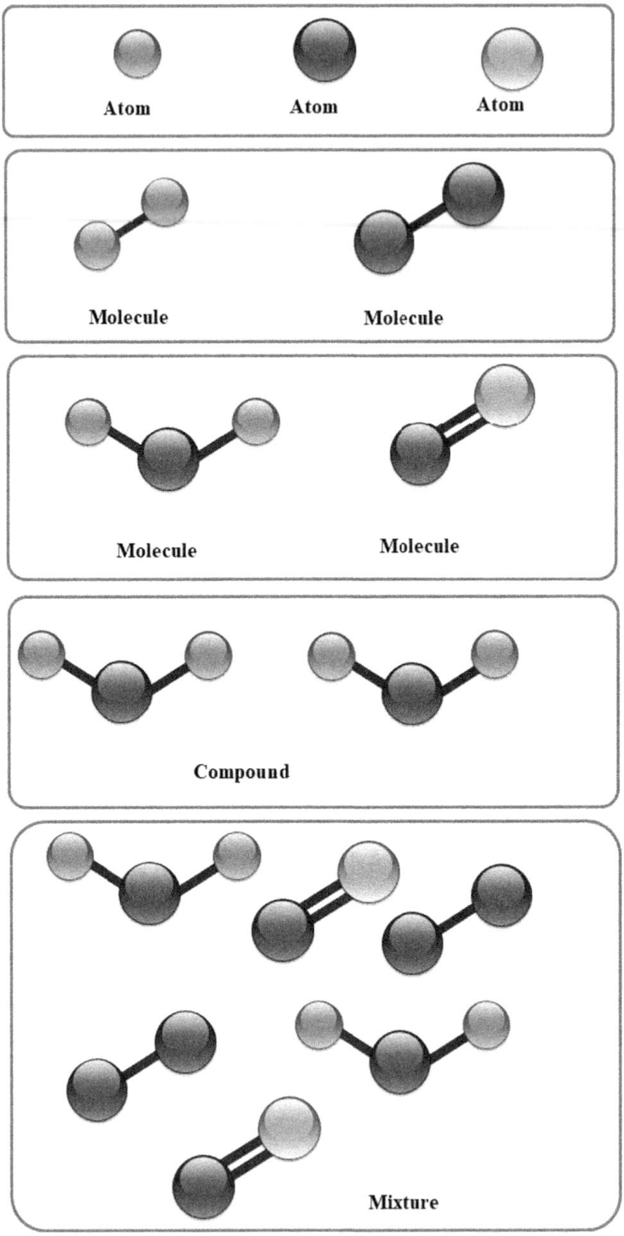
Atom
Atom
Atom
Molecule
Molecule
Molecule
Molecule
Compound
Mixture

FIGURE 1.2 Illustration of the differences between an atom, molecule, compound, and mixture.

refers to a matter which has a specific purpose. For example, construction material. A log of wood lying on the roadside is matter. But when picked with the intention of crafting something out of it, the log becomes a wooden material. Basically, such matter is raw material that becomes some useful object after processing.

Materials can be natural or made. Further, they can be made from natural resources by, for instance, mechanical processes or synthesized chemically from precursors by a chemical reaction. As an unlimited number of precursors are available, an unlimited number of materials can be prepared by various chemical processes and methods.

1.4 Chemical formulae

Chemical symbols are used to represent the elements. For example, **C** stands for carbon, **O** for oxygen, **S** for sulfur, and **Na** for sodium. On the other hand, for a molecule, the chemical symbols of the atoms are used to write the molecular formula. For example, the formula for carbon monoxide is **CO**. This molecular formula tells us that each molecule of carbon monoxide consists one carbon atom bonded to one oxygen atom. If we are referring to more than one formula, then the plural, 'formulae,' is used.

Numbers in a formula are used to denote when a molecule contains more than one atom of an element; the numbers are written below the element's symbol. For example, CO_2 is the formula for carbon dioxide. It tells you that each molecule has one carbon atom and two oxygen atoms. Some formulae are more complicated. For example, the formula for sodium sulfate is Na_2SO_4. This compound contains two sodium atoms (Na_2), one sulfur atom (S), and four oxygen atoms (O_4). To determine the formula of a new material or compound, it is necessary to understand how this compound is formed, i.e., the precursors and chemical reaction involved and write a balanced chemical equation of the reaction used to prepare this material.

1.5 Chemical reactions

Chemical reactions involve breaking apart the molecules and combining the parts to form bigger molecules. An element changes to a different element not through chemistry, but only through nuclear fission or fusion. A chemical reaction of a substance involves rearrangement of the atom's molecular or ionic structure and is distinct from a change in the physical form or a nuclear reaction. A chemical reaction can also be defined as a process where one or more substances, the reactants, are converted to one or more different substances, the products. Substances can be chemical elements or compounds. A chemical reaction rearranges the constituent atoms of the reactants to create different substances as products.

2. Bulk materials and nanomaterials

2.1 Definitions

Bulk materials are particles whose size exceeds 100 nm in all dimensions. In such materials, physical properties are independent of size. The prefix 'nano' refers to the Greek prefix meaning 'dwarf' or an object with very small size, and it depicts 1000 millionth of a meter (1 nm = 10^{-9} m). With NMs, different physical properties can depend on their size and shape. NMs have particle sizes measurable in nanometers (nms) (Fig. 1.3). NMs are materials or chemical substances that are manufactured and used at a very small scale, i.e., 1−100 nm in at least one dimension. The British Standards Institution (BSI) has proposed some definitions for the scientific terms related to NMs (Jeevanandam et al., 2018).

Fig. 1.4 lists some important definitions of some terms used in nanotechnology-related studies such as nanoparticle, nanostructure, and nanomaterial. Nanofiber, for instance, is a term used to refer to those fibers with diameters in the nanometer range. Nanofibers can be generated from different sources such as polymers and hence have different physical properties and application potentials. Nanofibers are materials with two similar exterior nanoscale dimensions and a third larger dimension.

Nanoparticulate materials became important when researchers realized that size influences the physicochemical properties of a substance, including its chemical, electrical, mechanical, and optical properties. Consequently, NMs have attracted considerable interest owing to their unique properties (Gleiter, 2000). Nanoparticulate materials can be potentially applied for numerous applications including water treatment plants, oil refineries, petrochemical industries, industrial processes, catalytic processes, buildings and building materials, diagnostics, and drug delivery. There are numerous methods for producing NPs including chemical precipitation, condensation, pyrolysis, hydrothermal synthesis attrition, and ion implantation (Pokropivny & Skorokhod, 2007; Mulvaney, 2015). In brief, the main difference between NMs and bulk materials is that the predominant thermal, mechanical, optical, electric, and magnetic properties of the material change at a larger rate in nanoscale compared to that in bulk scale. This is because at the nanoscale, the materials have more interfaces thus increasing the overall surface area by increasing the interacting faces, which leads to change in the properties. This change in properties can be considered as an enhancement since they can be used for several novel applications in the field of electronics, storage devices, and defense equipment.

Any material with dimensions in nanoscale regime does not solely qualify as an NM. Essentially, it has to exhibit distinct properties arising from 'quantum confinement effects' to be classified as NM. The 'quantum confinement effect' arises from the discretization of electronic energy states,

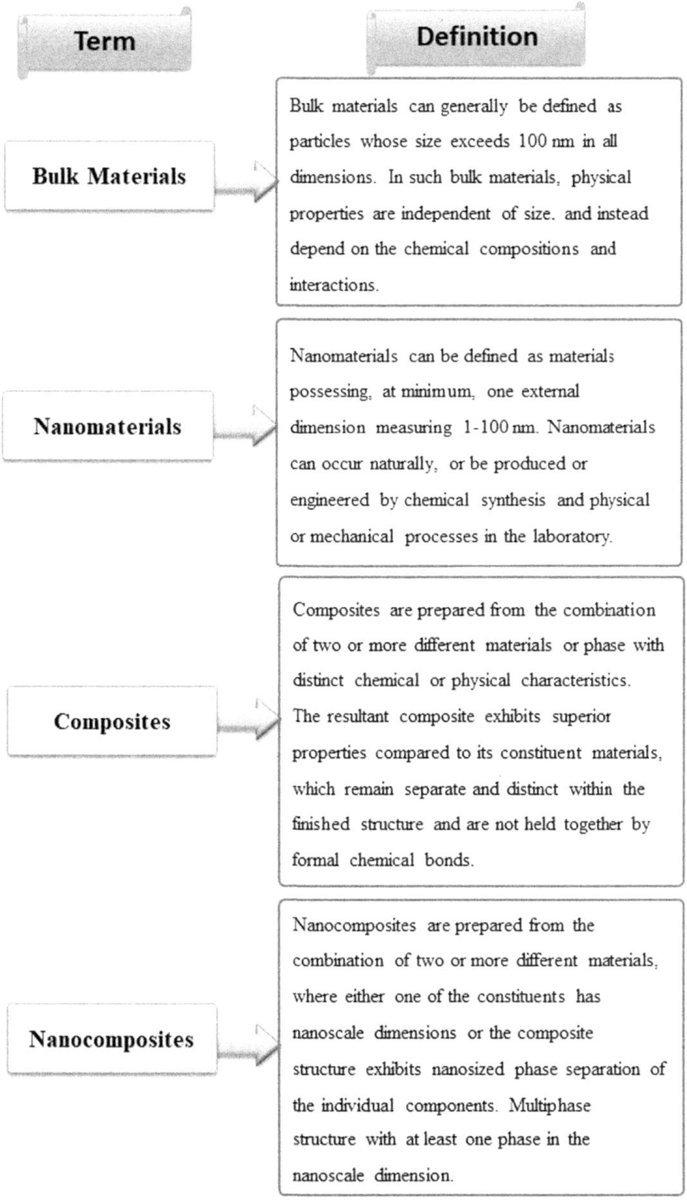
Term
Definition
Bulk Materials
Bulk materials can generally be defined as particles whose size exceeds 100 nm in all dimensions. In such bulk materials, physical properties are independent of size, and instead depend on the chemical compositions and interactions.
Nanomaterials
Nanomaterials can be defined as materials possessing, at minimum, one external dimension measuring 1-100 nm. Nanomaterials can occur naturally, or be produced or engineered by chemical synthesis and physical or mechanical processes in the laboratory.
Composites
Composites are prepared from the combination of two or more different materials or phase with distinct chemical or physical characteristics. The resultant composite exhibits superior properties compared to its constituent materials, which remain separate and distinct within the finished structure and are not held together by formal chemical bonds.
Nanocomposites
Nanocomposites are prepared from the combination of two or more different materials, where either one of the constituents has nanoscale dimensions or the composite structure exhibits nanosized phase separation of the individual components. Multiphase structure with at least one phase in the nanoscale dimension.

FIGURE 1.3 Some materials, related terminologies, and their definitions.

confining the electrons in a material within a very small space. The excitonic mechanisms that govern these effects are among the other sections that have

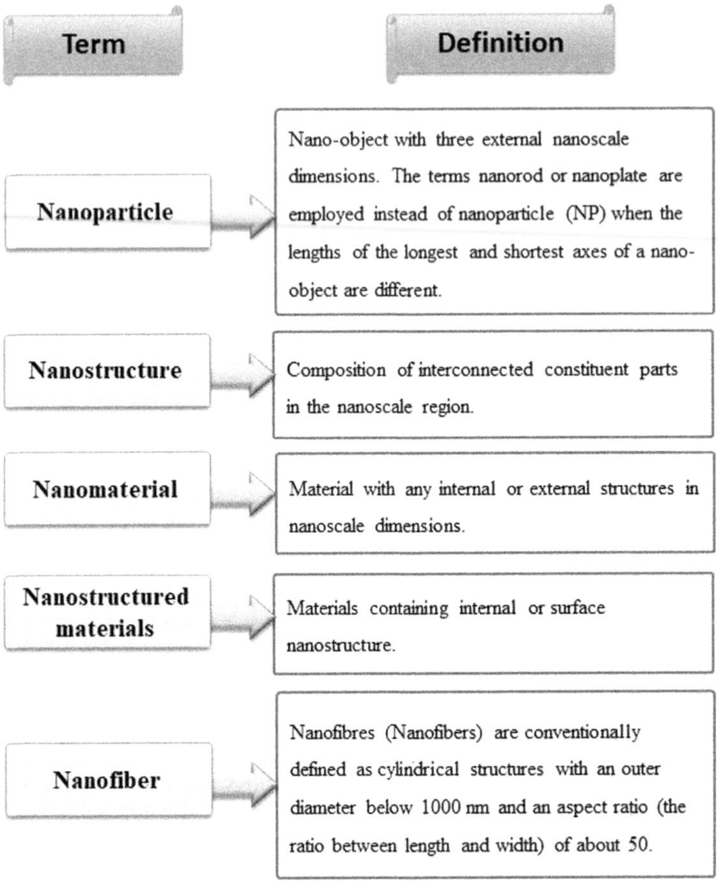
Term
Definition
Nanoparticle
Nano-object with three external nanoscale dimensions. The terms nanorod or nanoplate are employed instead of nanoparticle (NP) when the lengths of the longest and shortest axes of a nano-object are different.
Nanostructure
Composition of interconnected constituent parts in the nanoscale region.
Nanomaterial
Material with any internal or external structures in nanoscale dimensions.
Nanostructured materials
Materials containing internal or surface nanostructure.
Nanofiber
Nanofibres (Nanofibers) are conventionally defined as cylindrical structures with an outer diameter below 1000 nm and an aspect ratio (the ratio between length and width) of about 50.

FIGURE 1.4 Definitions of some terms used in nanoscience.

not been elucidated in detail here. Deriving from the fact of confinement of electron movement in a material, different materials exhibit different size limits within the nm frame below, wherein they begin to show NM-like properties. This limit is called the 'Bulk to Nano transition limit.' For example, the nanoproperties of gold nanoparticles start from a cluster size below 1000 atoms only (Please look for the size in nm online), whereas the same happens for copper nanoparticles from a cluster size of only 100 atoms or less.

NMs show high diversity, differing in chemical composition, size, shape, crystallinity, and surface modification (Schwirn et al., 2014) (Fig. 1.5). Considering this plurality of forms of the materials, the physical and chemical

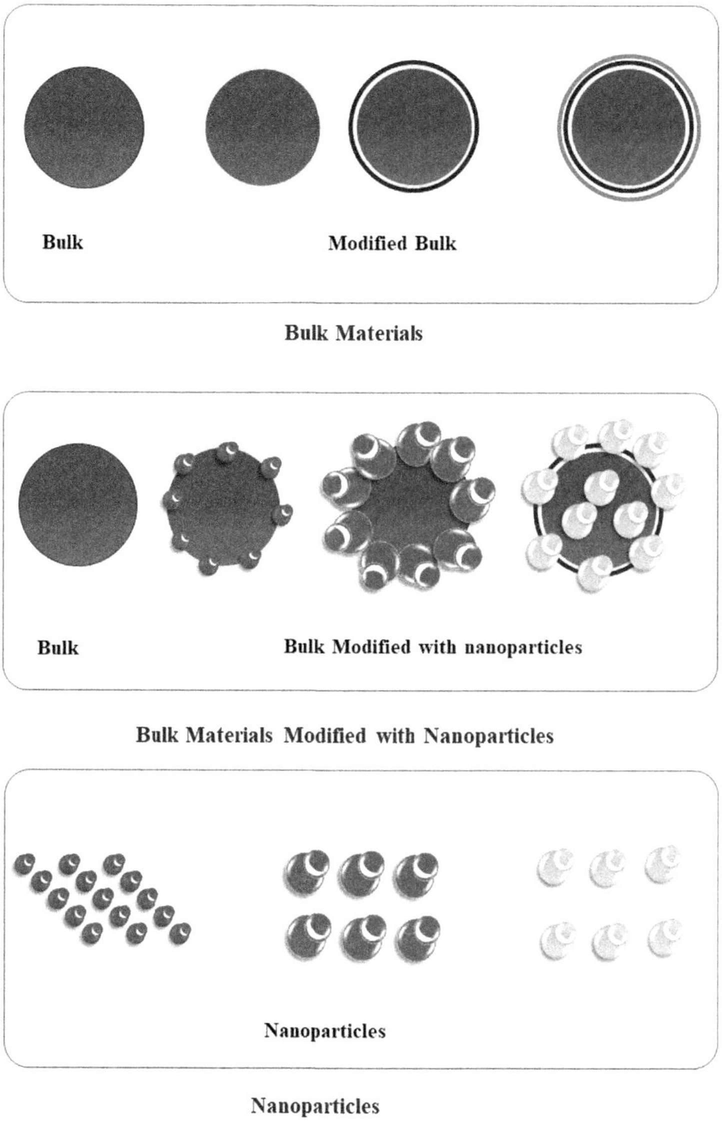
Bulk
Modified Bulk
Bulk Materials
Bulk
Bulk Modified with nanoparticles
Bulk Materials Modified with Nanoparticles
Nanoparticles
Nanoparticles

FIGURE 1.5 Illustration of the different forms of a substance: bulk materials and NMs.

characteristics change. Thus, a synthetic approach must be adopted to adequately develop a material with the targeted properties.

2.2 Differences between bulk and nanomaterials

NMs behave very differently from regular bulk materials for different reasons including surface area to volume ratio. For materials with side lengths, radii, and heights in nanoscale regime (less than one micron), NMs' surface areas play larger roles than the volume they occupy. With their surfaces dictating the materials properties, markedly different behaviors can be observed on NMs' part compared to bulk materials as depicted in Fig. 1.6. Additionally, NMs display unique properties as shown in Fig. 1.7. Some very cool concepts including superparamagnetism, surface plasmon resonance, and self-assembly may occur. These NMs with nanoscale properties have great potential to be applied in several applications including environment, water, photovoltaic energy storage, and supercomputing. All this is made possible due to NMs being extremely small particles having a large surface area to volume ratio.

What are the differences between bulk and nanomaterials?

The basic differences can be observed in the following distinguishing features:

- **Quantum arrangement**: The quantum arrangement defines the surface states of the material, which, in turn, defines how the particle is shaped, to further define the quantum states in the particle. This has a direct influence on the optical, electrical, thermal, structural, and mechanical properties of the nanostructured materials. In bulk materials, atoms are organized in an "infinite" structure, be it monocrystal, polycrystal with texture, or amorphic things with only a local structure. Nevertheless, most atoms have "infinite" number of atoms in some vicinity. Only very few are at the surface. Whereas in NMs, every constituent has a small number of neighbors. Consequently, it is somewhere in between. Molecules have discrete quantum levels, while bulk has bands (an average of infinite discrete quantum levels). Nano almost has a discrete subset of subbands, which are quantum states not found in bulk or molecules. It must be noted that quantum states determine how things behave on the macroscale.
- **Small size**: Having a size comparable to the free-electron path has several consequences. In molecules, there are no "free" electrons. And in bulk, every "free" electron is bounced at scattering centers. If the material becomes very small, it can have a "free" electron that travels through the entire structure without bouncing any scattering center (Poole and Owens, 2005). If the electrons move freely in space with modulated fields, this leads to new electronic properties. But here again, these new electrical properties change the game, and not the field.
- **Surface to volume ratio**: In bulk materials, the number of constituents at the surface is a tiny fraction, whereas for NMs, the ratio is very high. It affects the rate at which chemical reactions take place. Therefore, the reactivity of NMs is different.

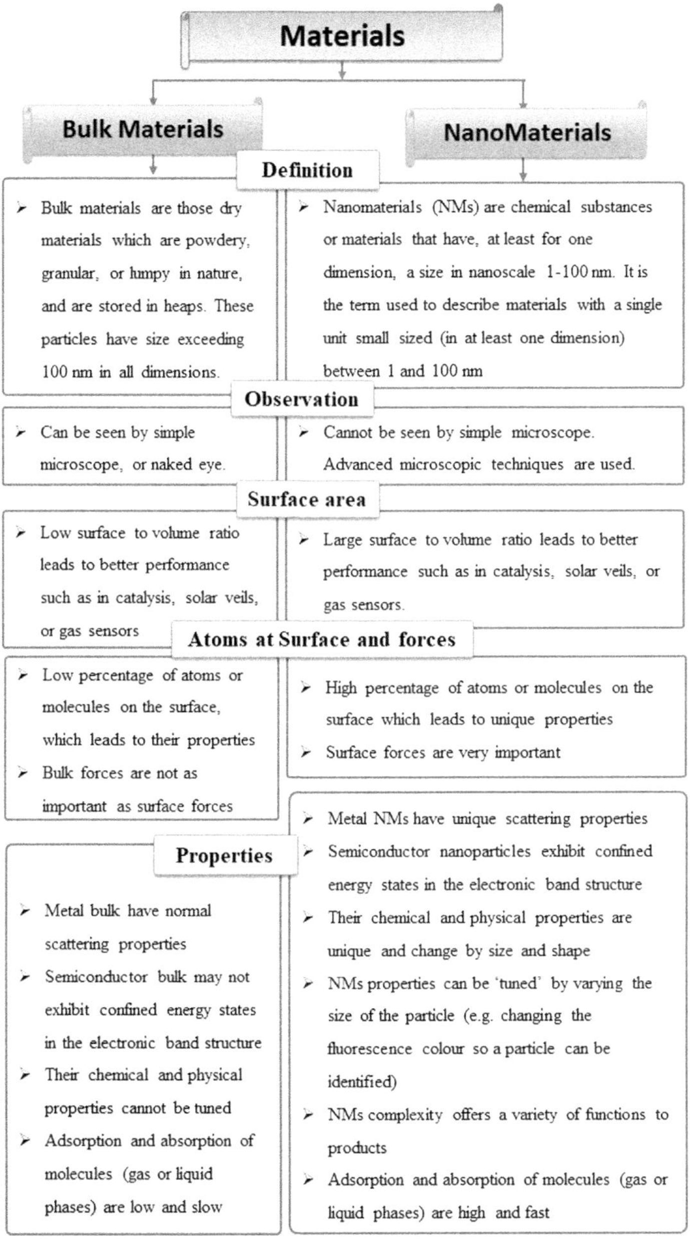
Materials
Bulk Materials
NanoMaterials
Definition
Bulk materials are those dry materials which are powdery, granular, or lumpy in nature, and are stored in heaps. These particles have size exceeding 100 nm in all dimensions.
Nanomaterials (NMs) are chemical substances or materials that have, at least for one dimension, a size in nanoscale 1-100 nm. It is the term used to describe materials with a single unit small sized (in at least one dimension) between 1 and 100 nm
Observation
Can be seen by simple microscope, or naked eye.
Cannot be seen by simple microscope. Advanced microscopic techniques are used.
Surface area
Low surface to volume ratio leads to better performance such as in catalysis, solar veils, or gas sensors
Large surface to volume ratio leads to better performance such as in catalysis, solar veils, or gas sensors.
Atoms at Surface and forces
Low percentage of atoms or molecules on the surface, which leads to their properties
Bulk forces are not as important as surface forces
High percentage of atoms or molecules on the surface which leads to unique properties
Surface forces are very important
Properties
Metal bulk have normal scattering properties
Semiconductor bulk may not exhibit confined energy states in the electronic band structure
Their chemical and physical properties cannot be tuned
Adsorption and absorption of molecules (gas or liquid phases) are low and slow
Metal NMs have unique scattering properties
Semiconductor nanoparticles exhibit confined energy states in the electronic band structure
Their chemical and physical properties are unique and change by size and shape
NMs properties can be 'tuned' by varying the size of the particle (e.g. changing the fluorescence colour so a particle can be identified)
NMs complexity offers a variety of functions to products
Adsorption and absorption of molecules (gas or liquid phases) are high and fast

FIGURE 1.6 Key differences between bulk and nanomaterials.

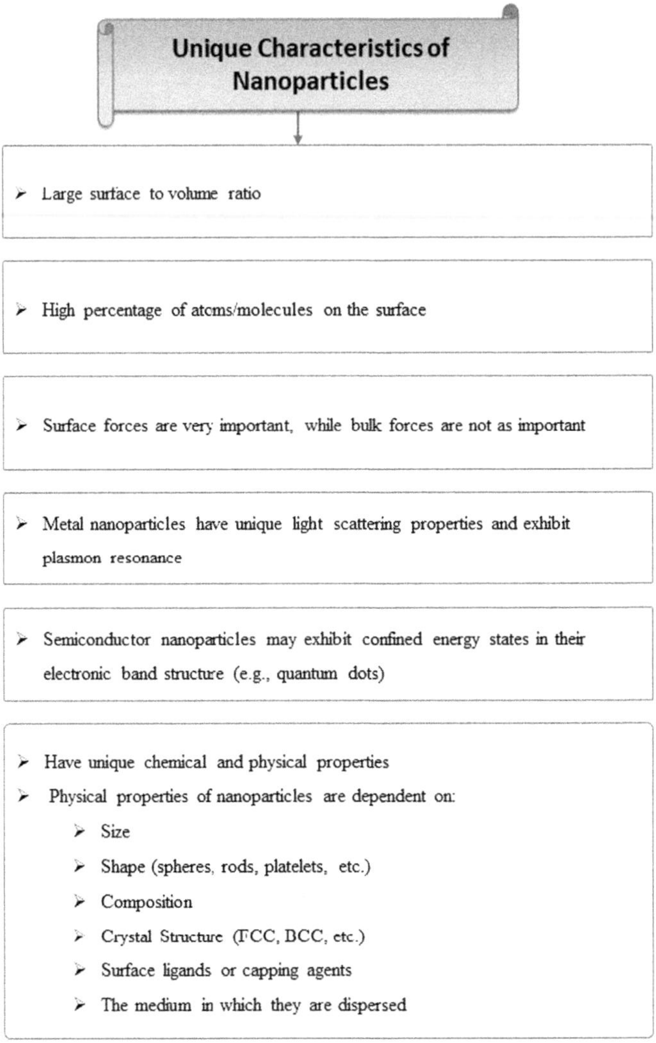
Unique Characteristics of Nanoparticles
➢ Large surface to volume ratio
➢ High percentage of atcms/molecules on the surface
➢ Surface forces are very important, while bulk forces are not as important
➢ Metal nanoparticles have unique light scattering properties and exhibit plasmon resonance
➢ Semiconductor nanoparticles may exhibit confined energy states in their electronic band structure (e.g., quantum dots)
➢ Have unique chemical and physical properties
➢ Physical properties of nanoparticles are dependent on:
➢ Size
➢ Shape (spheres, rods, platelets, etc.)
➢ Composition
➢ Crystal Structure (FCC, BCC, etc.)
➢ Surface ligands or capping agents
➢ The medium in which they are dispersed

FIGURE 1.7 Unique characteristics of nanoparticles.

The difference in the bulk material's physical properties when its dimension is reduced might be very important. For example, silver is highly diamagnetic in bulk crystalline form, but if reduced to clusters of 13 or 14 atoms, its properties will be changed (Pereiro et al., 2007; Pereiro and Baldomir, 2007). Titanium dioxide nanoparticles have been found to be more efficient in killing germs when walls are painted with it. NMs are found to be useful for supercapacitors, some new batteries, robotics, and even computing, among other things. There are several examples of NMs, and new ones are being discovered, which will bring about a revolution in engineering, medicines, and other walks of life. Nevertheless, nm is still larger when compared with an atomic scale, as depicted in Fig. 1.8.

Does picotechnology succeed nanotechnology?

It seems there is plenty of room for the development of picotechnology.

2.3 Types of nanomaterials

The size of the structure below which a material starts behaving like an NM is different for different materials. Thus, it is wrong to generalize that when the dimensions of a material fall under nanoscale regime, they automatically become NMs (Buzea et al., 2007).

Nanoscale materials range from 1 to 100 nms, and they can be classified as in Fig. 1.9, since they are presented in the following geometries:

- ➢ Type one, 0D: 0-dimensional NMs that have all of the three dimensions in nanoscale range. Examples are heterogeneous nanoparticles arrays, uniform particles arrays (quantum dots), core–shell quantum dots, onions, hollow spheres and nanolenses.
- ➢ Type two, 1D: one-dimensional NMs that have one of the three dimensions in nanoscale range. Examples are nanotubes, nanowires, nanorods, nanobelts, nanoribbons, hierarchical nanostructures, montmorillonite clay and nanographene platelets.
- ➢ Type three, 2D: two-dimensional NMs that have two of the three dimensions in nanoscale range. Examples are graphene, Mxenes, branched structures, nanoprisms, nanoplates, nanosheets, nanowalls, and nanodisks.
- ➢ Type four, 3D: Formed by the arrangement of multiple 0D, 1D, or 2D materials to form the 3D structure. Examples are nanoballs (dendritic structures), nanocoils, nanocones, nanopillers, nanoflowers, graphite, and multi-nanolayers.

NMs can be further categorized into two, namely, natural and artificial/synthetic NMs. On the basis of applications, these synthetic NMs can be categorized as functional materials (at least one dimension is in nm range and has very specific properties in comparison to raw material) and nonfunctional materials.

Most common examples of NMs include fullerene, carbon nanotubes, graphene, and quantum dots. In nature, colors of insects are often not pigments at all, but very tiny structures that scatter light selectively to give an effect of color. Fig. 1.10 shows a classification of NMs based on their properties.

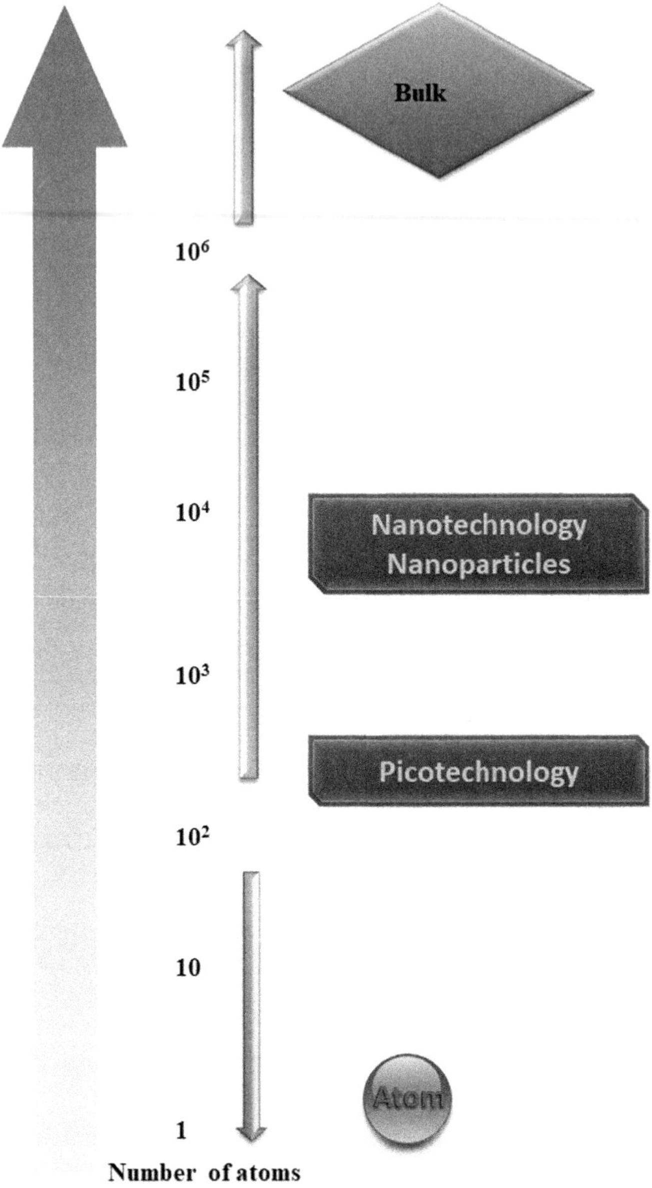
Bulk
10^6
10^5
10^4
Nanotechnology
Nanoparticles
10^3
Picotechnology
10^2
10
Atom
1
Number of atoms

FIGURE 1.8 Molecules, nanoparticles, and bulk materials can be almost distinguished by the number of atoms comprising each type of material.

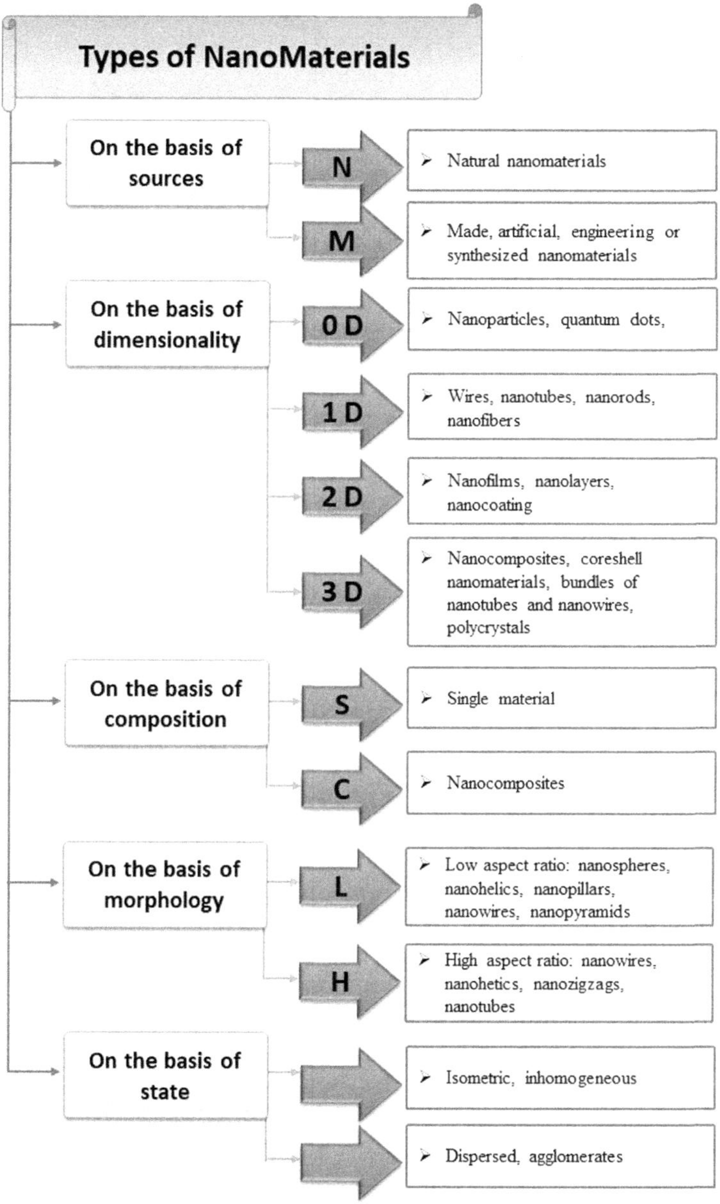
Types of NanoMaterials
On the basis of sources
N
➢ Natural nanomaterials
M
➢ Made, artificial, engineering or synthesized nanomaterials
On the basis of dimensionality
0 D
➢ Nanoparticles, quantum dots,
1 D
➢ Wires, nanotubes, nanorods, nanofibers
2 D
➢ Nanofilms, nanolayers, nanocoating
3 D
➢ Nanocomposites, coreshell nanomaterials, bundles of nanotubes and nanowires, polycrystals
On the basis of composition
S
➢ Single material
C
➢ Nanocomposites
On the basis of morphology
L
➢ Low aspect ratio: nanospheres, nanohelics, nanopillars, nanowires, nanopyramids
H
➢ High aspect ratio: nanowires, nanohetics, nanozigzags, nanotubes
On the basis of state
➢ Isometric, inhomogeneous
➢ Dispersed, agglomerates

FIGURE 1.9 Typical types of NMs.

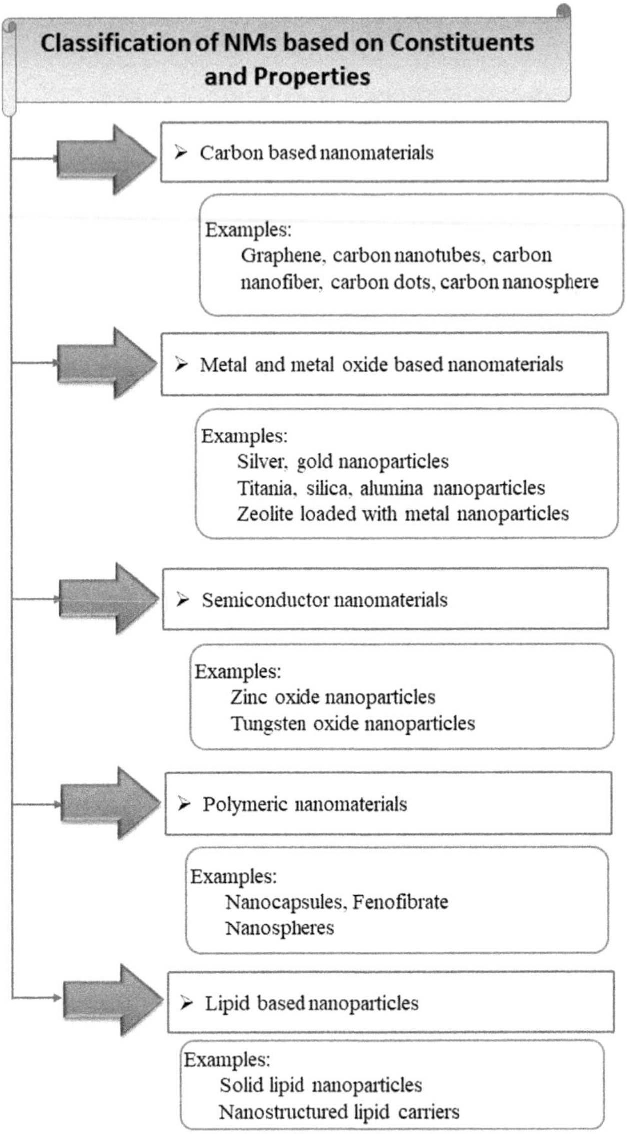
Classification of NMs based on Constituents and Properties
Carbon based nanomaterials
Examples:
Graphene, carbon nanotubes, carbon nanofiber, carbon dots, carbon nanosphere
Metal and metal oxide based nanomaterials
Examples:
Silver, gold nanoparticles
Titania, silica, alumina nanoparticles
Zeolite loaded with metal nanoparticles
Semiconductor nanomaterials
Examples:
Zinc oxide nanoparticles
Tungsten oxide nanoparticles
Polymeric nanomaterials
Examples:
Nanocapsules, Fenofibrate
Nanospheres
Lipid based nanoparticles
Examples:
Solid lipid nanoparticles
Nanostructured lipid carriers

FIGURE 1.10 Classification of NMs based on their properties.

3. Science of nanomaterials

3.1 Chemistry of materials

Chemistry is the science that systematically studies the following:

(i) the composition, properties, and activity/behavior of matter, which is defined as anything that has mass and volume, takes up space, and is made up of particles (Brown et al., 2018);
(ii) the elements that makeup matter to the compounds composed of atoms, molecules, and ions, their composition, structure, properties, and behavior; and
(iii) the changes chemicals undergo during a reaction with other substances (Carsten, 2001; Armstrong, 2012).

Chemistry is a branch of science that deals with the study of the chemical composition, properties, and structure of various compounds and its constituents. It deals with the transformations and energy released and absorbed during the chemical reactions. Chemistry provides an understanding of the many changes of the matter from one form to another. Chemistry involves the ability to describe ingredients in a material and how they change when the material is used. A great challenge in chemistry involves the development of a coherent explanation of the complex behavior of materials: why they appear as they do, what gives them their enduring properties, and how interactions among different substances can bring about the formation of new substances and the destruction of reactants. Furthermore, chemistry concerns with the utilization of natural substances and creation of artificial ones. Fig. 1.11 displays how chemistry is important in different fields.

3.2 Science of nanomaterials

The science of NMs involves the study of nanostructures and NMs and their properties. This science is cross-disciplinary, where scientists from a range of fields including chemistry, physics, and materials science study it for a better understanding. This science involves the study of properties of matter at the nanoscale, with particular focus on the unique, size-dependent properties of solid-state materials. New methods of synthesis are required to make materials at the nanoscale regime, employing both bottom-up and top-down techniques. It is equally important that new characterization approaches are needed (Weiss, 2014; Feynman, 1992).

3.3 Nanoscience and nanotechnology

Before the term nanotechnology was used, concepts and ideas related to nanoscience and nanotechnology started with a talk titled, "There's Plenty of Room at the Bottom," by physicist, Richard Feynman, at an American

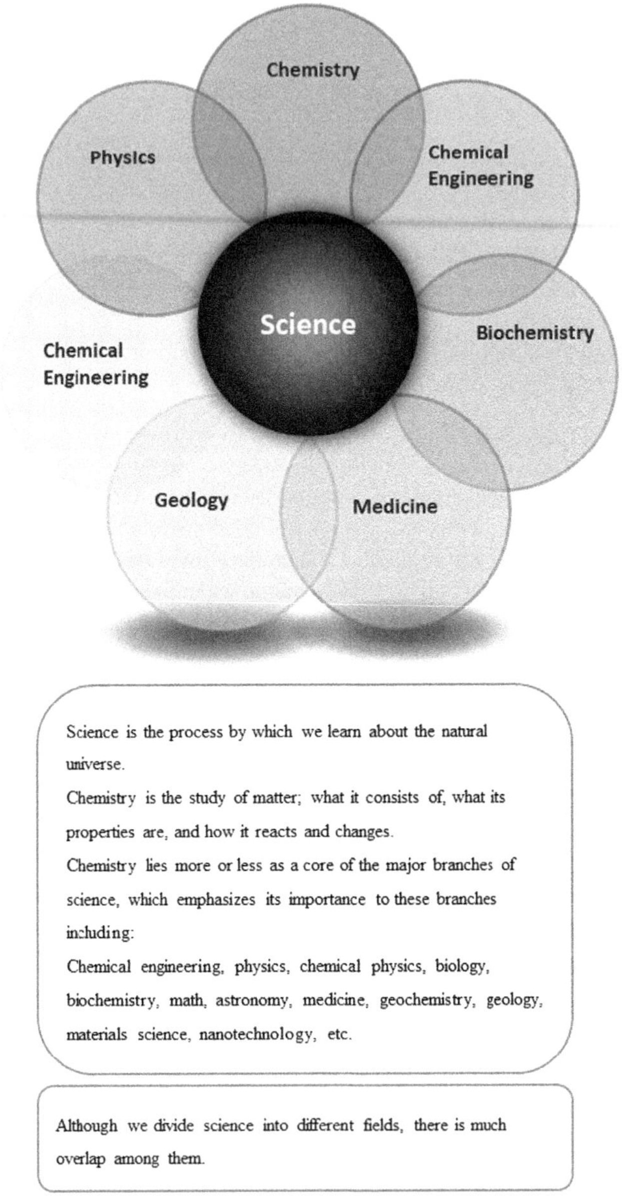
Chemistry
Physics
Chemical Engineering
Science
Biochemistry
Chemical Engineering
Geology
Medicine
Science is the process by which we learn about the natural universe.
Chemistry is the study of matter; what it consists of, what its properties are, and how it reacts and changes.
Chemistry lies more or less as a core of the major branches of science, which emphasizes its importance to these branches including:
Chemical engineering, physics, chemical physics, biology, biochemistry, math, astronomy, medicine, geochemistry, geology, materials science, nanotechnology, etc.
Although we divide science into different fields, there is much overlap among them.

FIGURE 1.11 The relationships between some of the major branches of science. Chemistry lies more or less in the middle, which emphasizes its importance in the different branches of science.

Physical Society meeting at the California Institute of Technology (CalTech) on December 29, 1959 (Feynman, 1992). Feynman described a process where scientists could manipulate and control individual atoms and molecules.

Following which, Professor Norio Taniguchi coined the term nanotechnology (Taniguchi, 1974). Modern nanotechnology began in 1981 with the development of the scanning tunneling microscope that could "see" individual atoms.

Nanoscience is the study of the composition, properties, and activity of matter in nanoscale (Fig. 1.12). It involves the manipulation and engineering of NMs and their structures (Weiss, 2012). In nanoscience, atomic physics converges with the physics and chemistry of complex systems. Important properties of materials, including electrical, optical, thermal, and mechanical properties, are determined by the way the molecules and atoms assemble on the nanoscale into larger structures. Moreover, in nm-sized structures, these properties often appear different than on macroscale, as quantum mechanical effects become important (Damasceno et al., 2012).

Nanotechnology applies nanoscience to use new NMs and nanosize components in useful products. Ultimately, nanotechnology provides an ability to design custom-made materials and products with new enhanced properties, nanoelectronic components, and types of smart materials and sensors, as well as interfaces between electronics and other systems.

Nanoscience and nanotechnology are two terms commonly used when dealing with NMs. Fig. 1.13 depicts the distinction between the two. While nanoscience studies NMs and their properties, nanotechnology uses NMs to create something new or different. Nanoscience converges physics, materials science, and biology to deal with the manipulation of materials at atomic and molecular scales, whereas nanotechnology aims to observe, measure, manipulate, assemble, control, and manufacture matter at the nm scale. Both involve the ability to see and control individual atoms and molecules. Nanotechnology is an advanced area of research that allows the production of a wide class of materials, including particulate materials with at least one dimension of less than 100 nms.

During synthesis, the properties of the NMs can be controlled by controlling some conditions, as physical properties of nanoparticles are dependent on the following:

- Size
- Shape (spheres, rods, platelets, etc.)
- Composition—crystal structure; face-centered cubic (FCC), body-centered cubic (BCC), etc.
- Surface ligands or capping agents
- Medium in which they are dispersed

By controlling these features, NMs with the following unusual properties can be obtained:

➢ Lowered phase transition temperatures
➢ Increased mechanical strength
➢ Different optical properties

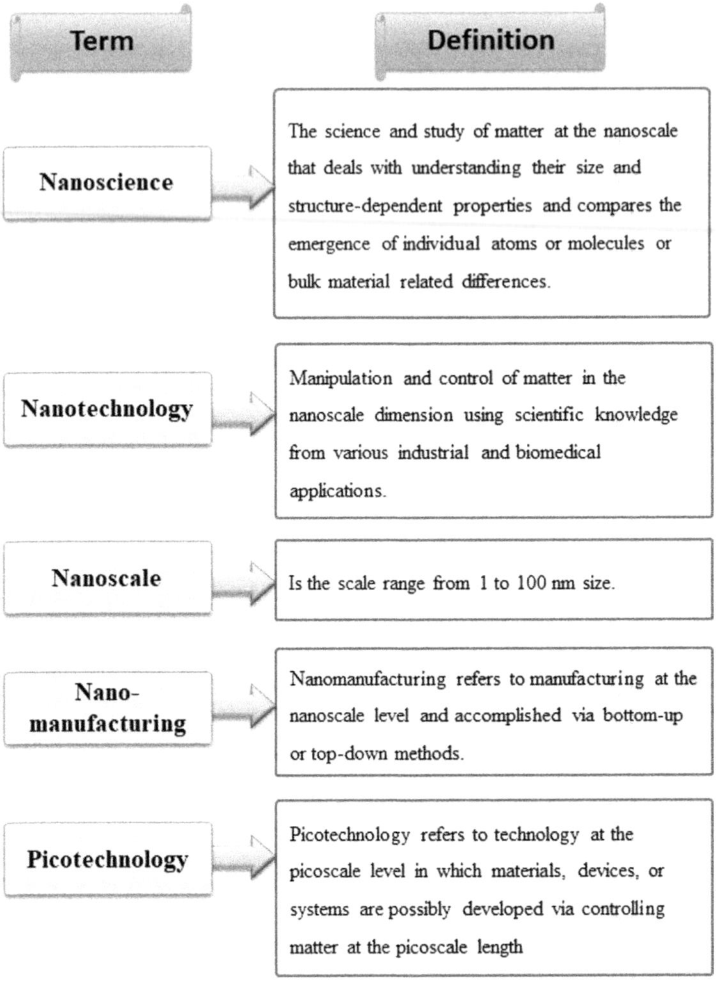
Term
Definition
Nanoscience
The science and study of matter at the nanoscale that deals with understanding their size and structure-dependent properties and compares the emergence of individual atoms or molecules or bulk material related differences.
Nanotechnology
Manipulation and control of matter in the nanoscale dimension using scientific knowledge from various industrial and biomedical applications.
Nanoscale
Is the scale range from 1 to 100 nm size.
Nano-manufacturing
Nanomanufacturing refers to manufacturing at the nanoscale level and accomplished via bottom-up or top-down methods.
Picotechnology
Picotechnology refers to technology at the picoscale level in which materials, devices, or systems are possibly developed via controlling matter at the picoscale length

FIGURE 1.12 Definitions of some terms used in nanoscience and nanotechnology.

- ➢ Altered electrical conductivity
- ➢ Magnetic properties
- ➢ Self-purification and self-perfection

4. Surface science and interface

4.1 Surface science

Surface science is the study of the physical and chemical phenomena that take place at the interface of two phases; for instance, at interfaces between two

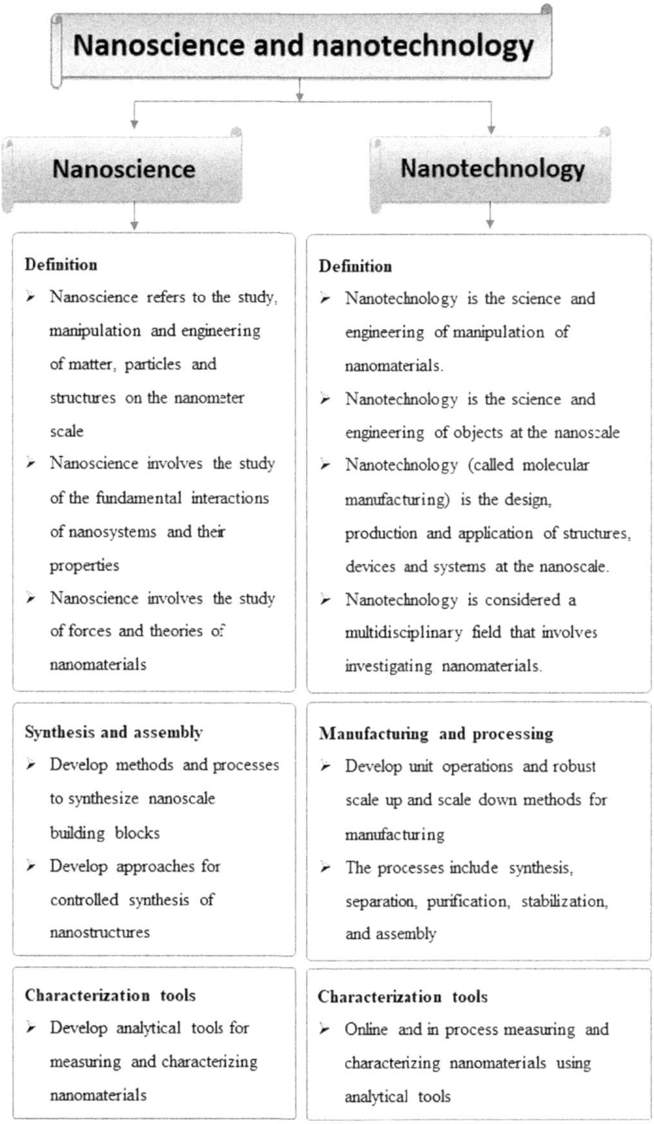
Nanoscience and nanotechnology
Nanoscience
Nanotechnology
Definition
Nanoscience refers to the study, manipulation and engineering of matter, particles and structures on the nanometer scale
Nanoscience involves the study of the fundamental interactions of nanosystems and their properties
Nanoscience involves the study of forces and theories of nanomaterials
Definition
Nanotechnology is the science and engineering of manipulation of nanomaterials.
Nanotechnology is the science and engineering of objects at the nanoscale
Nanotechnology (called molecular manufacturing) is the design, production and application of structures, devices and systems at the nanoscale.
Nanotechnology is considered a multidisciplinary field that involves investigating nanomaterials.
Synthesis and assembly
Develop methods and processes to synthesize nanoscale building blocks
Develop approaches for controlled synthesis of nanostructures
Manufacturing and processing
Develop unit operations and robust scale up and scale down methods for manufacturing
The processes include synthesis, separation, purification, stabilization, and assembly
Characterization tools
Develop analytical tools for measuring and characterizing nanomaterials
Characterization tools
Online and in process measuring and characterizing nanomaterials using analytical tools

FIGURE 1.13 General illustration of nanoscience and nanotechnology.

solids, liquids, or gases, as well as between solid and liquid, liquid and gas, and gas and solid. It is a research field that borders between chemistry and solid-state physics. The science and technology of interacting surfaces in relative motion is known as tribology. Some related practical applications are grouped together as surface engineering. Surface science includes surface chemistry and surface physics (Prutton, 1994). It is of particular importance to the fields of heterogeneous catalysis, electrochemistry, and geochemistry. Surface science deals with several phenomena including adhesion, adsorption, friction, lubrication, and heterogeneous catalysis (Christmann et al., 1974). Additionally, it is important for the production of semiconductor devices, fuel cells, self-assembled monolayers, biomaterials, and pharmaceuticals (Kolasinski, 2002).

Surface science relates to interface and colloid science (Luklema, 2005). Interfacial chemistry and physics are common subjects for both, nevertheless the methods are different. Additionally, interface and colloid science involves the study of macroscopic phenomena that occur in heterogeneous systems as a result of the peculiarities of interfaces.

4.2 Surface chemistry

This branch of chemistry studies chemistry at material surface, along with the properties of the surfaces and the chemical changes occurring at the surface. Simply put, it deals with types of surface phenomena.

It can be defined as the study of chemical reactions at interfaces. It is closely related to surface engineering, which aims to modify the chemical composition of a surface by incorporating selected elements or functional groups that produce various desired effects or improvements in the properties of the surface or interface. Further, surface chemistry is the study of processes that occur at the interfaces between gases and condensed phases such as liquids or solids.

It is the study of chemical reactions at surfaces and interfaces. Appreciating the interaction of molecules and atoms with the surfaces and each other while on the surface is crucial to understand desirable chemical reactions or process, including those involved in heterogeneous catalysis, and undesirable ones, such as those involved in corrosion chemistry. Surface chemistry overlaps with electrochemistry. It is of particular importance to the field of heterogeneous catalysis (Imbihl et al., 1982; Somorjai, 1994; Woodruff & Delchar, 1994).

Surface chemistry is crucial to understand the adhesion of gas or liquid molecules to the surface, known as adsorption. This can happen either due to chemisorption or physisorption, which are both included in surface chemistry. The behavior of a solution-based interface is affected by the surface charge, dipoles, energies, and their distribution within the electrical double layer.

4.3 Surface physics

Surface physics is the study of physical changes that occur at interfaces. Surface physics overlaps with surface chemistry. It investigates several fields including surface diffusion, surface reconstruction, surface phonons and plasmons, epitaxy and surface-enhanced Raman scattering, emission and tunneling of electrons, spintronics, and self-assembly of nanostructures on surfaces (Ibach, 2006).

There are several techniques to study the processes at surfaces. The examples of such techniques include scanning probe microscopy, X-ray photoelectron spectroscopy (XPS), surface X-ray scattering, and surface-enhanced Raman spectroscopy.

4.4 Surface in geometry

In geometry, surface is defined as a two-dimensional collection of points (flat surface), a three-dimensional collection of points whose cross section is a curve (curved surface), or the boundary of any three-dimensional solid. Generally, a surface is a continuous boundary that divides a three-dimensional space into two regions.

4.5 Difference between surface and interface

A surface is the shell of a macroscopic object (e.g., the inside of the material) in contact with its environment (the outside environment) (Fig. 1.14). An interface is the boundary between two phases. The surface of a material determines some of its chemical and physical properties including its optical appearance, stickiness, chemical reactivity, wetting behavior, and frictional behavior, among others.

4.5.1 Morphology (structure) of a surface

The structure or morphology of a surface is a macroscopic or ensemble property that defines its form and shape. The structure of a surface is given by the atomic and molecular composition and arrangement of the atoms in space. The topography of a surface is its profile determined by valleys, planes, and hills (Figs. 1.15 and 1.16). Several techniques and instruments, such as atomic force microscopy (AFM) and scanning electron microscopy (SEM), can be used to characterize the materials surface.

An interface is the separating layer between two condensed phases (usually of molecular dimensions). While adhesion can be defined as the attractive interactions between two different media, cohesion is defined as the attractive interactions within a phase (solid or liquid) (Fig. 1.17). At the border of a solid or liquid in contact with liquid or vapor, usually, there is no abrupt change in density, but a more or less continuous transition from high to low density. The interface either comprises evaporating bulk material or condensing material from the gas phase (Jonas and Vamvakaki, 2010; Hendon et al., 2016).

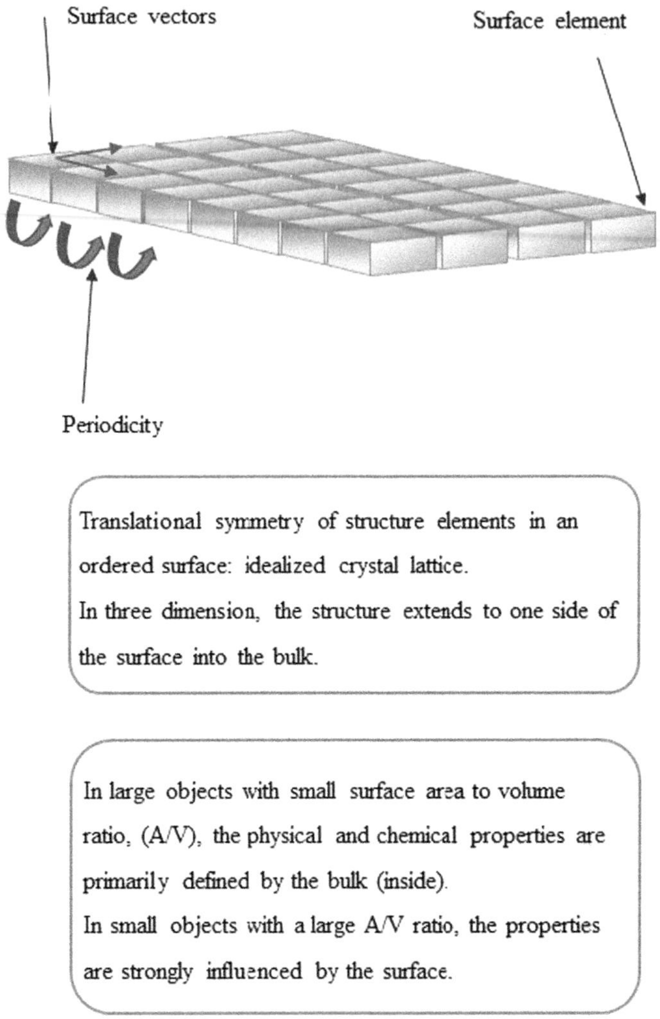
Surface vectors
Surface element
Periodicity
Translational symmetry of structure elements in an ordered surface: idealized crystal lattice.
In three dimension, the structure extends to one side of the surface into the bulk.
In large objects with small surface area to volume ratio, (A/V), the physical and chemical properties are primarily defined by the bulk (inside).
In small objects with a large A/V ratio, the properties are strongly influenced by the surface.

FIGURE 1.14 Schematic representation of a surface.

4.6 Examples of surface and interface

There are different surface and interface scenarios including solid–liquid interface and solid–vapor interface.

4.6.1 Scenario I

The first scenario covered is the solid and liquid interface, where a liquid dissolves surface atoms, which leads to surface charges (Fig. 1.18).

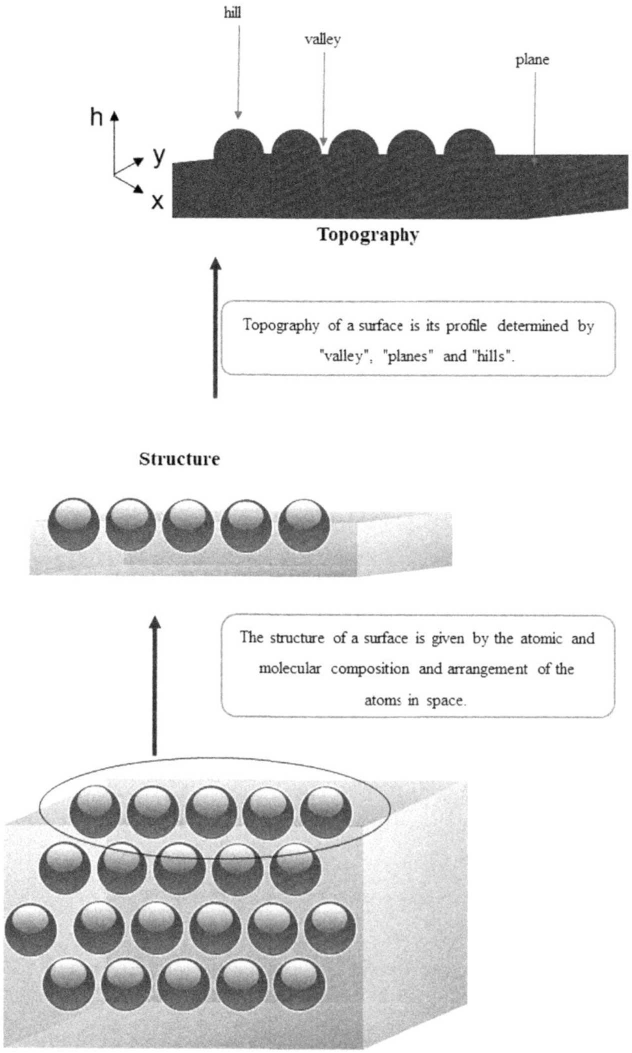
hill
valley
plane
h
y
x
Topography
Topography of a surface is its profile determined by "valley", "planes" and "hills".
Structure
The structure of a surface is given by the atomic and molecular composition and arrangement of the atoms in space.

FIGURE 1.15 An illustration depicting the difference between a surface and topography.

Furthermore, liquid molecules at the interface can be of a much higher order than those in the bulk or away from the interface.

4.6.2 Scenario II

The second scenario covered is a solid interacting with another solid. Here, when two crystalline solids are in atomic contact, the different lattice constants

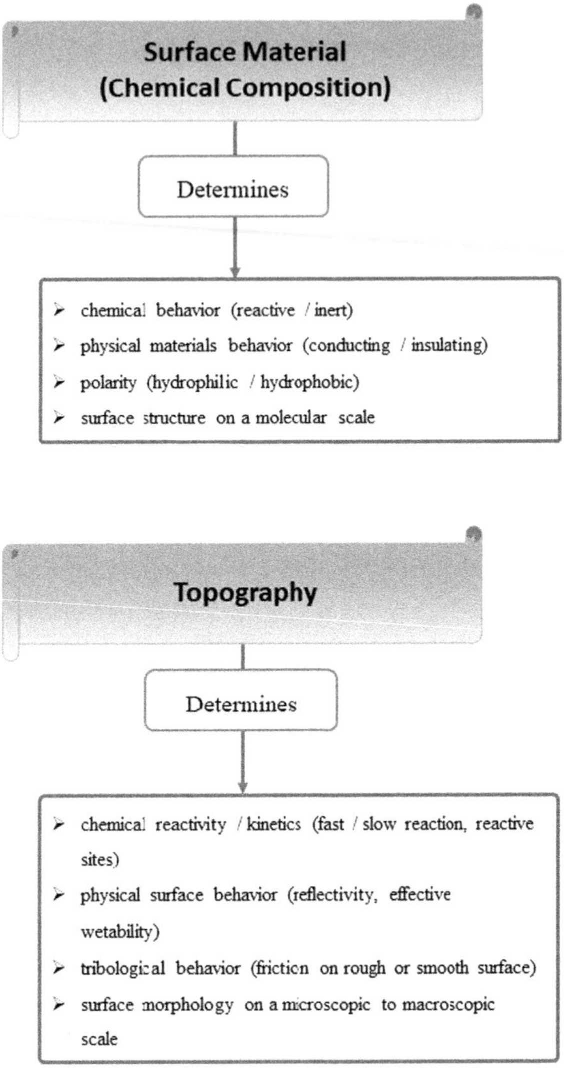
Surface Material
(Chemical Composition)
Determines
➢ chemical behavior (reactive / inert)
➢ physical materials behavior (conducting / insulating)
➢ polarity (hydrophilic / hydrophobic)
➢ surface structure on a molecular scale
Topography
Determines
➢ chemical reactivity / kinetics (fast / slow reaction, reactive sites)
➢ physical surface behavior (reflectivity, effective wetability)
➢ tribological behavior (friction on rough or smooth surface)
➢ surface morphology on a microscopic to macroscopic scale

FIGURE 1.16 Comparison between chemical composition and topography.

may generate strain at the interface (Fig. 1.19). If both materials react together, a new compound will be formed in contact region or interface, as interdiffusion may occur at higher temperatures.

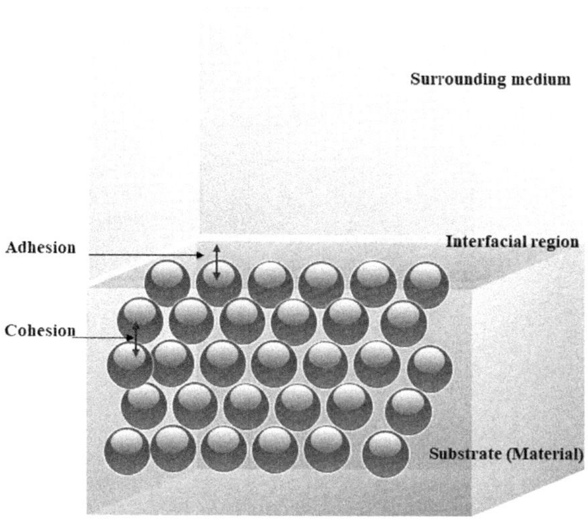
Surrounding medium
Adhesion
Interfacial region
Cohesion
Substrate (Material)

FIGURE 1.17 An illustration depicting the difference between an interface of a material and the surrounding.

4.6.3 Scenario III

The third scenario covered is a liquid interacting with another liquid (Fig. 1.20): both liquid phases are highly mobile, and the shape of interface is controlled by surface tension. Depending on solubility, the liquid molecules may migrate from one phase to the other through control by chemical potential or partition coefficient.

4.6.4 Scenario IV

The fourth scenario covered is the solid and vapor interface (Fig. 1.21). Usually, solids are highly immobile, and crystalline solids are highly ordered and structured. Generally, there is no evaporation of surface atoms and molecules, only lateral diffusion, which highly depends on the temperature. However, depending on the type of gas or vapor, there might be some interactions in long-term processes.

4.6.5 Scenario V

The fifth scenario covered is the liquid and gas interface (Fig. 1.22), where the liquid phase is highly mobile and disordered. Constant evaporation and recondensation at interface surface depends on the nature of the liquid and gas and on parameters such as temperature and interferences.

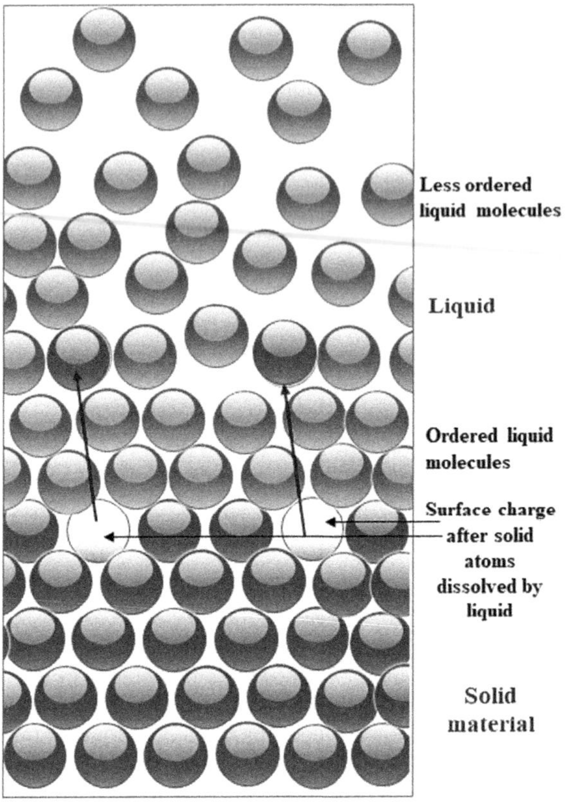
Less ordered
liquid molecules
Liquid
Ordered liquid
molecules
Surface charge
after solid
atoms
dissolved by
liquid
Solid
material

FIGURE 1.18 An illustration of solid–liquid surface and interface scenario.

5. Application of surface science

Surface science is important in many critical chemical processes because owing to unbalanced forces, otherwise known as residual forces, the atoms at the surface display exceptional properties that are different from those in the bulk. While atoms in the bulk are surrounded in all directions, thus exhibiting balanced forces of attraction, the forces of attraction at the surface are unbalanced, as they are attracted from the inside. The residual forces, or attraction forces acting outside, are not balanced, and they provide unique properties to the surface (Fig. 1.23). Surface science involves the interaction of surfaces of one material with the particles of another phase. Surface science plays an active role in various fields including adsorption, catalysis, colloids, electrodes, and separation.

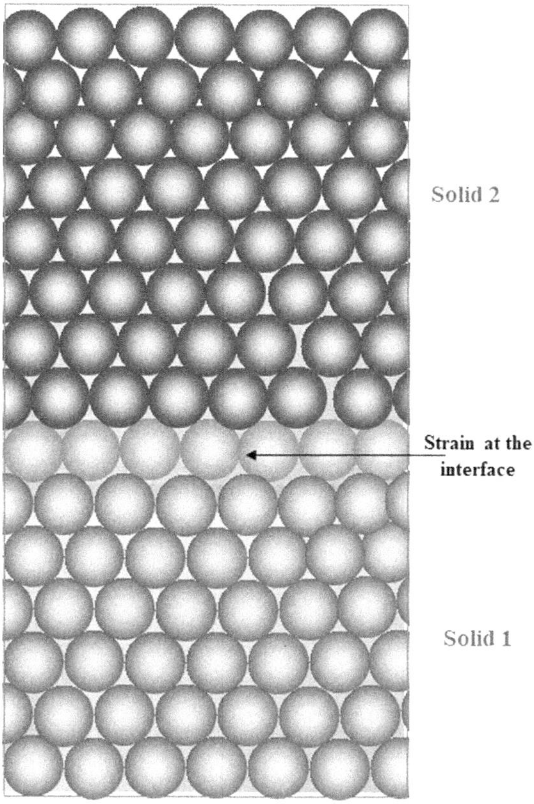
Solid 2
Strain at the interface
Solid 1

FIGURE 1.19 An illustration of solid–solid surface and interface scenario.

5.1 Adsorption

Adsorption refers to the adhesion of atoms, ions, or molecules from a gas, liquid, or dissolved solid to the surface of a material. It is considered a surface phenomenon that creates a film of the adsorbate on the surface of the adsorbent (solid material, for instance). However, adsorption differs from absorption, as the latter is the process where a fluid is dissolved by a liquid or solid (absorbent).

Adsorption can either be due to chemisorption or physisorption, and the strength of molecular adsorption to a material surface is critically important to the catalyst's performance. Surface science allows one to understand the mechanisms of adsorption and the manner to control its equilibrium and reversibility.

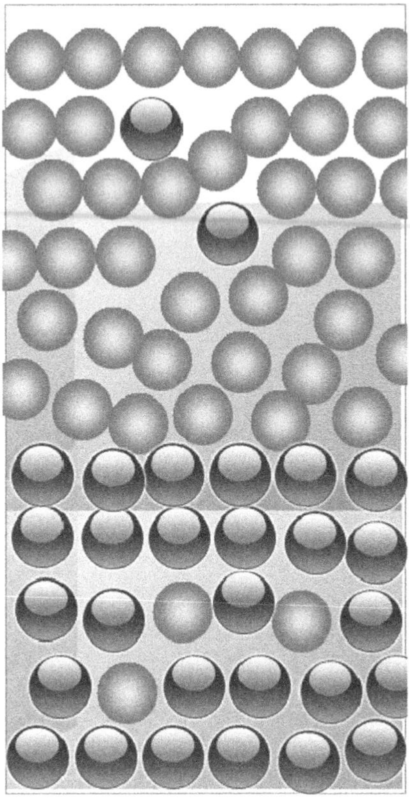
Liquid 2
Interface is controlled by surface tension
Liquid 1

FIGURE 1.20 An illustration of liquid–liquid surface and interface scenario.

5.2 Colloid

Colloid is a mixture in which one substance of microscopically dispersed insoluble particles is suspended throughout another substance. Colloid science deals with systems consisting of large molecules or small particles. The dispersed substance alone is called a colloid, while the overall mixture is called a colloidal suspension although a narrower sense of the word suspension is distinguished from colloids by larger particle size. Unlike a solution, whose solute and solvent constitute one single phase, a colloid has a dispersed phase (the suspended particles) and a continuous phase (the medium of suspension). To qualify as a colloid, the mixture must be one that does not settle or takes a considerably long time to settle. Surface science allows one to understand the mechanisms of colloid formation and the way to control its stability.

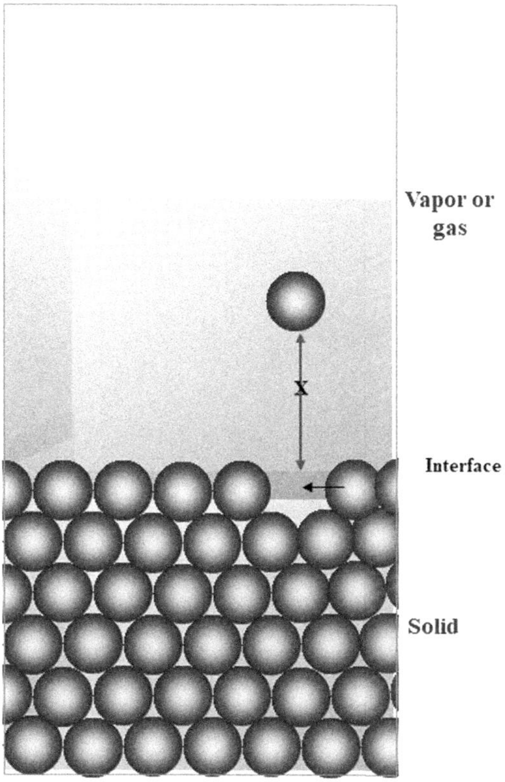
Vapor or
gas
X
Interface
Solid

FIGURE 1.21 An illustration of solid–liquid surface and interface scenario.

5.3 Emulsion

Surface science is crucial to understand the formation of emulsion, which is a mixture of two or more liquids that are normally immiscible (unmixable or unblendable). Although the terms colloid and emulsion are sometimes used interchangeably, emulsion should be used when both phases, dispersed and continuous, are liquids. In an emulsion, one liquid (the dispersed phase) is dispersed in the other (the continuous phase). Some examples of emulsions include vinaigrettes, homogenized milk, cutting fluids for metal working, petroleum emulsions, water-in-oil emulsion, and oil-in-water-in-oil.

Both colloidal and surface chemistry have applications and uses in relation to water, oil, pharmaceutical and chemical industries, nanotechnology, minerals, ceramics, technology and study of the behavior, control and manipulation of fluids, and microfluidics. Although research on surface chemistry of nanobiomaterials has been significant in recent years, there hasn't been any scientometric study on the topic.

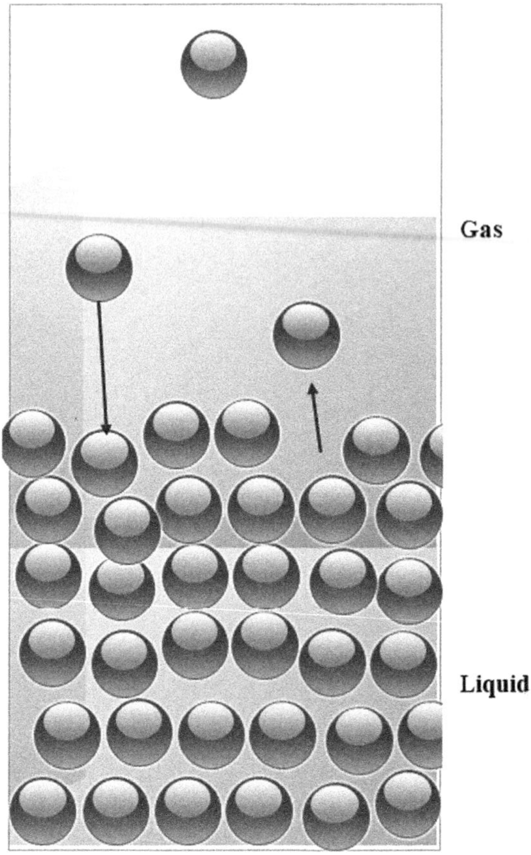
Gas
Liquid

FIGURE 1.22 An illustration of solid–liquid surface and interface scenario.

5.4 Other applications

There are several other applications of surface science in various fields (Bauer, 2014). Surface science allows one to understand, investigate, and develop more theories in many areas of real applications (Aballe et al., 2004; Tromp, 1993; Ibach, 2006; Bauer, 1990). The following are some of the other selected fields that apply surface science:

≻ Catalysis: adsorption, the adhesion of gas or liquid molecules to the surface, can be due to chemisorption or physisorption, and the strength of molecular adsorption to a catalyst surface is significant to the catalyst's performance
≻ Design of catalysts (industrial production of NH_3, e.g., for car exhaust)
≻ Inhibition of corrosion, for example, in ships, cars, buildings, or pipelines

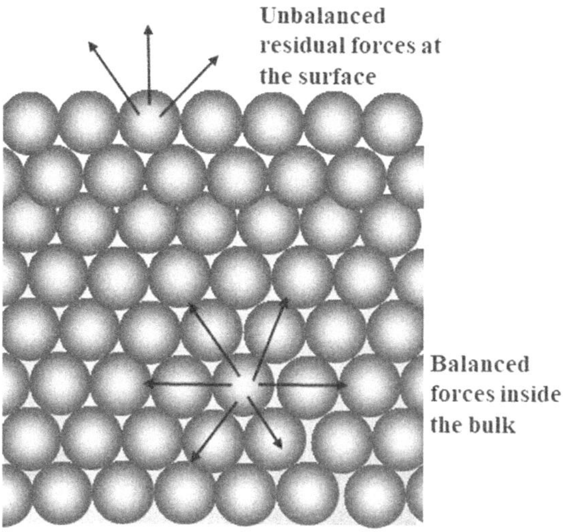
Unbalanced
residual forces at
the surface
Balanced
forces inside
the bulk

FIGURE 1.23 An illustration of the forces at the bulk and surface of a material.

➢ Modification of surface properties, which is very important in the design of the products listed below:
  - Friction: includes design of tires, bearings, and accessories for many equipment
  - Wear: for example, polymer lenses of glasses
  - Stickiness: for example, frying pan and adhesive tape
  - Wetting, condensation: for example, scuba diving goggles, outdoor gear, and inkjet printing
  - Antireflection: for example, picture frames and displays
  - Color such as paint, coating, and surface enhancements
➢ Microfluidics
➢ Sensors including chemical, physical, and biological
➢ Chip manufacturing/microelectronics
➢ Hard disks that are antifriction and ultrasmooth
➢ Biological surfaces such as biocompatibility and patterned cell growth

6. Theories of surface science

Theoretical surface science aims to understand the fundamental concepts that govern the geometric and electronic structure of surfaces and processes occurring on these surfaces such as gas–surface scattering, reactions at surfaces, and growth of surface layers. Some chemical reactions occur on catalyst surfaces, especially those occurring in the chemical industry. A catalyst, a car exhaust catalyst, for example, is used to increase the output of a chemical reaction and convert hazardous waste into less harmful products.

Theory and experiment equally contribute to the scientific progress of surface science. Particularly, computational surface science may act as a virtual chemistry and physics lab at surfaces. Computational experiments may add relevant information to the research process.

Theoretical concepts, microscopic approach, and computational tools could be necessary to understand and describe surface science. Based on the fundamental theoretical entity, the Hamiltonian, a hierarchy of theoretical methods, can be considered to describe surface processes through statistical and thermodynamic approaches (Gros et al., 1999; Groß, 2007, 2009).

6.1 The Hamiltonian

The theoretical description starts with the definition of the system under consideration and a determination of the fundamental interactions present in the system. Such information is contained in the Hamiltonian, which is the central quantity for any theoretical treatment. All physical and chemical properties of any system can be derived from its Hamiltonian. With regard to electrons, atoms, and other microscopic particles in surface science, laws of quantum mechanics, which require the solution of Schrodinger equation, provide proper descriptions (Groβ, 2009).

The fundamental particles in solid-state physics and chemistry are the nuclei and electrons that interact with each other through electrostatic forces. By considering valence electrons alone, relativistic effects are usually neglected, and by treating core and valence electrons on the same footing, any magnetic effects are neglected. By altering this, a system of nuclei and electrons is described by the nonrelativistic Schrodinger equation with a Hamiltonian. Simply put, when relativistic and magnetic effects are neglected, the Hamiltonian describes a system of nuclei and electrons in the following manner (Groβ, 2007):

$$H = T_{\mathrm{nucl}} + T_{\mathrm{el}} + V_{\mathrm{nucl-nucl}} + V_{\mathrm{nucl-el}} + V_{\mathrm{el-ek}}$$

where T_{nucl} and T_{el} are the kinetic energies of the nuclei and electrons. The other terms describe the electrostatic interaction between the positively charged nuclei and electronics. More information can be obtained at Groβ (2009).

Research on surface science focuses on electronic structure calculations and electronic and geometric structure of surfaces to further understand the interaction between atoms and molecules and surfaces. This allows more development of new materials with required surfaces toward advanced applications.

References

Aballe, L., Barinov, A., Locatelli, A., Heun, S., Kiskinova, M., 2004. Tuning surface reactivity via electron quantum confinement. Phys. Rev. Lett. 93, 196103, 4 pages.

Armstrong, J., 2012. General, Organic, and Biochemistry: An Applied Approach. Brooks/Cole, ISBN 978-0-534-49349-3, p. 48.

Bauer, E., 1990. Low energy electron microscopy. In: Vanselow, R., Howe, R. (Eds.), Chemistry and Physics of Solid Surfaces VIII. Springer, Berlin, pp. 267–288.

Bauer, E., 2014. Applications in surface science. In: Surface Microscopy with Low Energy Electrons. Springer, New York, NY.

Brown, T.L., LeMay Jr., Eugene, H., Bursten, B.E., Murphey, C.J., Woodward, P.M., Stoltzfus, M.W., Lufaso, M.W., 2018. Introduction: matter, energy, and measurement. In: Chemistry: The Central Science, fourteenth ed. Pearson, New York, ISBN 9780134414232, pp. 46–85.

Buzea, C., Pacheco, I., Robbie, K., 2007. Nanomaterials and nanoparticles: sources and toxicity. Biointerphases 2 (4), MR17–MR71. https://doi.org/10.1116/1.2815690 arXiv:0801.3280.

Carsten, R., 2001. Chemical Sciences in the 20th Century: Bridging Boundaries. Wiley-VCH, ISBN 3-527-30271-9, pp. 1–2.

Christmann, K., Schober, O., Ertl, G., Neumann, M., June 1, 1974. Adsorption of hydrogen on nickel single crystal surfaces. J. Chem. Phys. 60 (11), 4528–4540.

Damasceno, P.F., Engel, M., Glotzer, S.C., 2012. Crystalline assemblies and densest packings of a family of truncated tetrahedra and the role of directional entropic forces. ACS Nano 6, 609–614.

Demtröder, W., 2002. Atoms, Molecules and Photons: An Introduction to Atomic- Molecular- and Quantum Physics, first ed. Springer, ISBN 978-3-540-20631-6, pp. 39–42.

Feynman, R.P., 1992. There's plenty of room at the bottom. J. Microelectromech. Syst. 1 (1), 60–66.

Gleiter, H., 2000. Nanostructured materials: basic concepts and microstructure. Acta Mater. 48, 1–29.

Gros, A., Scheffler, M., Mehl, M.J., Papaconstantopoulos, D.A., 1999. Ab initio based tight-binding Hamiltonian for the dissociation of molecules at surfaces. Phys. Rev. Lett. 82, 1209.

Groß, A., 2007. Introduction to theoretical surface science. In: Experiment, Modeling and Simulation of Gas-Surface Interactions for Reactive Flows in Hypersonic Flights, pp. 3.1–3.22. Educational Notes RTO-EN-AVT-142, Paper 3. Neuilly-sur-Seine, France: RTO.

Groß, A., 2009. Theoretical Surface Science A Microscopic Perspective. Springer, ISBN 978-3-540-68966-9.

Hendon, C.H., Hunt, S.T., Milina, M., Butler, K.T., Walsh, A., Roman-Leshkov, Y., 2016. Realistic surface descriptions of heterometallic interfaces: the case of TiWC coated in noble metals. J. Phys. Chem. Lett. 7, 4475.

Ibach, H., 2006. Physics of Surfaces and Interfaces. Springer, Berlin, ISBN 978-3540347095, p. 171.

Imbihl, R., Behm, R.J., Christmann, K., Ertl, G., Matsushima, T., 1982. Phase transitions of a two-dimensional chemisorbed system-H on Fe(110). Surf. Sci. 117, 257–266.

Jeevanandam, J., Barhoum, A., Chan, Y.S., Dufresne, A., Danquah, M.K., 2018. Review on nanoparticles and nanostructured materials: history, sources, toxicity and regulations. Beilstein J. Nanotechnol. 9, 1050–1074. https://doi.org/10.3762/bjnano.9.98.

Jonas, U., Vamvakaki, M., 2010. From fluidic self-assembly to hierarchical structures—superhydrophobic flexible interfaces. Angew. Chem. Int. Ed. 49 (27), 4542–4543.
Kolasinski, K.W., 2002. Surface Science: Foundations of Catalysis and Nanoscience. Wiley, Chichester, ISBN 0471492450.
Lewis, G.N., 1916. The atom and the molecule. J. Am. Chem. Soc. 38 (4), 762–786. https://doi.org/10.1021/ja02261a002.
Luklema, J., 2005. Fundamentals of Interface and Colloid Science, pp. 1–5.
Mulvaney, P., 2015. Nanoscience vs nanotechnology—defining the field. ACS Nano 9 (3), 2215–2217.
Pereiro, M., Baldomir, D., 2007. Structure and static response of small silver clusters to an external electric field. Phys. Rev. 75 (3), 033202.
Pereiro, M., Baldomir, D., Arias, J.E., 2007. Unexpected magnetism of small silver clusters. Phys. Rev. 75 (6), 063204.
Pokropivny, V.V., Skorokhod, V.V., 2007. Classification of nanostructures by dimensionality and concept of surface forms engineering in nanomaterial science. Mater. Sci. Eng. C 27, 990–993.
Poole Jr., C.P., Owens, F.J., 2005. Introduction to Nanotechnology, first ed., ISBN 978-0471079354.
Prutton, M., 1994. Introduction to Surface Physics. Oxford University Press, ISBN 978-0-19-853476-1.
Pullman, B., 1998. The Atom in the History of Human Thought. Oxford University Press, Oxford, England, ISBN 978-0-19-515040-7, pp. 31–33.
Schwirn, K., Tietjen, L., Beer, I., 2014. Why are nanomaterials different and how can they be appropriately regulated under REACH? Environ. Sci. Eur. 26, 4. https://doi.org/10.1186/2190-4715-26-4.
Somorjai, G.A., 1994. Introduction to Surface Chemistry and Catalysis. Wiley, New York, ISBN 0471031925.
Stern, D.P., 2005. The Atomic Nucleus and Bohr's Early Model of the Atom. NASA/Goddard Space Flight Center. Archived from the original on 20 August 2007.
Taniguchi, N., 1974. On the basic concept of 'nano-technology'. In: Proceedings ICPE; Tokyo, Part II, Japan Society of Precision Engineering, "Nano-technology" mainly consists of the processing of separation, consolidation, and deformation of materials by one atom or one molecule.
Tromp, R.M., 1993. Surface stress and interface formation. Phys. Rev. B 47, 7125–7127.
Weiss, P.S., 2012. New tools lead to new science. ACS Nano 6, 1877–1879.
Weiss, P.S., 2014. Mesoscale science: lessons from and opportunities for nanoscience. ACS Nano 8, 11025–11026.
Woodruff, D.P., Delchar, T.A., 1994. Modern techniques of surface science. In: Cambridge Solid State Science Series, second ed. Cambridge University Press, Cambridge, UK, ISBN 0521424984.

Chapter 2

Adsorption technology and surface science

1. Introduction

Adsorption is defined as a process in which adsorbate (a component or components) from a gas or a liquid phase are attached to the surface of a solid phase (adsorbent) when the adsorbate is brought in contact with the solid phase (Fig. 2.1). Adsorption is an important aspect of many diverse processes in the chemical and process industries, including several chemical and biochemical reactions, purification and filtration, gas and liquid processing, and catalysis, to name a few (Slejko, 1985; Suzuki, 1990). Adsorption is a surface phenomenon and complex process. This necessitates careful planning and execution of its operations.

Adsorption technology is one of the most important technologies in areas such as catalysts, water purification, and surface modification and design. It is based on the accumulation of concentration at a surface and is the consequence of interactive forces of physical attractions between the porous solids surface and components molecules removed from the bulk phase. Thus, adsorption technology is considered a promising means to purify and separate gases and liquids.

Affinity with polar substances such as alcohols and water is determined by surface polarity. Polar adsorbents are considered hydrophilic. Silica gel, porous alumina, aluminosilicates, and zeolites are examples of this type of adsorbent (Serrano, 2007; Wu et al., 2021). Conversely, nonpolar adsorbents, such as carbonaceous adsorbents, polymer adsorbents, and silicalite, are generally hydrophobic and have more affinity with oil or hydrocarbons than water.

For a large adsorption capacity, a large specific surface area is preferred. However, when a significant internal surface area is created in a small volume, a huge number of tiny pores between the adsorption surfaces are necessarily created. The size of micropores determines the accessibility of adsorbate molecules to the internal adsorption surface. As a result, the pore size

Surface Science of Adsorbents and Nanoadsorbents. https://doi.org/10.1016/B978-0-12-849876-7.00006-3

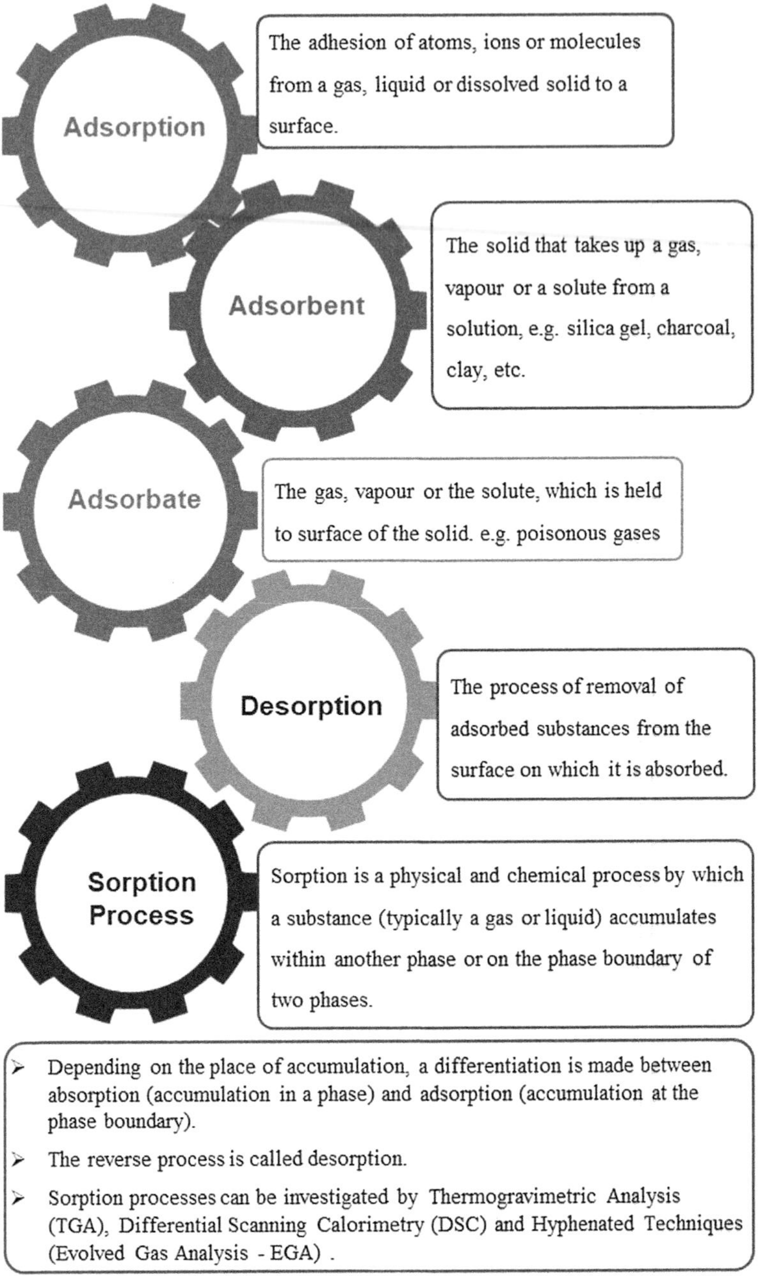
Adsorption
The adhesion of atoms, ions or molecules from a gas, liquid or dissolved solid to a surface.
Adsorbent
The solid that takes up a gas, vapour or a solute from a solution, e.g. silica gel, charcoal, clay, etc.
Adsorbate
The gas, vapour or the solute, which is held to surface of the solid. e.g. poisonous gases
Desorption
The process of removal of adsorbed substances from the surface on which it is absorbed.
Sorption Process
Sorption is a physical and chemical process by which a substance (typically a gas or liquid) accumulates within another phase or on the phase boundary of two phases.
➢ Depending on the place of accumulation, a differentiation is made between absorption (accumulation in a phase) and adsorption (accumulation at the phase boundary).
➢ The reverse process is called desorption.
➢ Sorption processes can be investigated by Thermogravimetric Analysis (TGA), Differential Scanning Calorimetry (DSC) and Hyphenated Techniques (Evolved Gas Analysis - EGA) .

FIGURE 2.1 Important terms in the adsorption process; adsorbent, adsorbate, desorption, and sorption.

distribution of micropores is another crucial feature for determining adsorbent adsorptivity. For certain uses, materials like zeolite and carbon molecular sieves can be precisely designed and tailored with precise pore size distributions (Fig. 2.2). Adsorption technology allows efficient production of such pure products from ore sources. This chapter discusses adsorption phenomena and some related terminologies.

2. Adsorption and absorption

Adsorption differs from absorption, where one substance is absorbed into the physical structures of a substance (absorbent) (Fig. 2.3). The absorbate is the material that is absorbed into another substance, while the absorbent is the substance that absorbs the absorbate. When an organic molecule enters a solid particle (such as a soil particle), the organic molecule becomes the absorbate, and the soil particle becomes the absorbent. The absorbent could be a gas, a liquid, or a solid, while the absorbate could be an atom, an ion, or any molecule (Fig. 2.4). Generally, absorbate and absorbent are present in two different phases although not always. As an example, the absorption property of chemicals is used on different occasions. Liquid–liquid extraction works on this premise. When two liquids exist in the same container, solutes can be removed from one to the other if the solute is more absorbed in one than the other. The absorbent must have a porous structure or adequate room to accommodate the absorbate in order to absorb. In addition, the absorbate molecule shall be of a size that allows it to fit inside the structure of the absorbent. Additionally, the absorption process is aided by the attraction forces between two components. Energy, like matter, is susceptible to absorption (into substances) (Table 2.1). Spectrophotometry is based on the absorption of light by atoms, molecules, and other species.

3. Adhesion

The tendency of dissimilar particles or surfaces to attach to one another is known as adhesion, whereas the tendency of similar or identical particles or surfaces to cling to one another is known as cohesion (Maeda et al., 2002; Popov et al., 2017). Many factors influence the strength of attachment between an adhesive and its substrate, including the method through which it happens and the surface area over which the two materials come into contact. When opposed to materials that do not wet each other, wet materials have a higher contact area.

The following mechanisms are proposed to explain adhesion:

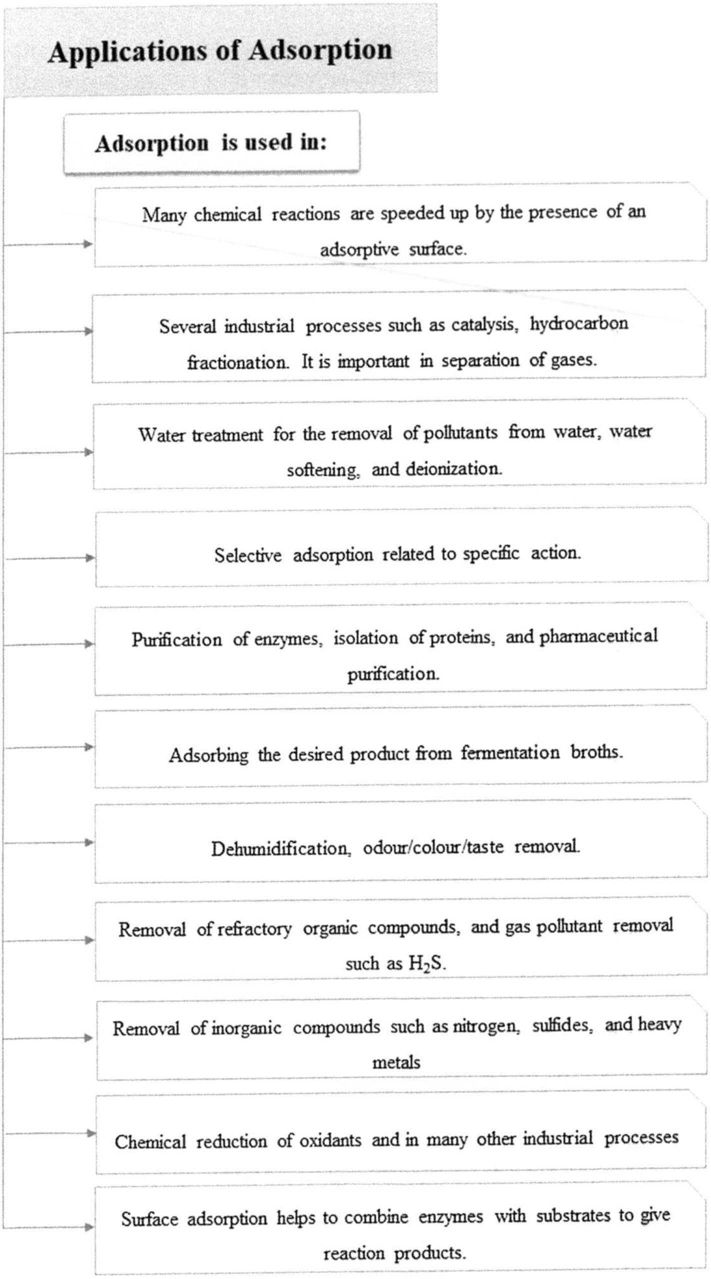
Applications of Adsorption
Adsorption is used in:
Many chemical reactions are speeded up by the presence of an adsorptive surface.
Several industrial processes such as catalysis, hydrocarbon fractionation. It is important in separation of gases.
Water treatment for the removal of pollutants from water, water softening, and deionization.
Selective adsorption related to specific action.
Purification of enzymes, isolation of proteins, and pharmaceutical purification.
Adsorbing the desired product from fermentation broths.
Dehumidification, odour/colour/taste removal.
Removal of refractory organic compounds, and gas pollutant removal such as H_2S.
Removal of inorganic compounds such as nitrogen, sulfides, and heavy metals
Chemical reduction of oxidants and in many other industrial processes
Surface adsorption helps to combine enzymes with substrates to give reaction products.

FIGURE 2.2 Examples of the applications and uses of adsorption process.

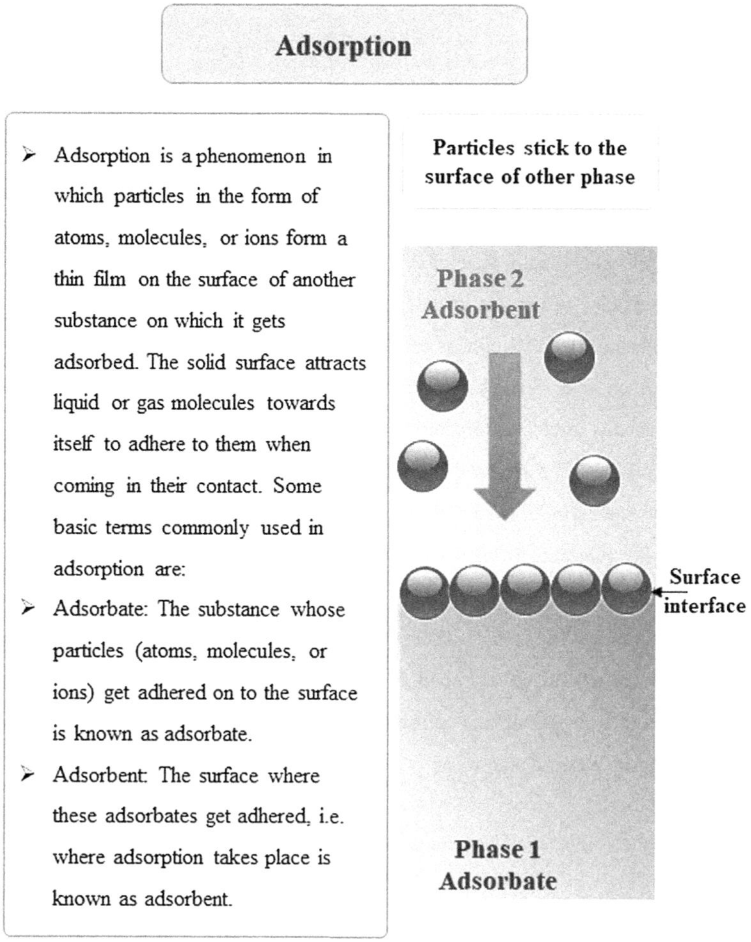
Adsorption
➢ Adsorption is a phenomenon in which particles in the form of atoms, molecules, or ions form a thin film on the surface of another substance on which it gets adsorbed. The solid surface attracts liquid or gas molecules towards itself to adhere to them when coming in their contact. Some basic terms commonly used in adsorption are:
➢ Adsorbate: The substance whose particles (atoms, molecules, or ions) get adhered on to the surface is known as adsorbate.
➢ Adsorbent: The surface where these adsorbates get adhered, i.e. where adsorption takes place is known as adsorbent.
Particles stick to the surface of other phase
Phase 2
Adsorbent
Surface interface
Phase 1
Adsorbate

FIGURE 2.3 An illustration of the adsorption with surface interface between adsorbent phase and adsorbate phase.

- Mechanical Adhesion: When an adhesive pushes its way into the microscopic pores of two materials, for example, mechanical interlocking occurs.
- Chemical Adhesion: At the joint, two materials may combine to form a composite.
- Dispersive Adhesion: Two materials are kept together by "Van der Waals forces," which is also known as adsorption. Electron motions or displacements inside the molecules cause these weak but frequent interactions between molecules of the materials.

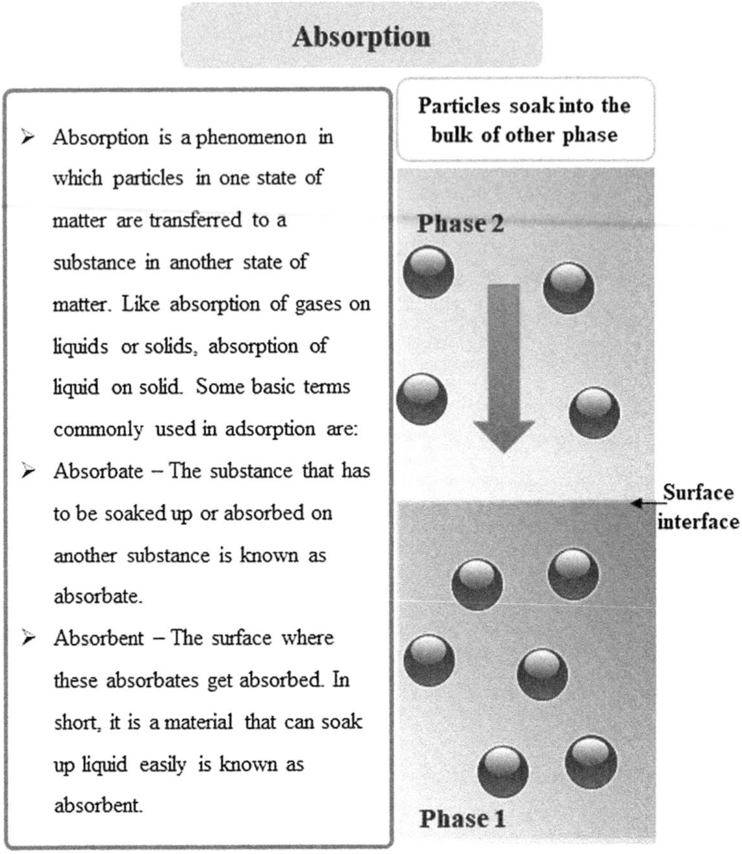
Absorption
➢ Absorption is a phenomenon in which particles in one state of matter are transferred to a substance in another state of matter. Like absorption of gases on liquids or solids, absorption of liquid on solid. Some basic terms commonly used in adsorption are:
➢ Absorbate – The substance that has to be soaked up or absorbed on another substance is known as absorbate.
➢ Absorbent – The surface where these absorbates get absorbed. In short, it is a material that can soak up liquid easily is known as absorbent.
Particles soak into the bulk of other phase
Phase 2
Surface interface
Phase 1

FIGURE 2.4 An illustration of absorption.

- Electrostatic Adhesion: Some conducting materials may allow electrons to travel through, resulting in an electrical charge difference at the junction. This forms a capacitor-like structure, as well as an attractive electrostatic force between the materials.
- Diffusive Adhesion: Diffusion may cause some materials to mix at the joint. Diffusive adhesion occurs when both materials' molecules are mobile and soluble in each other.

4. Tribology

Tribology is the study of the interactions of surfaces in motion. It entails the study and application of friction, lubrication, and wear principles. Complex

TABLE 2.1 Comparison between adsorption and absorption.

Term	Adsorption	Absorption
Definition	Adsorption is a process where substances such as gas, liquids, or dissolved solids adhere to the surface of another material, which could be solid or liquid.	Absorption is defined as a process where any substance (ions and molecules) is taken or absorbed by another substance, particularly a liquid or solid substance. It takes place by the process of diffusion or osmosis.
Phenomenon location	It's a surface phenomenon in which molecules just adhere to the adsorbent's surface. For example, silica adsorbs water vapor.	It is a bulk phenomenon in which absorbate molecules enter the absorbent. For example, anhydrous calcium chloride absorbs water.
Principle	Substances are absorbed onto an adsorbent's surface because the adsorbent contains unoccupied spaces that encourage particle attachment to the gaps.	Because of the availability of space and the nature of the particle, substances are absorbed into an absorbent.
Rate of process	The rate of adsorption may increase slowly achieving equilibrium. The rate of adsorption is rapid at the initial stage.	The absorption occurs at a uniform rate. The rate of absorption remains constant or uniform throughout the process.
Bonding	Van der Wall's forces or covalent bonds keep adsorbed elements linked to the adsorbent.	The absorbed components do not interact chemically with the absorbent and remain in the absorbent.
Concentration change	The concentration varies on the bulk of the adsorbent. The surface of the adsorbent has a higher concentration of adsorbate than the rest of the adsorbent.	After absorption, the absorbate concentration in the absorbent is homogeneous. The concentration does not change throughout the medium. It remains throughout the process.

Continued

TABLE 2.1 Comparison between adsorption and absorption.—cont'd

Term	Adsorption	Absorption
Heat exchange	Adsorption is primarily an exothermic process in which the surface's energy diminishes, resulting in a drop in the surface's residual forces.	Absorption is primarily an endothermic process since energy is supplied from the outside of the surface, and the absorbent's overall energy increases as a result of absorption.
Temperature effect	The adsorption depends on temperature.	There is no effect on temperature.
Separation	Adsorbent materials can be separated by passing a new substance through the adsorbent's surface and replacing the previously adsorbed material.	Absorbed materials can be separated into different phases based on their chemical interaction with the phases.
Examples	The principle of adsorption is used in separation methods such as adsorption chromatography to separate mixtures. Air conditioning, water purification, synthetic resin, and coolers are other common examples of adsorption.	Cold storage, ice production, turbine inlet cooling, and chillers are common examples of absorption.

Sorption is a generic term that refers to adsorption or absorption processes.

tribological interactions influence any product where one material slides over or scrapes against another (Majmuder et al., 2007). Tribology is most commonly associated with mechanical bearings, but it also applies to goods like hip implants, hair conditioners, lipsticks, powders, and lip gloss.

The production of compacted oxide layer glazes has been seen to defend against wear in high-temperature sliding wear where traditional lubricants cannot be used. In the manufacturing industry, tribology is crucial. Friction increases tool wear and the amount of force required to work a piece in metal-forming activities. This leads to higher costs due to the need for more frequent tool replacement, a loss of tolerance when tool dimensions move, and the need for more force to shape a piece. Tool wear is almost eliminated by a layer of lubricant that eliminates surface contact and reduces the required power by one-third.

5. Adsorption terms

Adsorption is where a solute, substance, molecule, liquid, or gas (a substance in solution) binds (adheres) to the surface of another substance or to a solid material or liquid (adsorbent), to form a film of molecules or atoms (adsorbate) (Bruch et al., 2007; Rouquerol et al., 1999, 2007; Yang, 2003) (Fig. 2.5). Adsorption occurs in many natural chemical, physical, and biological systems. It's crucial in several kinds of catalysis, especially when gases adsorb on metal surfaces. As a result of the decreased activation, this response is made easier. It occurs as a result of attractive interactions between the adsorbent's surface and the species being adsorbed. All of the constituent atoms of an adsorbent's bonding requirements (ionic, covalent, or metallic bonding) are fulfilled by other atoms in the material's bulk. However, atoms on the adsorbent's surface

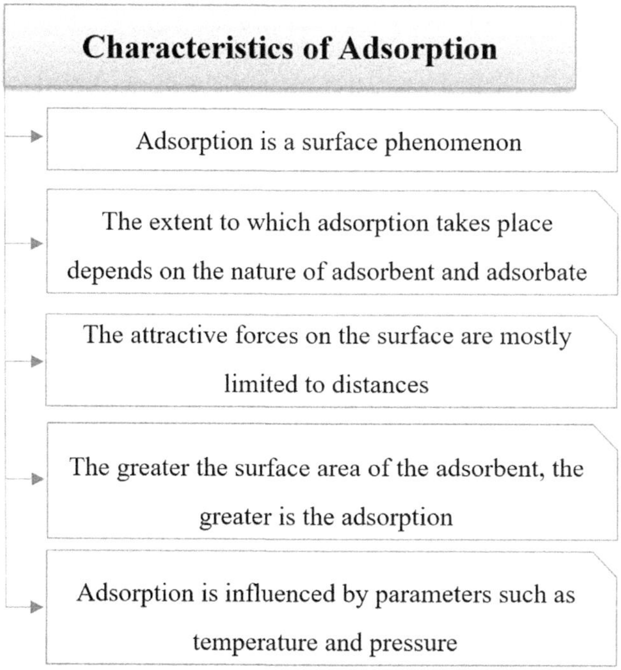
Characteristics of Adsorption
Adsorption is a surface phenomenon
The extent to which adsorption takes place depends on the nature of adsorbent and adsorbate
The attractive forces on the surface are mostly limited to distances
The greater the surface area of the adsorbent, the greater is the adsorption
Adsorption is influenced by parameters such as temperature and pressure

FIGURE 2.5 Some of the key characteristics of the adsorption process.

are not completely surrounded by other adsorbent atoms, and so can attract adsorbate molecules (Fig. 2.6). The particular nature of the bonding is determined by a variety of factors specific to the species involved (Saleh, 2015).

6. Classification of adsorption process

The adsorption process is commonly classified into physisorption (physical adsorption), chemisorption (chemical adsorption), and ion exchange adsorption.

Physical adsorption, also known as physisorption, happens when gas molecules accumulate onto the solid surface because of a weak Van der Waals force. Chemical adsorption, also known as chemisorption, takes place when gas atoms or molecules are bound to the surface of a solid by chemical

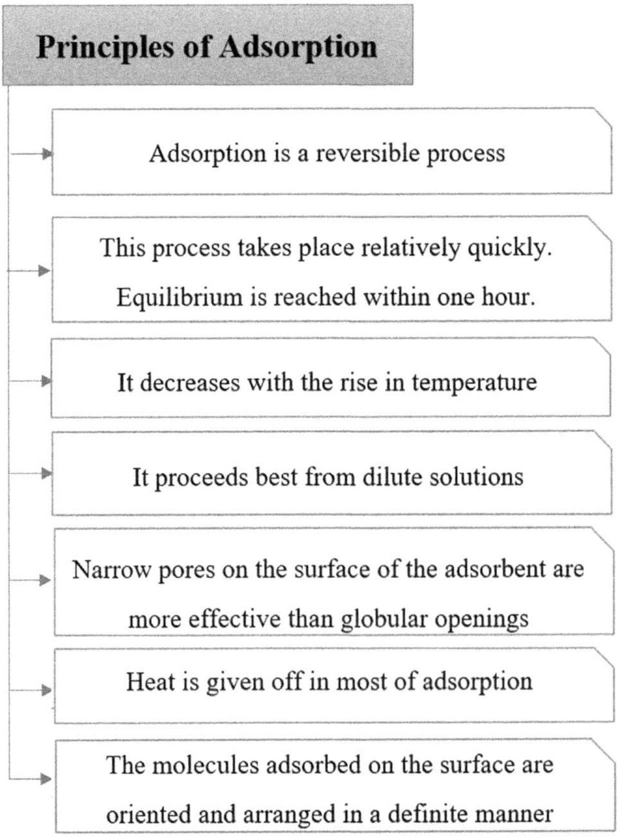
Principles of Adsorption
Adsorption is a reversible process
This process takes place relatively quickly.
Equilibrium is reached within one hour.
It decreases with the rise in temperature
It proceeds best from dilute solutions
Narrow pores on the surface of the adsorbent are
more effective than globular openings
Heat is given off in most of adsorption
The molecules adsorbed on the surface are
oriented and arranged in a definite manner

FIGURE 2.6 Critical principles of the adsorption process.

bonding. Chemical bonding can be either ionic or covalent. Chemisorption is also known as activated adsorption since it requires a high activation energy. It's difficult to tell the difference between these two processes when they happen at the same time (Saleh, 2016, 2018, 2021). When a physisorption occurs at a low temperature, it may transition to chemisorption as the temperature rises. Van der Waals force, for example, adsorbs dihydrogen on nickel first. Hydrogen atoms are formed when hydrogen molecules disintegrate, and chemisorption holds them on the surface (Fig. 2.7).

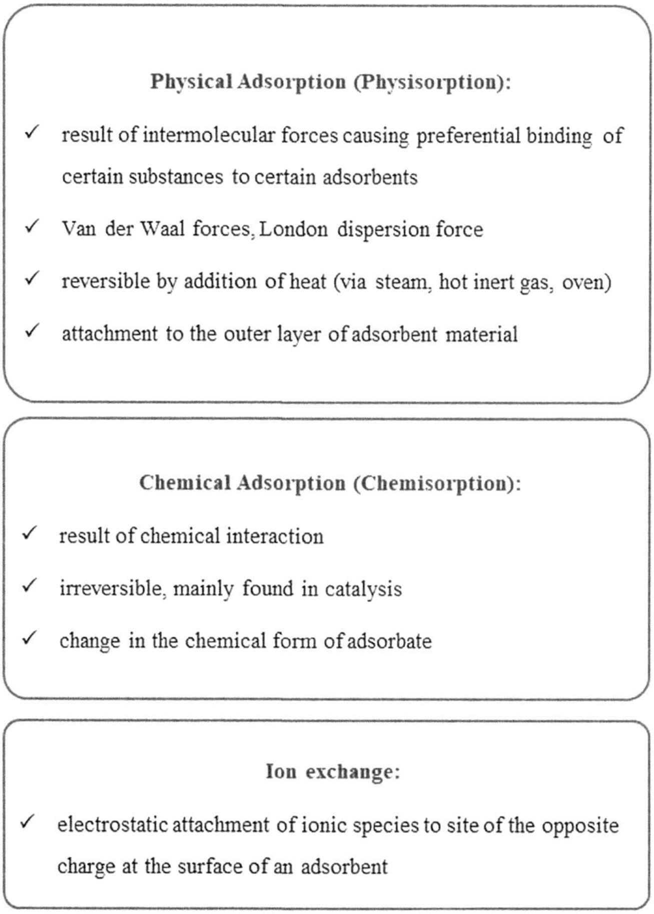
Physical Adsorption (Physisorption):
✓ result of intermolecular forces causing preferential binding of certain substances to certain adsorbents
✓ Van der Waal forces, London dispersion force
✓ reversible by addition of heat (via steam, hot inert gas, oven)
✓ attachment to the outer layer of adsorbent material
Chemical Adsorption (Chemisorption):
✓ result of chemical interaction
✓ irreversible, mainly found in catalysis
✓ change in the chemical form of adsorbate
Ion exchange:
✓ electrostatic attachment of ionic species to site of the opposite charge at the surface of an adsorbent

FIGURE 2.7 Definitions of the types of adsorption; physical adsorption (physisorption), chemical adsorption (chemisorption) and ion exchange.

6.1 Physisorption

Physisorption (physical adsorption) (Table 2.2), which is characterized by weak interactions, binds hydrogen as a result of Van der Waals force and is hydrophobic in nature (Fig. 2.8). Such an adsorption process usually decreases with an increase in temperature. The adsorbed molecules remain unbroken.

Characteristics of physisorption include the following:

➢ Lack of specificity: Because Van der Waals force is universal, an adsorbent's surface does not display any preference for a particular gas.

TABLE 2.2 Difference between physical and chemical adsorption.

Term	Physical adsorption	Chemical adsorption
Name	It is called physisorption.	It is called chemisorption.
Forces	The forces operating are weak Van der Waals forces.	The operating forces are chemical bonds, covalent or ionic.
Heat	The adsorption heat is low, approximately 20–40 $kJ\ mol^{-1}$.	The adsorption heat is high, approximately 40–400 $kJ\ mol^{-1}$.
Process direction	It is a reversible process; desorption occurs by an increase in temperature or decrease in pressure.	It is an irreversible process. Efforts to free the adsorbed gases provide various compounds.
Activation energy	No activation energy is necessary.	Activation energy is necessary.
Effect of temperature	It takes place at a low temperature and may decrease with increasing the temperature.	It may increase with increasing temperature.
Effect of pressure	It enhances with increasing the pressure.	It enhances with increasing the pressure.
Specific in nature	It is not specific.	It is specific since it takes place by the possible chemical bonds.
Effect of surface area	It increases with the increase in adsorbent surface area.	It increases with the increase in surface area of the adsorbent and of the active sites for forming bonds.
Layers	It forms multimolecular layers.	It forms a monomolecular layer.

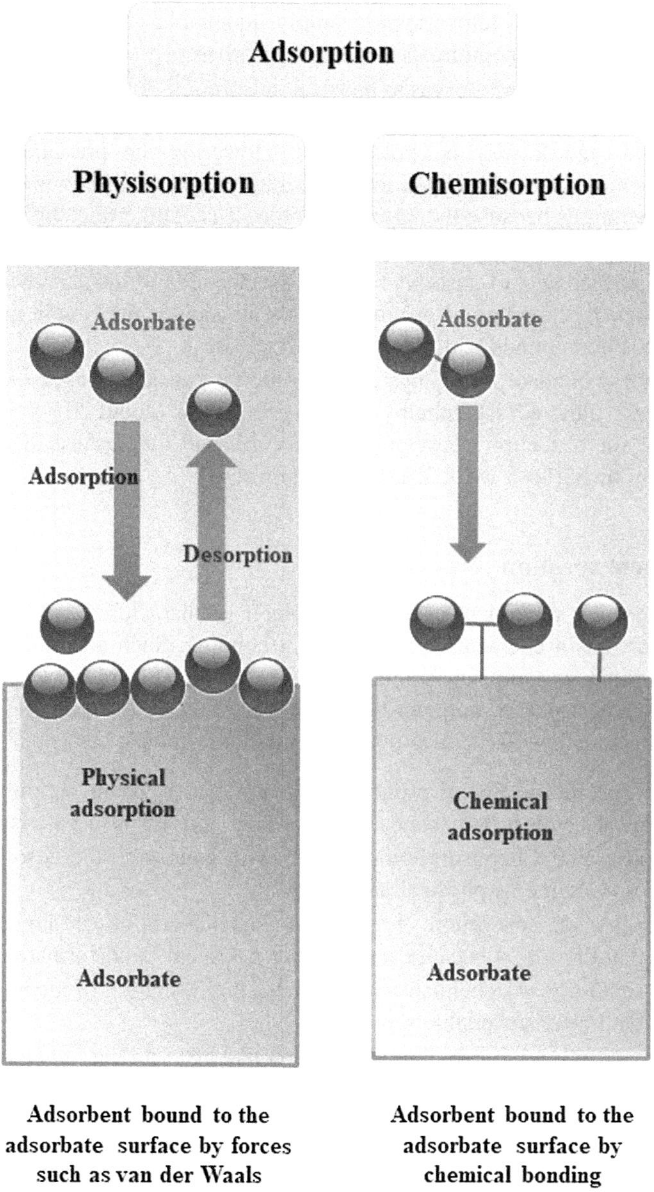
Adsorption
Physisorption
Chemisorption
Adsorbate
Adsorption
Desorption
Physical adsorption
Adsorbate
Adsorbate
Chemical adsorption
Adsorbate
Adsorbent bound to the adsorbate surface by forces such as van der Waals
Adsorbent bound to the adsorbate surface by chemical bonding

FIGURE 2.8 Illustration of the difference between physical adsorption (physisorption), and chemical adsorption (chemisorption).

➢ Nature of adsorbate: The quantity of gas absorbed by a material is measured by the gas's composition. Because the Van der Waals force is greater at critical temperatures, simply liquefiable gases (i.e., those with high critical temperature) are quickly adsorbed.

- Reversible nature: A gas's physical adsorption by a solid is usually reversible. As a result, as the volume of the gas lowers, more gas is adsorbed, and it may be evacuated by lowering the pressure. Physical adsorption occurs rapidly at low temperatures and declines with increasing temperatures because the adsorption process is exothermic (Le-Chatelier's principle).
- The surface area of adsorbent: As the surface area of the adsorbent grows, so does the extent of adsorption. Adsorbents include finely split metals and porous compounds with enormous surface areas.
- Enthalpy of adsorption: Physical adsorption is considered as an exothermic process, although its enthalpy of adsorption is low (about 20–40 kJ mol^{-1}) since the attraction between gas molecules and the surface of a solid is solely owing to a weak Van der Waals force.

6.2 Chemisorption

Chemisorption (chemical adsorption), which is characterized by strong interactions, has a characteristic of covalent bonding. Such adsorption process usually increases with an increase in temperature. The molecules that have been adsorbed may or may not be broken up.

Characteristics of chemisorption include the following:

- High specificity: Chemisorption is highly selective, occurring only when a chemical connection between the adsorbent and the adsorbate is possible.
- Surface area: Chemisorption enhances with increasing the adsorbent surface area, same as physical adsorption.
- Enthalpy of adsorption: Enthalpy of chemisorption is high (around 80–240 kJ mol^{-1}) because it comprises chemical bond formation.
- Irreversibility: As chemisorption includes the formation of a compound, it is a naturally irreversible process.
- Chemisorption is an exothermic reaction that takes a long time to complete at low temperatures because of the large activation energy. Adsorption, like other chemical changes, often increases as temperature rises.
- Physisorption of a gas adsorbed at low temperatures can transition to chemisorption at high temperatures. Chemisorption is typically aided by a high pressure environment.

6.3 Ion exchange

Ion exchange comprises an exchange of ions (adsorbates) between a liquid and solid phase (adsorbent). It is a chemical process that removes dissolved ionic contaminants from the water. Such an adsorption process usually increases with an increase in temperature.

Adsorption is a surface effect where a solid holds molecules of a fluid as a thin film on its surface. Whereas ion exchange involves an interchange or substitution of ions throughout a pair of fluids across a boundary. Ion exchange resins can be used to soften the water by substituting Ca^{2+} ion with Na^{+} or H^{+} ion and also to remove heavy metal ions from water.

- The following are the examples of equilibrium of the cation exchange processes:
  - Homovalent exchanges: The exchange of Cobalt(II) and Calcium(II) ions on calcium bentonite (Nagy et al., 2016).
  - Heterovalent exchanges: The exchange of Strontium(II) and Sodium(I) ions on sodium bentonite. The exchange of Cesium(I) and rare-earth elements (Pr(III), Nd(III), and Dy(III)) cations on Pr(III) bentonite, Nd(III) bentonite, and Dy(III) bentonite produced from calcium bentonite).

During ion exchange, ions of positive (cations) or negative (anions) charge in the liquid solution replace dissimilar and displaceable ions of the same charge contained in the solid ion exchanger. The ion exchanger comprises insoluble, immobile, and permanently bound co-ions of the opposite charges. For example, the process of water softening using ion exchange (Fig. 2.9).

$$Ca^{2+}(aq) + 2NaR(s) \leftrightarrow CaR_2(s) + 2Na^{+}(aq)$$

Certain adsorbates are mostly selectively transported from the fluid phase to the surface of insoluble, hard particles suspended in a vessel or packed in a column in sorption processes such as ion exchange, adsorption, and chromatography.

7. Factors affecting degree of adsorption

The degree of adsorption is influenced by several factors, some of which are enumerated below.

7.1 Surface area

Although all of the forces operating between the particles inside the adsorbent are mutually balanced, the particles on the surface are not surrounded on all sides by atoms or molecules of their sort, resulting in unbalanced or residual attractive forces. The forces of the adsorbent attract the adsorbate particles to its surface. The extent of adsorption increases as the surface area per unit mass of the adsorbent increases at a certain temperature and pressure.

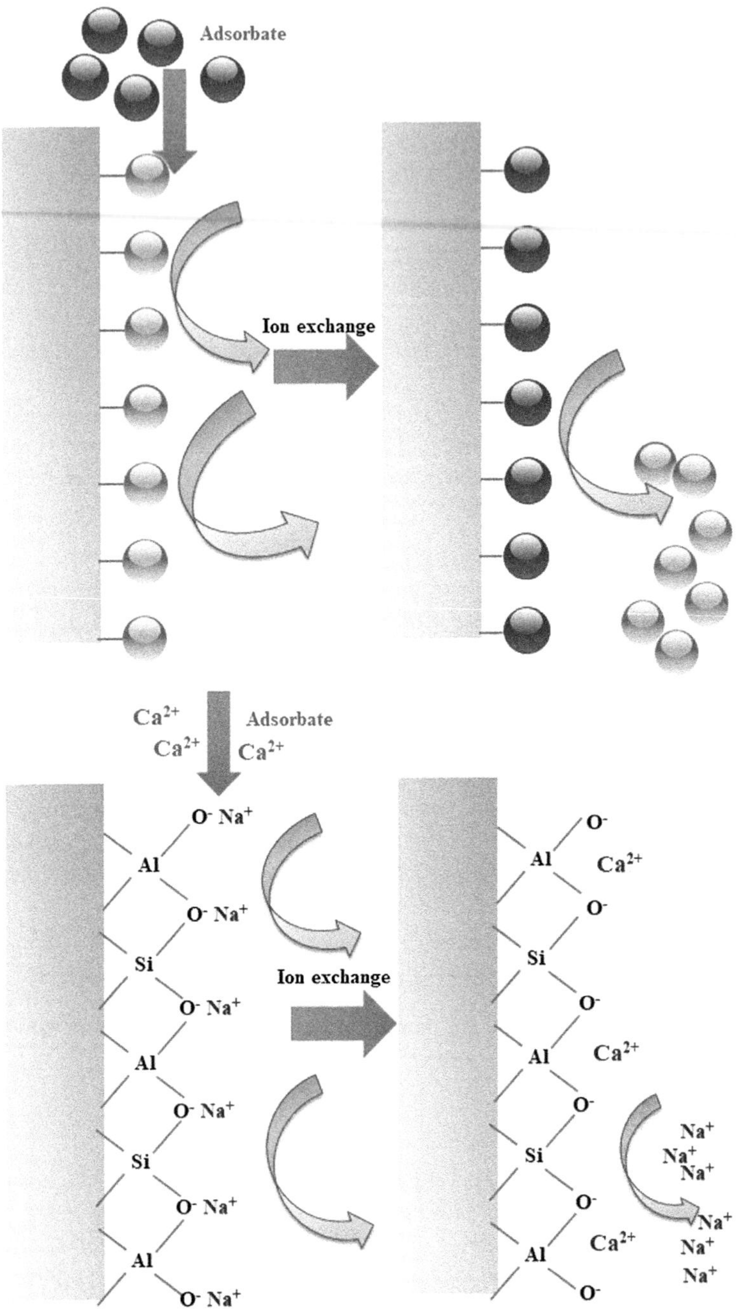
Adsorbate
Ion exchange
Ca2+
Ca2+
Adsorbate
Ca2+
O- Na+
Al
O- Na+
Si
O- Na+
Al
O- Na+
Si
O- Na+
Al
O- Na+
Ion exchange
O-
Al
Ca2+
O-
Si
O-
Al
Ca2+
O-
Si
O-
Al
Ca2+
O-
Na+
Na+
Na+
Na+
Na+
Na+

FIGURE 2.9 Ion exchange adsorption process.

7.2 Heat of adsorption

The heat of adsorption is an important factor when featuring adsorption. During adsorption, the residual forces of the surface decrease, i.e., the surface energy decreases, which appears as heat. Thus, adsorption is invariably an exothermic process. Hence, ΔH of adsorption is constantly in the negative. The molecules' freedom of movement is reduced when a gas is adsorbed. This relates to a decrease in the entropy of the gas as a result of adsorption, i.e., S is negative. Adsorption results in a decrease in the system's enthalpy and entropy. The thermodynamic condition for a process to be spontaneous is that it occurs at a constant temperature and pressure, ΔG is negative, i.e., Gibbs energy must decrease. Based on, $\Delta G = \Delta H - T\Delta S$, ΔG shall be mostly negative if ΔH is appropriately negative as $T\Delta S$ is positive. As a result, in a spontaneous adsorption process, the combination of these two elements causes G to be negative. Enthalpy becomes less negative as adsorption progresses, eventually approaching $T\Delta S$, and ΔG becomes zero. Thus, an equilibrium is achieved.

7.3 Solubility of adsorbate

The lesser the extent of adsorption, the greater the solubility and the stronger the solute–solvent connection.

7.4 Other factors

In brief, among others (Fig. 2.10), the following are some factors affecting the degree of adsorption:

- **(i)** Certain properties, including chemical structure and polarity, and nature, for instance, the physiochemical nature, which also includes surface functional groups, of the adsorbent and adsorbate.
- **(ii)** A mixture of adsorbates, where the compounds can increase adsorption, act independently, or might interfere with each another.
- **(iii)** Conditions including temperature and pressure: usually, adsorption is improved at decreasing temperature and increasing pressure.
- **(iv)** Generally, larger surface areas of adsorbents and high attractive force between the surface of the adsorbent and the adsorbate are preferred for the adsorption process.
- **(v)** The size and volume of the pore of the adsorbent.
- **(vi)** Other factors include temperature, pressure, and nature of the adsorbate and adsorbent.
- **(vii)** The presence of other solutes in the solution.

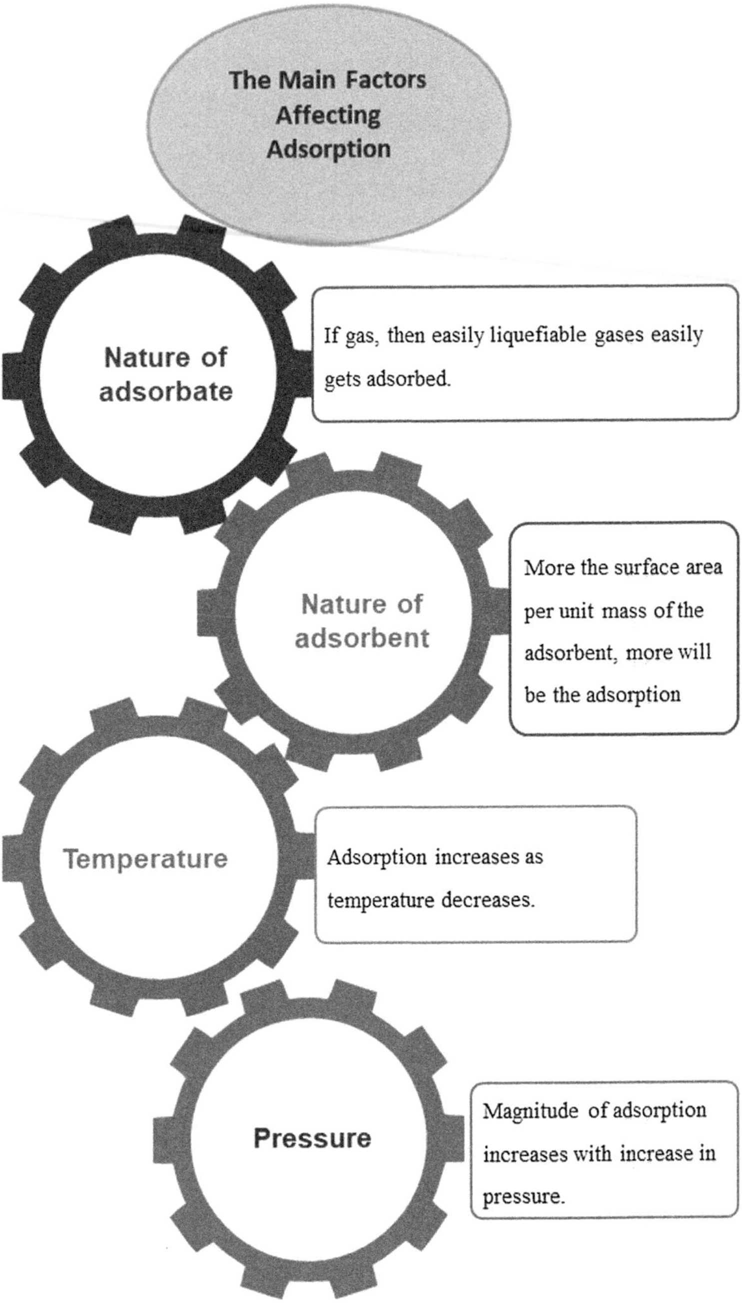
The Main Factors Affecting Adsorption
Nature of adsorbate
If gas, then easily liquefiable gases easily gets adsorbed.
Nature of adsorbent
More the surface area per unit mass of the adsorbent, more will be the adsorption
Temperature
Adsorption increases as temperature decreases.
Pressure
Magnitude of adsorption increases with increase in pressure.

FIGURE 2.10 Main factors affecting the rate of adsorption.

8. Requirements for sorbents

A material must fulfill certain specific properties to be selected as an adsorbent. Generally, these requirements include the following:

- ➢ High selectivity to assist better separation
- ➢ High capacity to reduce the amount of sorbent required
- ➢ Favorable kinetic and transport properties for rapid sorption
- ➢ Thermal and chemical stability
- ➢ Mechanical and hardness strengths to prevent crushing and erosion
- ➢ Free-flowing tendency to make filling and emptying vessels easier
- ➢ High resistance to fouling for long life
- ➢ No tendency to promote undesirable chemical reactions
- ➢ Regeneration capability
- ➢ Reasonable cost

9. Interactions

The following are the types of interactions (adsorption forces) between adsorbent and adsorbate:

- Dipole–dipole interactions
- London or Van der Waals forces
- Coulomb-unlike charges
- Hydrogen bonding
- Point charge and a dipole
- Point charge neutral species
- Covalent bonding with reaction
- London dispersion or Van der Waals force is often predominant

10. BET theory

Brunauer–Emmett–Teller (BET) theory was developed to elucidate the physical adsorption of gas molecules on solid surfaces and serve as the basis for a significant analysis method for measuring the specific surface area of materials. The BET equation (Brunauer, Emmett, and Teller who developed the theory), which was reported in 1938 (Brunauer et al., 1938), continues to be widely used technique in determining the number of molecules or atoms of a gas required to form a monolayer or multilayers of adsorbed molecules on the adsorbent surface (Thommes et al., 2015; Lowell et al., 2004). Physical adsorption, or physisorption, is a term used to describe the phenomenon. BET theory can be used for multilayer adsorption systems and typically utilizes as adsorbate, gases (mostly nitrogen) that do not react chemically with nanomaterial surface (adsorbents) to measure specific surface area. When nitrogen is employed, standard BET analysis is performed at the boiling temperature of nitrogen (77 K). Furthermore, to quantify the surface area at various temperatures and measuring scales, probing adsorbates are used with a reduced frequency. Argon, carbon dioxide, and water have all been used as adsorbates. Specific surface area may be a scale-dependent attribute for which no precise values can be determined. So, The amount of specific surface area calculated using BET theory is governed by the adsorbate molecules and their adsorption cross-section (Hanaor et al., 2014; Galarneau et al., 2018; Sing, 1998; Rouquerol et al., 2007) (Fig. 2.11).

The BET equation defines the link between the quantity of gas molecules adsorbed (X) at a particular relative pressure (P/P_0) and the heat of adsorption (C), as per the following equation:

$$\frac{1}{X[(P_0/P)-1]} = \frac{1}{X_mC} + \frac{C-1}{X_mC}\left(\frac{P}{P_0}\right)$$

where X_m is the gas monolayer, X is the number of molecules of the used gas adsorbed at a relative pressure (P/P_0), and C refers to a second parameter related to the adsorption heat. The BET equation strictly defines a linear plot of $1/X\ [(P_0/P) - 1]$ versus P/P_0, that is restricted to a limited region of the adsorption isotherm, typically in the P/P_0 range of 0.050–0.35.

The surface area, SA, is defined from the intercept and slope as per the equation as:

$$\text{SA} = \frac{1}{\text{slope} + \text{intercept}} \cdot \text{CSA}$$

where the cross-sectional area of the adsorbate is CSA (Lowell et al., 2004).

BET theory concepts are an extension of the Langmuir theory (Langmuir, 1916, 1918), which is a theory for monolayer to multilayer sorption, and it includes the hypotheses as:

- ➢ Gas molecules physically adsorb in layers onto a solid.
- ➢ Gas molecules interact with the layers around them.
- ➢ Each layer is subjected to the Langmuir theory.
- ➢ The first layer's adsorption enthalpy is constant and higher than the second (and higher layers).
- ➢ Adsorption enthalpy for the second (and higher) layer is the same as liquefaction enthalpy.

11. Adsorption principles

According to adsorption principles, strong attractive surface forces trap gas (adsorbate) molecules in the pores of the adsorbent. The forces include Van der Waals force and electrostatic force, between the adsorbate molecules and atoms that compose the adsorbent surface. For similar pressure and

Term	Definition
Adsorption equilibria	If the adsorbent and adsorbate are contacted long enough, an equilibrium will be established between the amount of adsorbate adsorbed and the amount of adsorbate in solution.
Adsorption isotherm	The process of adsorption is studied through graphs known as adsorption isotherms.
Basic Adsorption isotherm	Graph between the amounts (in Y axis) of adsorbate (x) adsorbed on the surface of adsorbent (m) and pressure (P) (in X axis) at constant temperature.

- In adsorption, adsorbate gets adsorbed on adsorbent.
- The direction of the equilibrium shifts in the direction where tension can be relieved. Excess pressure on the equilibrium system causes the equilibrium to shift in the direction where the number of molecules decreases.
- Number of molecules decreases in forward direction. Therefore, forward direction of equilibrium will be favored.
- After saturation pressure Ps, adsorption does not occur anymore. Only limited vacancies are present on the surface of the adsorbent.
- At high pressure a stage is reached when all the sites are occupied and further increase in pressure does not cause any difference in adsorption process. At high pressure, Adsorption is independent of pressure.

FIGURE 2.11 Main steps in basic adsorption isotherms.

temperature circumstances, the molecular distance inside the pores of the adsorbent is substantially smaller than in the gaseous phase. Fig. 2.12 provides a graphic illustration of the same. As a result, the density of the adsorbate in the adsorbed phase becomes liquid-like (Suzuki, 1990).

Equilibrium and kinetics are essential for a successful adsorption process. When adsorbate particles take too long to reach the adsorbent particle's interior, they lead to a low throughput despite good adsorption capacity. Adsorbent capacity, on the other hand, is limited when a high volume of adsorbent is required for a specific amount of adsorbate particles throughput. As a result, a good adsorbent should have both strong adsorptive capacity and good kinetics (Do, 1998). The following aspects must be met in order to meet the two requirements:

- ➢ The adsorbent needs to have a lot of surface area or micropore volume.
- ➢ For the transfer of molecules to the interior, the adsorbent must have a somewhat wide pore network.

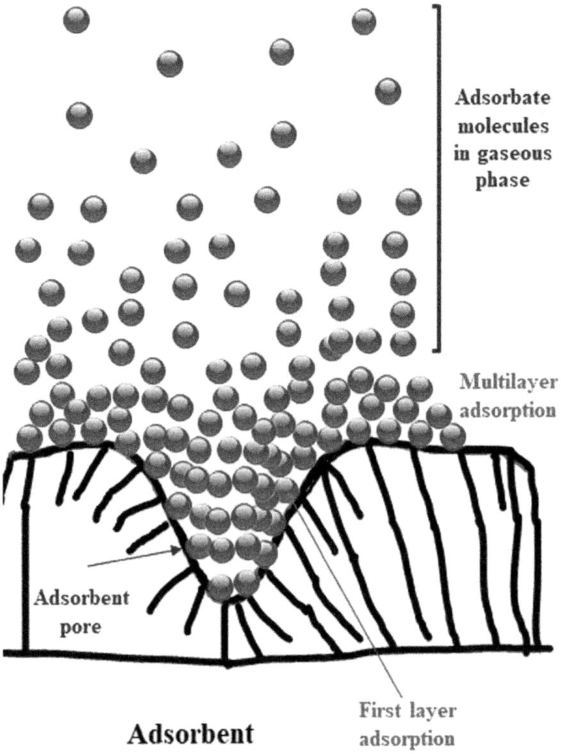
Adsorbate
molecules
in gaseous
phase
Multilayer
adsorption
Adsorbent
pore
Adsorbent
First layer
adsorption

FIGURE 2.12 An illustration of a classic adsorption process with a relative comparison of the adsorbate density between the surrounding phase (such as gas or liquid) and the adsorbed phase.

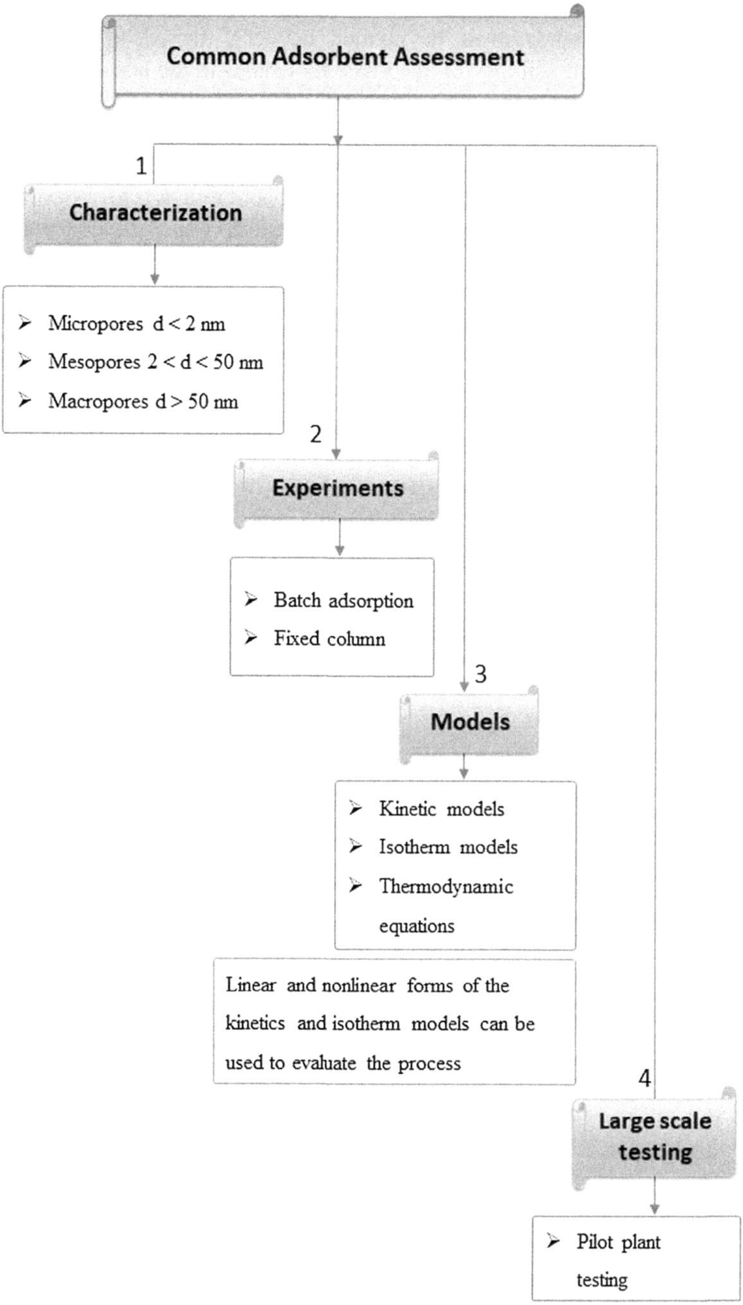
Common Adsorbent Assessment
1
Characterization
Micropores d < 2 nm
Mesopores 2 < d < 50 nm
Macropores d > 50 nm
2
Experiments
Batch adsorption
Fixed column
3
Models
Kinetic models
Isotherm models
Thermodynamic equations
Linear and nonlinear forms of the kinetics and isotherm models can be used to evaluate the process
4
Large scale testing
Pilot plant testing

FIGURE 2.13 Common steps to perform and describe the adsorption process of a system.

In order to provide a big surface area in a small volume, a significant number of micropores must be created between the adsorption surfaces (Fig. 2.13). The accessibility of adsorbate molecules to the adsorption surface is determined by the size of the micropore; consequently, the pore size distribution (PSD) of micropores is an important feature for defining adsorbent adsorptivity (Suzuki, 1990). An excellent adsorbent is said to have a combination of two pore ranges: micropore and macropore ranges (Do, 1998). In fine powders and crystalline adsorbents, macropores serve as diffusion pathways for adsorbate molecules from outside the granule to the micropores (Suzuki, 1990). International Union of Pure and Applied Chemistry (IUPAC) (Sing et al., 1985) recommended the classification of pore size as:

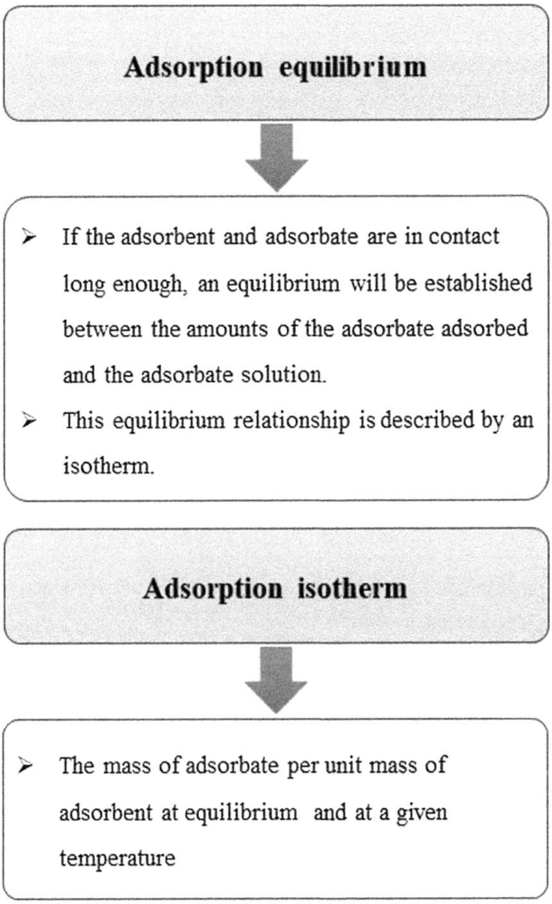
Adsorption equilibrium
➢ If the adsorbent and adsorbate are in contact long enough, an equilibrium will be established between the amounts of the adsorbate adsorbed and the adsorbate solution.
➢ This equilibrium relationship is described by an isotherm.
Adsorption isotherm
➢ The mass of adsorbate per unit mass of adsorbent at equilibrium and at a given temperature

FIGURE 2.14 Difference between adsorption equilibrium and adsorption isotherm.

- ≻ Micropores are used when pore size is $d < 2$ nm
- ≻ Mesopores are used when pore size is $2 < d < 50$ nm
- ≻ Macropores are used when pore size is $d > 50$ nm

12. Adsorption equilibrium

After a period of time, adsorbate molecules settle onto adsorbent surfaces, achieving adsorption equilibrium or state (Fig. 2.14). Equilibrium adsorbate uptake (C) is the quantity of adsorbate accumulated on the adsorbent surface at equilibrium conditions, and it is a function of equilibrium pressure (P) and equilibrium temperature (T). The adsorption isotherm at temperature is the change in equilibrium adsorbate uptake (C) against equilibrium pressure (P) while the temperature is kept constant (isothermal process) (T).

13. Conclusions

The phenomenon of higher concentration of any species of gas, liquid, or solid at the surface compared to the bulk of a material is termed adsorption. The process of removing adsorbed compounds from the surface on which they were absorbed is known as desorption. Of the several types of adsorbents, the commonly used ones include activated charcoal, silica, zeolites, alumina, and bentonite clay. Absorption is the process through which gas or liquid particles are evenly distributed throughout the bulk at a constant rate. Adsorption is a surface phenomenon that occurs when tensions on the surface of solids and liquids are imbalanced. At a given temperature and pressure, the extent of adsorption increases as the surface area per unit mass of adsorbent increases. Physisorption and chemisorption are the two types of adsorption. There are several industrial applications of adsorption including gas separation, water purification, masks making technology, separation technology, and isolation and purification of natural products.

References

Bruch, L.W., Cole, M.W., Eugene, Z., 2007. Physical Adsorption: Forces and Phenomena. Dover Publications, Mineola, NY, ISBN 978-0486457673.

Brunauer, S., Emmet, P.H., Teller, E., 1938. Adsorption of gases in multimolecular layers. J. Am. Chem. Soc. 60, 309–319.

Do, D.D., 1998. Adsorption Analysis: Equilibria and Kinetics. Imperial College Press, Singapore.

Galarneau, A., Mehlhorn, D., Guenneau, F., Coasne, B., Villemot, F., Minoux, D., Aquino, C., Dath, J.-P., 2018. Specific surface area determination for microporous/mesoporous materials: the case of mesoporous FAU-Y zeolites. Langmuir Am. Chem. Soc. 34 (47), 14134–14142.

Hanaor, D.A.H., Ghadiri, M., Chrzanowski, W., Gan, Y., 2014. Scalable surface area characterization by electrokinetic analysis of complex anion adsorption. Langmuir 30 (50), 15143–15152.

Langmuir, I., 1916. The constitution and fundamental properties of solids and liquids. J. Am. Chem. Soc. 38, 2221–2295.

Langmuir, I., 1918. The adsorption of gases on plane surfaces of glass, mica and platinum. J. Am. Chem. Soc. 40, 1361–1403.

Lowell, S., Shields, E., Martin, T., Matthias, T., 2004. Characterization of Porous Solids and Powders: Surface Area, Pore Size and Density, first ed. Springer, Dordrecht, The Netherlands.

Maeda, N., Chen, N., Tirrell, M., Israelachvili, J.N., 2002. Adhesion and friction mechanisms of polymer-on-polymer surfaces. Science 297 (5580), 379–382.

Majmuder, A., Ghatak, A., Sharma, A., 2007. Microfluidic adhesion induced by subsurface microstructures. Science 318 (5848), 258–261.

Nagy, N.M., Kovács, E.M., Kónya, J., 2016. Ion exchange isotherms in solid: electrolyte solution systems. J. Radioanal. Nucl. Chem. 308, 1017–1026.

Popov, V.L., Pohrt, R., Li, Q., 2017. Strength of adhesive contacts: influence of contact geometry and material gradients. Friction 5 (3), 308–325.

Rouquerol, F., Rouquerol, J., Sing, K., 1999. Adsorption by Powders and Porous Solids: Principles, Methodology, and Applications. Academic Press, San Diego, ISBN 0125989202.

Rouquerol, J., Llewellyn, P., Rouquerol, F., 2007. Is the bet equation applicable to microporous adsorbents? Stud. Surf. Sci. Catal. 160, 49–56.

Saleh, T.A., 2015. Isotherm, kinetic, and thermodynamic studies on Hg (II) adsorption from aqueous solution by silica-multiwall carbon nanotubes. Environ. Sci. Pollut. Control Ser. 22 (21), 16721–16731. https://doi.org/10.1007/s11356-015-4866-z.

Saleh, T.A., 2016. Nanocomposite of carbon nanotubes/silica nanoparticles and their use for adsorption of Pb (II): from surface properties to sorption mechanism. Desalination Water Treat. 57 (23), 10730–10744. https://doi.org/10.1080/19443994.2015.1036784.

Saleh, T.A., 2018. Simultaneous adsorptive desulfurization of diesel fuel over bimetallic nanoparticles loaded on activated carbon. J. Clean. Prod. 172, 2123–2132.

Saleh, T.A., 2021. Protocols for synthesis of nanomaterials, polymers, and green materials as adsorbents for water treatment technologies. Environ. Technol. Innovat. 24, 101821.

Serrano, P., Guillermo, C., Botas, J.A., Gutierrez, F.J., 2007. Characterization of adsorptive and hydrophobic properties of silicalite-1, ZSM-5, TS-1 and Beta zeolites by TPD techniques. Separ. Purif. Technol. 54 (1), 1–9.

Sing, K.S.W., 1998. Adsorption methods for the characterization of porous materials. Adv. Colloid Interface Sci. 76–77, 3–11.

Sing, K.S.W., Everett, D.H., Haul, R.A.W., Moscon, L., Pierotti, R.A., Rouguerol, J., Siemieniewska, T., 1985. Reporting physisorption data for gas/solid systems with special reference to the determination of surface area and porosity. Pure Appl. Chem. 57, 603.

Slejko, F.L., 1985. Adsorption Technology. Marcel Dekker, New York.

Suzuki, M., 1990. Adsorption Engineering. Elsevier Science Publishers, Tokyo.

Thommes, M., Katsumi, K., Neimark, A.V., Olivier, J.P., Rodriguez-Reinoso, F., Rouquerol, J., Sing, K.S.W., 2015. Physisorption of gases, with special reference to the evaluation of surface area and pore size distribution (IUPAC Technical Report). Pure Appl. Chem. 87, 1051–1069.

Wu, S., Wang, Y., Sun, C., Yang, M., Zhi, Y., 2021. Novel preparation of binder-free Y/ZSM-5 zeolite composites for VOCs adsorption. Chem. Eng. J. 417, 129172.

Yang, R.T., 2003. Adsorbents: Fundamentals and Applications. Wiley-Interscience, Hoboken, NJ, ISBN 0471297410.

Chapter 3

Kinetic models and thermodynamics of adsorption processes: classification

1. Adsorption reaction models and empirical models

1.1 Overview

The process of solute molecules attaching to the surface of an adsorbent is known as adsorption. It is a mass transfer process where adsorbate molecules transfer from their phase to the adsorbent. The process is conducted in a batch or column setup.

Two main processes are involved in adsorption, namely, physical (physisorption) and chemical (chemisorption) adsorption. While physisorption results from weak forces of attraction (van der Waals), chemisorption includes the formation of a strong bond between the solute and adsorbent and comprises the transfer of electrons. Adsorption kinetics is defined as a line (or curve) that measures the rate of retention or release of a solute (adsorbate) from, for instance, an aqueous media to solid-phase interface at a given adsorbent's dose, pH, flow rate, and temperature.

Popular design variables include quantifiable thermodynamic quantities like temperature equilibrium constant, as well as nonmeasurable counterparts like Gibbs free energy change, enthalpy, and entropy. These variables are used to analyze and predict adsorption process mechanisms. This chapter discusses the kinetics and thermodynamics involved in the adsorption process.

1.2 Steps in adsorption mass transfer

The adsorption kinetic research provides data on the adsorption rate, adsorbent performance, and mass transfer mechanisms. The kinetics of mass transfer in adsorption involves the following steps (Fig. 3.1):

- Step one is external diffusion, where the adsorbate transfers through the liquid film around the adsorbent. It involves a mass transfer of the solute

Surface Science of Adsorbents and Nanoadsorbents. https://doi.org/10.1016/B978-0-12-849876-7.00003-8

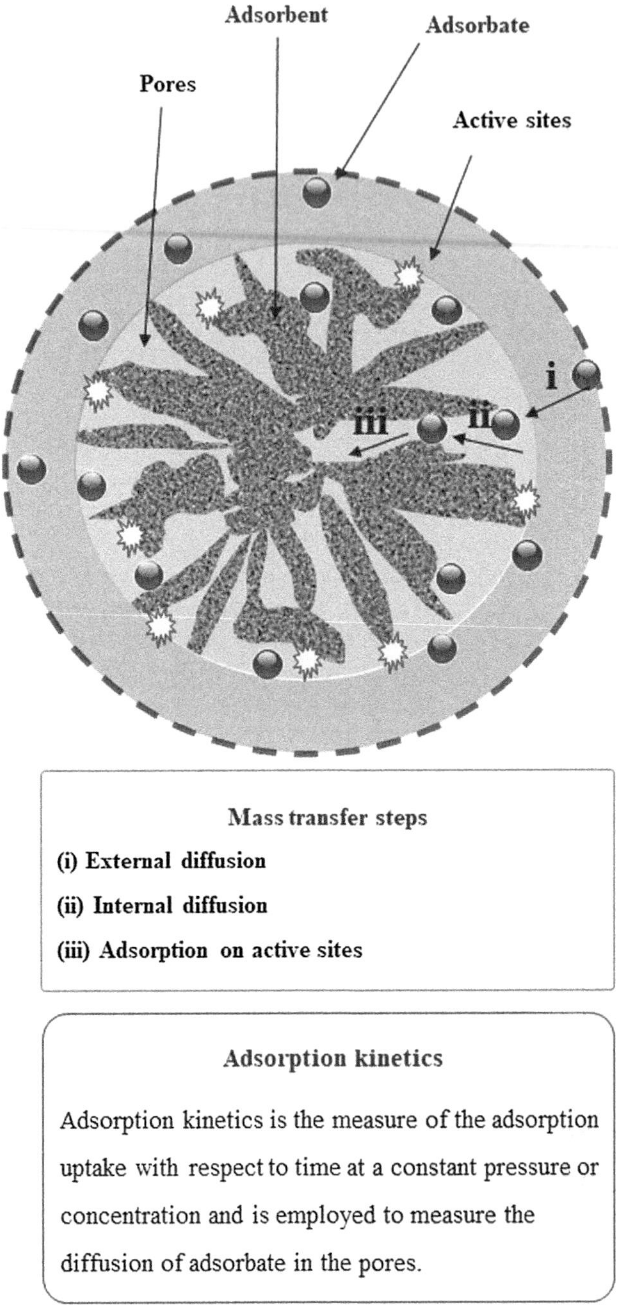

Adsorbent
Adsorbate
Pores
Active sites
i
ii
iii
Mass transfer steps
(i) External diffusion
(ii) Internal diffusion
(iii) Adsorption on active sites
Adsorption kinetics
Adsorption kinetics is the measure of the adsorption uptake with respect to time at a constant pressure or concentration and is employed to measure the diffusion of adsorbate in the pores.

FIGURE 3.1 Main steps in adsorption mass transfer.

from the bulk solution through stagnant film surroundings to the particle's external surface (external or film mass transport). The concentration difference between the bulk solution and the surface of the adsorbent is the driving force of the external diffusion.

➢ Step two is internal diffusion, which involves the adsorbate diffusing through the pores of the adsorbent. Internal or intraparticle diffusion occurs when an adsorbate hops from one available adsorption site to another in a series of adsorption–desorption reactions. Pore volume diffusion (diffusion in fluid-filled pores), surface diffusion (migration along the pore surface in which an adsorbate bounces from one available adsorption site to another in a sequence of adsorption–desorption processes), or a combination of both can produce this (Zhang et al., 2009).

➢ Step three is the adsorption of the adsorbate in the adsorbent's active sites or solute adhesion to the adsorbent's surface.

Furthermore, the following must be noted:

✔ Because the adsorption stage is normally much faster than the first two, the overall rate of adsorption is frequently controlled by the first or second step, whichever is slower, or a combination of both (Malash and El-Khaiary, 2010).

✔ Adsorption reaction models, such as first- and second-order kinetic equations, do not independently describe the processes outlined above (Qiu et al., 2009; Ocampo-Pérez et al., 2012).

✔ When using these kinetic models, it's common to assume that the total rate of adsorption is only governed by the adsorbate's adsorption rate on the adsorbent surface, and that intraparticle diffusion and external mass transport can be ignored (Ocampo-Pérez et al., 2012).

2. Mass transfer

Adsorption is a mass transfer process where adsorbate molecules transfer from the liquid phase to the solid adsorbent. Fig. 3.2 displays a general schematic illustration of the same where molecules from the adsorbate are adsorbed onto the porous adsorbent (Table 3.1). Although both adsorption and ion exchange follow similar steps and are described using adsorption diffusion models, ion exchange differs in the stoichiometric characters.

2.1 Stages in sorption process

Generally, adsorption involves the following consecutive stages:

- Bulk diffusion or convection in the bulk: transportation of the adsorbate molecules from the bulk to the adsorbents.

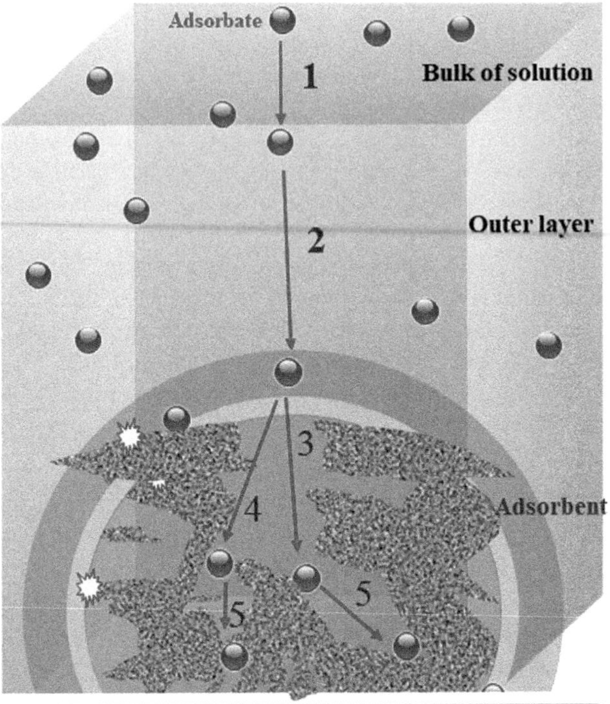
Adsorbate
1
Bulk of solution
Outer layer
2
3
4
Adsorbent
5
5

Diffusion and surface reaction steps:

1. At the bulk of the solution, bulk transport or bulk diffusion/convection.
2. At Outer layer or boundary layer: External mass transfer
3. At adsorbent particle; Film diffusion
4. At adsorbent particle; surface diffusion
5. At adsorbent particle; Adsorption

Some reaction kinetics:

(i) Pseudo-first-order: for system with negligible change in bulk solution concentration

(ii) Pseudo-second-order: typically used for surface reaction limited processes

(iii) Elovich: derived for adsorption on heterogeneous surfaces

FIGURE 3.2 A schematic illustration of the diffusion mechanisms involved in adsorption process.

TABLE 3.1 Common steps of sorption process.

	Adsorption of species from a fluid phase onto a porous solid consists, fundamentally, of the following three steps:	
	Step	**Description**
1	Film diffusion	Mass transfer of the adsorbate by diffusion from the bulk fluid phase through the boundary fluid film to the solid's external surface.
2	Intraparticle diffusion	Mass transfer of the adsorbate by diffusion into both the adsorbed phase and adsorbent's pores by consecutive rounds of pore diffusion and surface diffusion.
3	Interaction	Adsorption of the species on the active sites, pores, or surface of the solid through physical or chemical mechanisms.

- Molecular (or film) diffusion or external mass diffusion: transportation from the bulk solution to the adsorbent surface by diffusion through the boundary layer or liquid film surrounding the adsorbent particles (Table 3.2).
- Intraparticle diffusion or pore diffusion: internal diffusion of the adsorbate molecules from the exterior surface into the pores of the adsorbent, along

TABLE 3.2 Steps involved in adsorbate transport as well as pore and surface diffusion.

Step	Position	Phenomenon	Comment
1	Bulk of solution	Bulk transport or bulk diffusion/convection	It is a fast process for well-mixed systems.
2	Outer or boundary layer	External mass transfer	The thickness of the boundary layer depends on the agitation rate.
3	Adsorbent particle	Film diffusion	It is a slow process. Refer to Weber–Morris model for pore diffusion.
4		Porous diffusion/surface diffusion	It is a slow process and diffusion rate dependent.
5		Adsorption	It is a very fast process.

pore-wall surfaces, or both. It occurs slowly through pore volume diffusion and surface diffusion.

- Physical and/or chemical reaction: adsorption (solute attachment on the adsorbent surface sites) between the adsorbate and active sites of the adsorbent (Hai et al., 2017). Adsorption rate determines the adsorption of adsorbate molecules onto the adsorbent particles at the active sites. The adsorption mechanisms are evaluated to study and distinguish the respective transport mechanisms that limit the adsorption process (Bansal et al., 2009; Walter, 1984).

2.2 Why study kinetics for sorption process

Adsorption kinetics is an important parameter to consider during adsorption designing (Fig. 3.3). Kinetics determines the rate at which adsorption takes place. It is influenced by the contact time, solute concentration, and the surface complexity of the adsorbent. The suitability of any model relies on the error level in a correlation coefficient (R^2) and a sum of squared errors (SSEs).

Several adsorption kinetic models are discussed within this section including the pseudo-first-order model (Lagergren, 1898), pseudo-second-order model (), Elovich model (Elovich and Larinov, 1962), mixed-order model (Guo and Wang, 2019a), Ritchie's equation (Ritchie, 1977), and phenomenological mass transfer models (Table 3.3).

Moreover, there are other models (Qiu et al., 2009) developed for describing the kinetic of adsorption processes such as the film-pore mass transfer (FPMT) model (Guo and Wang, 2019a), Largitte double step model (Largitte and Pasquier, 2016a,b), Gaulke's unified kinetic model (Gaulke et al., 2016), Brouers–Sotolongo fractal kinetic model, and fractal-like adsorption kinetic model (El Boundati et al., 2019). Despite the fact that these models are rarely employed by academics, they are expected to aid in the investigation of adsorption mechanisms and the modeling of adsorption systems.

3. Mass transfer models

There are several models to describe mass transfer; some of which are discussed below.

3.1 Pseudo-first-order kinetic model

The Lagergren pseudo-first-order model assumes that the rate of change in solute uptake over time is proportional to the difference in saturation concentration and amount of solid absorption over time. This applies during the initial stages of an adsorption process. The model could represent the following three conditions:

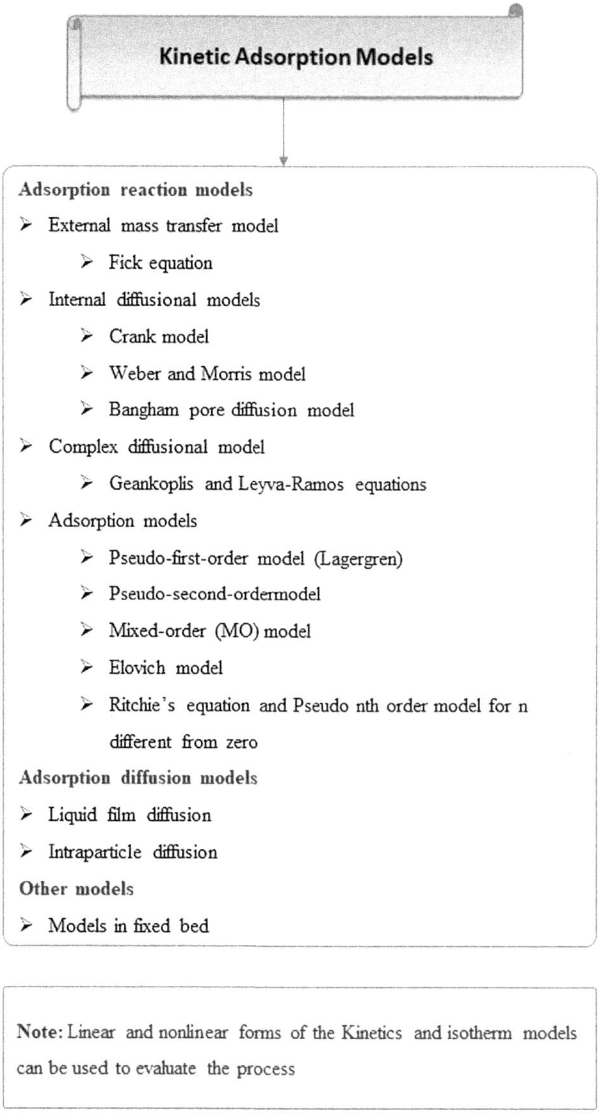
Kinetic Adsorption Models
Adsorption reaction models
External mass transfer model
Fick equation
Internal diffusional models
Crank model
Weber and Morris model
Bangham pore diffusion model
Complex diffusional model
Geankoplis and Leyva-Ramos equations
Adsorption models
Pseudo-first-order model (Lagergren)
Pseudo-second-ordermodel
Mixed-order (MO) model
Elovich model
Ritchie's equation and Pseudo nth order model for n different from zero
Adsorption diffusion models
Liquid film diffusion
Intraparticle diffusion
Other models
Models in fixed bed
Note: Linear and nonlinear forms of the Kinetics and isotherm models can be used to evaluate the process

FIGURE 3.3 Common models of kinetics used to describe the adsorption.

- First condition is that this model can be applied more visibly when the value of initial adsorbate concentration is high.
- Second condition is that the adsorption process is in its initial stage. When *t* approaches zero, the second kinetic model can be estimated to the first (Hu et al., 2018). For example, in the adsorption of lead ions onto peat, it

TABLE 3.3 Mathematical equations for some of the models used in kinetics modeling of adsorption experimental data.

	Linear form	Plot	Calculated coefficient
Reaction-based kinetic models			
Lagergren pseudo-first order	$\log(q_e - q_t) = \log q_e - \left(\frac{k_1 t}{2.303}\right)$ $\ln(q_e - q_t) = \ln q_e - k_1 t$	$\log(q_e - q_t)$ versus t	$k_1 = -2.303 \times$ slope; $q_e = 10^{\text{intercept}}$
Pseudo-second order	$\frac{t}{q_t} = \frac{1}{k_2 q_e^2} + \frac{t}{q_e}$	t/q_t versus t	$q_e = 1/\text{slope}$ $k_2 = \text{slope}^2/\text{intercept}$
Elovich model	$q_t = \frac{1}{\beta}\ln(\alpha\beta) + \frac{1}{\beta}\ln t$	q_t versus $\ln t$	Elovich differential equation can be solved assuming $\alpha\,\beta_t \gg 1$, and by applying the boundary conditions of $q_t = 0$ at $t = 0$ and $q_t = q_t$ at $t = t$
Diffusion-based kinetic models			
Weber–Morris intraparticle diffusion	$q_t = k_{id} t^{\frac{1}{2}} + C$	q_t versus $t^{1/2}$	$k_{id} = \text{slope}$
Boyd	$\beta_t = -\ln\left(\frac{\pi^2}{6}\right) - (\ln(1 - F(t))$		

where q_e: is the adsorption capacity at equilibrium (mg adsorbate/g adsorbent). q_t: is the adsorption capacities at time t (mg adsorbate/g adsorbent). k_1: is the rate constant of the Lagergren first-order kinetic model (min^{-1}). t: is the time of the experiment (min). k_2: is the rate constant of the pseudo-second order kinetic model (min^{-1}). k_{id}: is the intraparticle diffusion rate constant (mg adsorbate/g adsorbent $min^{1/2}$). C: is the plot's intercept of the M-W model (mg adsorbate/g adsorbent). α: is the initial adsorption rate (mg g^{-1} min^{-1}). β: is defined as desorption constant (g mg^{-1}) during any experiment.

was reported that the first model adequately represented the kinetic data for the first 20 min (Ho and Mckay, 1999).

- ≻ Third condition is that the adsorbent has few active sites. In this case, the external or internal diffusion is the rate-controlling step. To better understand this, the pseudo-first-order model can describe the adsorption of some metal ions or hydrophilic molecules onto microplastics (hydrophobic adsorbent) (Turner and Holmes, 2015; Guo et al., 2019a). The diffusion of hydrophilic adsorbates to the surface of the hydrophobic microplastics is difficult, which makes the external or internal diffusion the rate-limiting step. Conversely, the second model can describe the adsorption of hydrophobic organic molecules such as hydrocarbons (hexane, octane, or decane) and oil onto microplastics (Xu et al., 2019; Hu et al., 2017), as the diffusion of hydrophobic molecules to the surface of microplastics is easier compared to hydrophilic adsorbates. The adsorption onto adsorbent with active sites is the rate-controlling step. Consequently, the second model describes a condition where a few active sites exist into the adsorbent or a few adsorbate ions or molecules interact with the active sites of the adsorbent.

Therefore, it is suggested to acquire an accurate estimation of the kinetic parameters in the first-order kinetic model keeping in mind the following:

(i) A trial and error process to find the optimal q_e value (Ho and McKay, 1998).

(ii) An application of a nonlinear optimization method. A good guide for using the nonlinear technique can be used.

3.2 Pseudo-second-order kinetic model

This model was developed based on the assumption that the rate-limiting step is chemical adsorption or chemisorption, which predicts the behavior over the whole range of adsorption (Fig. 3.4). The equation of the pseudo-second-order kinetic model can be used to obtain related parameters (Blanchard et al., 1984).

In most cases, experimental data for adsorption kinetics fit well with the linear form of pseudo-second-order kinetic equation. However, caution is necessary when drawing conclusions based on the fitting to the linear form of the equation. For example, a plot of t/q_t versus t can be linear, and R^2 values for the linear form of the equation can be high ($R^2 > 0.99$). Nevertheless, significantly lower corresponding R^2 values for the nonlinear equation imply an inadequacy of the pseudo-second-order kinetic equation. In such a case, it is recommended to check if the adsorption process involves other possible mechanisms of interactions (aside from chemisorption) such as π−π

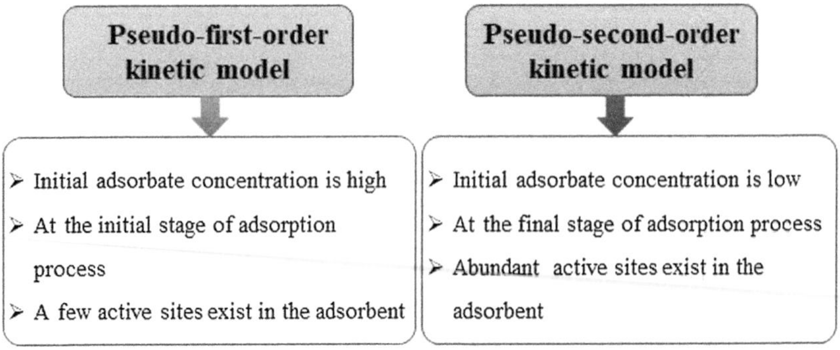
Pseudo-first-order kinetic model
Pseudo-second-order kinetic model
➢ Initial adsorbate concentration is high
➢ At the initial stage of adsorption process
➢ A few active sites exist in the adsorbent
➢ Initial adsorbate concentration is low
➢ At the final stage of adsorption process
➢ Abundant active sites exist in the adsorbent

FIGURE 3.4 The possible physical meanings of the pseudo-first-order and pseudo-second-order kinetic models.

interactions and pore filling. In general, the nonlinear equation provides kinetic model parameters that are more accurate than the linear equation (Tran et al., 2017; Lima et al., 2015).

This model could represent the following three conditions (Guo et al., 2019b):

- First condition is that this model can be applied more visibly when the value of the initial adsorbate concentration is low. This is because when the experimental data are fitted to this model equation, the R^2 decreases by increasing the initial concentration. For example, the results of the adsorption of methylene blue dye on molecular polyoxometalate indicated that the R^2 values of the second model decreased from 0.898 to 0.590 with the increase of C_0 from 140 to 300 mg L^{-1} (Sabarinathan et al., 2019).
- Second condition is at the final period of the adsorption. Compared to the first model, the second could better represent the lead ion adsorption onto peat for 20–90 min (Ho and Mckay, 1999). Nevertheless, most reported literature uses the pseudo-second-order kinetic equation from the first to final (equilibrium) stage of the sorption process. Some literature recommends the use of kinetic data for which a fractional uptake of lower than 85% can be fitted using this model (Simonin, 2016), as the kinetic data closer to or at equilibrium produced bias and unfairly promoted the second model.
- Third condition is that this model can be used when the adsorbents are abundant with active sites. Most literature reported that when adsorbents use plenty of functional groups or active sites for the adsorption of metal ions or molecules, the experimental data were better modeled by the second kinetic equation. For example, in the report on the adsorption of Pb(II) onto hydrochar and modified hydrochar, for hydrochar, the value of R^2 of the PSO model was 0.945, while for modified hydrochar, the value was 0.990 (Xia et al., 2019).

3.3 Mixed-order model

Generally, the first-order kinetic equation describes the diffusion step while the second-order kinetic equation describes the adsorption step on active sites. Thus, the mixed-order model signifies the overall adsorption process (Guo and Wang, 2019). The following conditions fulfill the assumption of the mixed-order model:

(i) An arbitrary stage of adsorption;
(ii) The rate-controlling step is diffusion or adsorption;
(iii) An arbitrary initial concentration of adsorbate in the media.

The mixed-order model is used to describe the adsorption of sulfamethazine on metal-organic frameworks (MOFs) and antibiotics on microplastic (Zhuang et al., 2020; Guo and Wang, 2019c). The mixed-order model is a differential equation, solved by the Runge–Kutta method, and be based on MATLAB or UI in Excel software.

3.4 Elovich model

The Roginsky–Zeldovich adsorption kinetic equation (generally known as Elovich equation) was initially established to define chemisorption kinetics of gas onto solids (Elovich and Larinov, 1962; Mclintock, 1967; Ngah et al., 2004). However, it has recently gained increasing attention to describe the adsorption of liquid–solid systems. This equation is widely used in adsorption kinetics, describing the chemisorption (chemical reaction) mechanism in nature. The approaching equilibrium parameter of the Elovich equation can be used to define adsorption kinetics characteristic curves.

The Elovich model is an empirical model with no clear physical implications. It's a popular way to simulate gas chemisorption on solids. The following are the basic assumptions of the Elovich model:

➢ The activation energy increases with adsorption time; and
➢ The surface of the adsorbent is heterogeneous.

This model has been used to simulate liquid-phase kinetics. For lead adsorption onto activated carbon, the Elovich equation was determined to provide the best fit (Largitte and Pasquier, 2016a,b). Elovich kinetics were also observed for dye adsorption on eggshell biocomposite beads (Elkady et al., 2011). This model was used to analyze adsorption results attained from the adsorption of the molecules over spent-a catalyst. The reported data revealed the adsorption process was chemisorption (Hussein et al., 2018). Adsorption of Zn^{2+}, Co^{2+}, Ni^{2+}, Cu^{2+}, and Fe^{3+}, by solvent-impregnated resins, is best defined by the Elovich equation (Juan and Chen, 1997).

3.5 Ritchie's equation

Ritchie's equation was developed to model the kinetic data of the adsorption of gases or liquids onto a solid (Ritchie, 1977). This model can be applied, for example, to the adsorption data of CO_2 onto any adsorbents such as piperazine-modified activated alumina (Fashi et al., 2018). Ritchie's equation can be also used to fit the kinetic data of the adsorption of organic molecules onto different adsorbents including the adsorption of methylene blue dye onto calcium alginate, ball-milled biochar, and their composites (Wang et al., 2018). Its physical meaning is that this adsorption is dominated by the adsorption on active sites. Thus, one ion or molecule as adsorbate can occupy '*n*' active sites. Instead of considering the desorption process, this model primarily represents the adsorption process with an order factor of one to two or greater than two. The model equation, on the other hand, is an empirical equation with no precise physical meanings.

3.6 Brouers–Sotolongo fractal kinetic model

The model was established specifically to describe the adsorption processes occurring at nanomaterials, which typically present a sole environment including heterogeneous surfaces and complex reactions (Al-Musawi et al., 2016). Reports show that adsorbents formed of nanomaterials have boundaries that efficiently separate mass and pore spaces (fractal surfaces) (Gaspard et al., 2006). The nanomaterials fractality, which may have resulted from the well-defined pore network formed during the synthesis and postsynthesis procedures, is a predictable property that might affect adsorbent adsorptive capabilities. The complicated nature of adsorption on nanomaterials and its impact on kinetics are rarely considered, and as a result, most kinetic models are used (Do and Wang, 1998).

This model was applied to study the adsorption of phosphorous onto granulated apatite filters in treatment wetlands (Delgado-González, 2021). Moreover, it was used for modeling Methylene blue adsorption on Agave Americana fiber using fractal kinetics. The Brouers–Sotolongo kinetic model well suited the experimental data, with no systematic variation of the global fractal time index, which fluctuates between 0.6 and 0.9 (Hamissa et al., 2014). The Brouers–Sotolongo fractal equation was used to model the kinetic data acquired from the sorption of antibiotics over the nanoadsorbents (Al-Musawi et al., 2016).

3.7 Pseudo-nth-order model

The pseudo-nth-order model is utilized for modeling the adsorption kinetic data of some adsorbates on solid adsorbents. For example, Pb(II) adsorption onto sulfuric acid–treated wheat bran (Özer, 2007). Adsorption processes with an order factor of one to two or greater than two are mostly represented by the model. In some cases, with an increase in initial concentration of the adsorbate, the *n* value decreases. For example, n value of tetracycline adsorption on chitosan adsorbent increased with increase in the initial concentration of tetracycline (Caroni et al., 2009). On the other hand, the n value was reported as 2.276 at low initial concentration of 17β-Estradiol (0.2 mg L^{-1}), while it decreased to 0.979 at high initial concentration of C_0 (6 mg L^{-1}) (Liu et al., 2019). Nevertheless, there is no correlation between n values and adsorption conditions, such as initial concentration, pH, and temperature. Thus, this model requires comprehensive investigation and more experiments to be conducted to obtain more data at different conditions rather than fitting the data. It is recommended to use the nonlinear regression method by Statistica 6.0 or SigmaPlot 11 software (Tseng et al., 2014).

4. External diffusion models

In external diffusion models, the adsorbate diffusion in a bounding liquid layer around the adsorbent is believed to be the slowest step. A number of equations were created to model the external mass transfer processes.

4.1 Frusawa and Smith model

It is an adsorption rate equation developed to describe the external diffusion processes (Frusawa and Smith, 1973; Özer et al., 2005). The Frusawa and Smith model assume external diffusion to be the slowest step, intraparticle diffusion to be negligible, and isotherm to be linear.

4.2 Mathews and Weber (M&W) model

The Mathews and Weber (M&W) model was developed to describe external diffusion (Mathews and Weber, 1977). It can be used by the nonlinear regression method. The M&W model can be used to describe the adsorption kinetic data of dyes and metals onto adsorbents such as biosorbent and minerals (Erdoğan and Ulku, 2012).

The basis of the homogeneous surface diffusion model (HSDM) (Mathews and Weber, 1976) included the influence of external mass transfer, and unsteady-state surface diffusion in the particle, as well as a nonlinear adsorption isotherm.

The following are the assumptions of the model:

- ➢ The adsorbent particle is presumed to be a homogenous solid sphere with surface diffusion transporting the adsorbate.
- ➢ Mass transport through film and surface diffusion alone is the rate-controlling process.
- ➢ A driving force is the resistance of a liquid film to mass transfer at the particle's outer surface.
- ➢ At the carbon particle's outer surface, an instantaneous equilibrium between the adsorbate and carbon particle takes place.

4.3 Phenomenological external mass transfer model

Film diffusion is assumed to be the slowest phase in this model, and equilibrium is achieved on the adsorbent's surface (Hines and Maddox, 1985; Ruthven, 1984a,b). The concentration gradient of adsorbent in the liquid film is the driving factor for external diffusion.

5. Internal diffusion models

Internal diffusion models, unlike external diffusion models, consider adsorbate diffusion within the adsorbent to be the slowest stage. Adsorption onto active sites and diffusion of adsorbate in the liquid film around the adsorbent are both immediate processes. Some of these include the Weber and Morris model, Boyd's intraparticle diffusion model, and phenomenological internal mass transfer model, and they are discussed in the following sections.

5.1 Boyd's intraparticle diffusion model

Boyd developed theoretical models for ion-exchange kinetics (Boyd et al., 1947). These kinetic models were discovered to be applicable to adsorption systems by the adsorption community, and Boyd's diffusion models have subsequently been used in various adsorption investigations, largely to establish the rate-controlling step (Castillejos et al., 2011; El-Khaiary and Malash, 2011).

Boyd's intraparticle diffusion equation has been developed to represent the internal mass transfer processes. This kinetic model was proposed to describe the diffusion of adsorbate through a bounding liquid film, assuming the concentration gradient as linear (Boyd et al., 1947). The Boyd equation can be used to identify an adsorbent's rate-controlling step and be expressed in the following manner:

$$F = \frac{q}{q_e} = 1 - \frac{6}{\pi^2} \times \sum_{n=1}^{\infty} \frac{1}{n^2} \times e^{-n^2 \times Bt}$$

$$F = 1 - \left(\frac{6}{\pi^2}\right) \exp(-Bt)$$

$$B = \frac{\pi^2 \times D_i}{R^2}$$

$$B_t = -\ln\frac{\pi^2}{6} - \ln(1 - F(t)) \quad \text{for } F(t) > 0.85$$

$$B_t = \left(\sqrt{\pi} - \sqrt{\pi - \frac{\pi^2 F(t)}{3}}\right)^2 \quad \text{for } F(t) \le 0.85$$

$$B_t = -0.4977 - \ln(1 - F)$$

where B_t is the function of F, which is the fraction of solute adsorbed at different times, t, and can be calculated. q_t and q_e are the amounts adsorbed on the adsorbent at any time, t, and at equilibrium state. B_t values at different contact times are determined and used to calculate the effective diffusion coefficient (D_2).

The film diffusion coefficient (D_1) values can be calculated from the slope of plots of qt/qe versus $t^{1/2}$ plots.

The following are the assumptions for these equations (Reichenberg, 1953; Viegas et al., 2014):

- ➢ The mathematical function of B_t is F, and the reverse is also true. If the diffusion coefficient D_i does not fluctuate with F over the range of values involved, the values of B_t can be plotted against experimental results of t, and a straight line (of B slope) passing through the origin shall be formed. F is solely dependent on D_i/R^2 for a given value of t. As a result, F is unaffected by the concentration of solute ions.
- ➢ The square of the particle radius is inversely proportional to B. Under particle diffusion, dF/dt and dq/dt are proportional to B for a given value of F, and the rate of exchange is inversely proportional to the square of the particle radius for the F values.
- ➢ Reichenberg (1953) obtained the approximations for $F(t) > 0.850$ and $F(t) \leq 0.850$.

When experimental data obey the Boyd's intraparticle diffusion model, B is a constant. Pore diffusion determines the rate of mass transfer if the plot in Fig. 3.5A is a straight line that passes through the origin (or particle diffusion mechanism). If, on the other hand, the plot is linear but does not pass through the origin or is nonlinear, it is said to be nonlinear (Fig. 3.5B and C), film diffusion or external mass transport is then a dominating factor (Sharma and Das, 2012).

Boyd's plots are typically multilinear, and statistical tools combined with traditional graphical analysis are frequently required to establish the number and location of breakpoints as well as to reduce the subjectivity of linear segments obtained only by visual analysis. For this, the piecewise linear regression (PLR) statistical method can be used in a Microsoft Excel spread sheet (Malash and El-Khaiary, 2010) by considering one, two, three, or four linear segments. Then, the regression is obtained. Following which, segmented line with confidence intervals of the linear segments is checked. Statistical indicators are used to evaluate the goodness of fit.

Examples:

Experimental data obtained for the adsorption kinetics of metronidazole molecules on microporous activated carbon were fitted to this model using the linear form. The calculated values of R^2 ranged between 0.9471 and 0.9885, which indicates good fitting. This model was applied for the adsorption of phenol onto polymeric adsorbent (NDA-100) (Meng, 2005). This model was

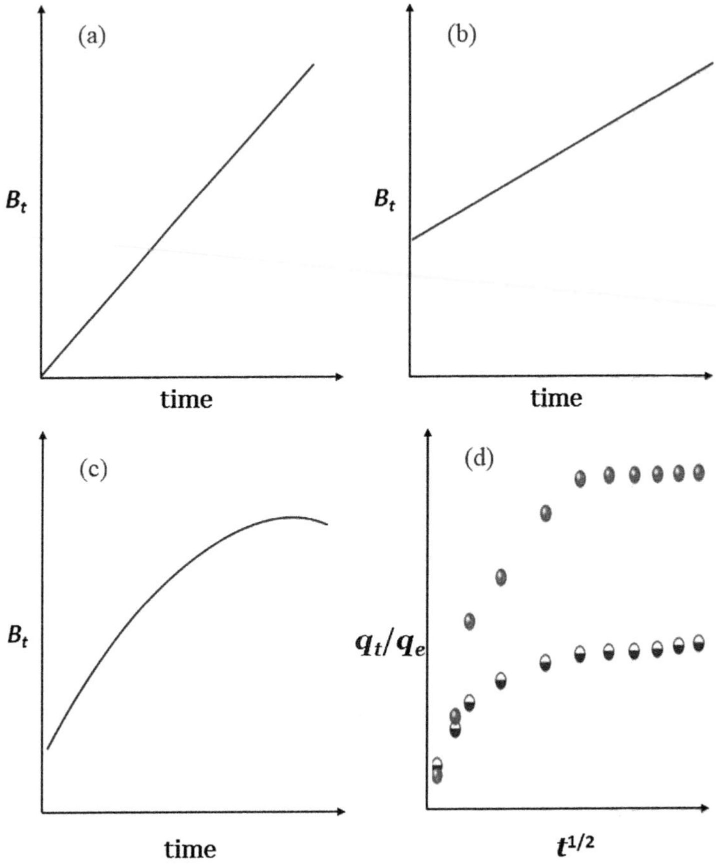
(a)
B_t
time
(b)
B_t
time
(c)
B_t
time
(d)
q_t/q_e
$t^{1/2}$

FIGURE 3.5 Typical characteristic curves based on the Boyd Model; (a) Linear Boyd model with line passes through the origin, (b) Plot of Boyd model with line that does not pass through the origin, (c) Nonlinear Boyd model, (d) intraparticle diffusion model.

also used to show that external mass transfer is the rate-determining stage in the adsorption of anionic and cationic dyes on carbonaceous particles generated from Juglansregia shell biomass (Crini et al., 2007).

When the experimental data obey Boyd's intraparticle diffusion model, *B* is a constant. If the straight line passes through zero points, this indicates that the sorption process is mostly controlled by intraparticle diffusion.

For example, experimental data of lead adsorption on mansonia wood sawdust were fitted to Boyd's model, which indicated that the adsorption process was controlled by intraparticle diffusion (Ofomaja, 2010). On the other hand, the plot of B_t versus t didn't pass through (0,0), which indicated that the adsorption of phenol onto poly(methyl methacrylate) was not controlled by intraparticle diffusion (Muhtaseb et al., 2011).

5.2 Weber and Morris model

This model is used to describe the intraparticle diffusion process (Weber and Morris, 1963). If the experimental data fit this model, then intraparticle diffusion is the controlling process. Else, multiple processes control the adsorption. This model can be used to determine the rate-controlling step. For example, experimental data of adsorption of phenol onto poly(methyl methacrylate) fit well to this model (Muhtaseb et al., 2011).

Weber–Morris discovered that the adsorption of adsorbates changes proportionally with the square root of time ($t^{1/2}$) in most adsorption processes (Weber and Morris, 1962)

$$q_t = k_{id}t^{\frac{1}{2}} + C$$

K_{id} is intraparticle diffusion constant (mg g^{-1} $min^{1/2}$)
C is the thickness of the boundary layer, from the intercept.

If the plot of q_t versus $t^{1/2}$ is linear, and it passes through the origin, then intraparticle diffusion alone is the rate-determining step. Intercept (C) provides a sign about the thickness of the boundary layer if simultaneously the film diffusion likewise occurs (Brouers, 2014).

The intraparticle diffusion model accords the best with experimental data in the kinetics of nitrate adsorption onto modified wheat residues, showing that intraparticle diffusion was the key rate-determining phase during the adsorption process (Wang et al., 2007). Arsenic removal by granular ferric hydroxide (GFH) fits to intraparticle diffusion, which is a significant mass transport process in the adsorption of arsenic ions over a packed-bed column (Badruzzman et al., 2004). In some cases, this model revealed sorption processes to not only be controlled by intraparticle diffusion, but also by film diffusion (Lazaridi and Asouhidou, 2003).

5.3 Phenomenological internal mass transfer model

Internal diffusion is assumed to be the slowest process in the model, and equilibrium is achieved at the liquid–solid interface. The equilibrium phenomenon is described by the adsorption isotherm (Crank, 1979).

6. Pore volume and surface diffusion model

The fundamental assumptions of this model include the following (Leyva-Ramos and Geankoplis, 1985; Souza et al., 2017):

(i) The adsorbent particle is spherical.
(ii) The convective mass transport in the adsorbent pores is negligible.

(iii) The solution is mostly homogeneous.
(iv) The adsorption on active sites is mostly simultaneous.

7. Models for adsorption onto active sites

The models for adsorption onto active sites assume adsorption onto the active sites to be the slowest step and diffusion process to be negligible. Langmuir kinetic model, phenomenological adsorption onto active sites model, and Ritchie's equation can describe this phenomenon.

The controlling mechanism can be a single step or a series of stages. When compared to intraparticle diffusion, the adsorption step in porous materials is naturally significantly faster. The equilibrium between the fluid and the solid is considered to be immediate (Schwaab et al., 2017).

8. Biot number

8.1 Calculation of biot number

Following is the determination of the adsorption system's exterior and internal diffusion coefficients, or the mass transfer coefficient (k_f) and particle surface diffusion (D_s). The surface diffusion modified Biot number (Bi) is obtained as follows:

$$Bi = \frac{k_f \cdot d_p \cdot C_0}{2 \cdot \rho_p \cdot D_s \cdot q_0}$$

where C_o refers to the initial liquid-phase concentration, and q_o is the equilibrium solid-phase concentration (Roy et al., 1993; Erosa et al., 2001).

8.2 Why calculate biot number

The Biot number is the ratio of the rate of diffusion within the particle to the rate of transport over the liquid layer. It establishes a criterion for determining whether external diffusion or surface diffusion is dominant.

➢ **Case 1**

For $Bi \ll 1$, external mass transport resistance is considered the mass transfer-controlling step (Boschi et al., 2011; Erosa et al., 2001; Traegner and Suidan, 1989). Nevertheless, the Bi limit, displaying internal diffusion's supremacy, is not universally accepted.

➢ **Case 2**

For $Bi \gg 1$ (Boschi et al., 2011), others use $Bi \gg 100$ (Prasad and Srivastava, 2009; Traegner and Suidan, 1989). For the latter, internal and exterior mass transfer are both crucial for the adsorption rate, as shown by Bi values between 1 and 100.

---

9. Adsorption process and model evaluation

Model performance indicators are used to evaluate the models. The kinetic models are typically applied to plot curves to obtain the associated parameters (slopes and intercepts) (Fig. 3.6). Therefore, the models are typically subjected to accuracy (goodness-of-fit), with the use of the performance indicators for determining the kinetic model best describing the interaction between the solutes and the adsorbent (George et al., 2018). Some indicators or error functions used for this are discussed in the following sections.

9.1 Coefficient of correlation

Coefficient of correlation or determination (R^2) denotes the variance in the mean. Its value is used for analyzing the degree of fitting of a kinetic model with experimental data. This indicator is used as a criterion of goodness-of-fit.

Examples:

- ≫ A higher R^2 for the pseudo-second-order kinetic (PSO) model compared to that of the pseudo-first-order kinetic (PFO) model for the sorption of examined concentration of adsorbate on adsorbent indicates that the kinetic data were best described by the pseudo-second-order kinetic model.
- ≫ High R^2 of the pseudo-second-order kinetic model for the amoxicillin at various concentrations over the carbon adsorbents indicates that the kinetic data were best defined with the model of pseudo-second-order kinetic (Moussavi et al., 2013).
- ≫ R^2 of the pseudo-second-order kinetic model was higher compared to that of the first-order model for the sorption of loperamide on lignocellulosic-

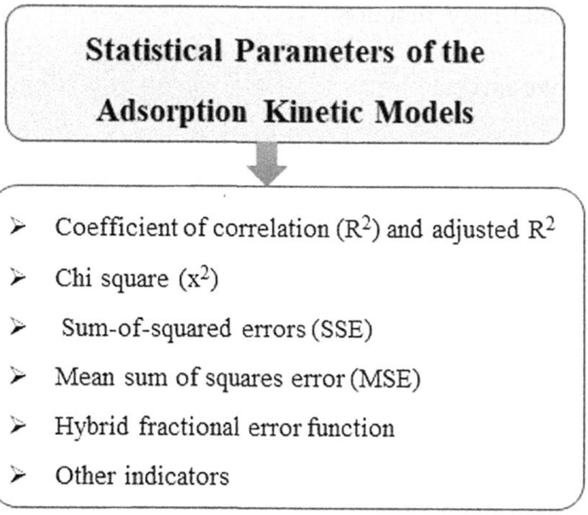
Statistical Parameters of the
Adsorption Kinetic Models
Coefficient of correlation (R2) and adjusted R2
Chi square (x2)
Sum-of-squared errors (SSE)
Mean sum of squares error (MSE)
Hybrid fractional error function
Other indicators

FIGURE 3.6 Parameters and indicators for model validity evaluation.

alumina. Thus, the pseudo-second-order model fitted more than the pseudo-first-order model (Velinov et al., 2019).

9.2 Chi-square

Chi-square (χ^2) function is used to decide the best fit for adsorption results. Chi-square or Pearson's Chi-square test is calculated by the sum square change between experimental results and computed results, with each square difference being divided by its corresponding value (Velinov et al., 2019). The calculated Chi-square is compared with the statistical tabulated value, and a conclusion to accept the hypothesis is obtained if the calculated value is lower.

Chi-square was used to evaluate the modeling of sorption kinetics of experimental results. As reported by El Bardiji et al. (2020), the *f* fractal-like multiexponential (*f*-mexp) equation fits the experimental results well over the sorption range. Other works reported on the use of the model to assess the goodness-of-fit of the sorption kinetic results for arsenic ions adsorption over $MnFe_2O_4$ adsorbent (Podder and Majumder, 2017).

9.3 Other indicators

Some other statistical indicators that can be used to define the model that best fits the adsorption experimental results include the following:

- ✓sum of squared errors (SSE)
- ✓mean sum of squared error (MSE)
- ✓hybrid fractional error function
- ✓average relative error
- ✓Spearman's correlation coefficient
- ✓nonlinear chi-square test
- ✓hybrid fractional error function
- ✓Marquardt's percent standard deviation (MPSD)
- ✓standard deviation of relative errors

10. Adsorption thermodynamics

In adsorption, the adsorbate molecules are more stabilized onto the surface of the adsorbent than in the media, due to the reduction in the energy level of the adsorbate molecules which accumulate in the pores of the adsorbent with a phase transformation. The adsorbed phase is the transformed phase of the adsorbate molecules, and it is treated as a thermodynamically identifiable phase, despite the fact that the precise location of the phase boundary is unknown (Ruthven, 1984a,b). Consequently, the assumption is that, unlike the gaseous phase, the thermodynamic states of the adsorbed phase are dependent on adsorption uptake as well as pressure and temperature. The evaluation of the thermodynamic quantities of adsorbed phase, such as specific heat capacity, the heat of adsorption, internal energy, entropy, and enthalpy, are important for thermodynamic analysis of an adsorption process. Nevertheless, this research does not study the thermodynamic formulations of these adsorbed phase quantities.

The temperature dependences of adsorption are investigated in equilibrium experiments. Various explanations of the behavior of adsorption systems have been proposed, which indicate the exothermic or endothermic nature of the process. Adsorption thermodynamic parameters should be evaluated (Thaligari et al., 2016; Saha and Cowdhury, 2011) to create the feasibility and spontaneity of such a process. Thus, thermodynamic parameters including change in enthalpy (ΔH°), Gibbs free energy change (ΔG°), change in entropy (ΔS°), adsorption potential (A), isosteric heat of adsorption (ΔH_X), hopping number (n), activation energy (E_a), sticking probability (S^*), and adsorption density (ρ) are determined using experimental data obtained from adsorption procedures (Ebelegi et al., 2020).

10.1 Gibbs free energy of change (ΔG°)

As one of the important parameters, Gibbs free energy of change (ΔG°) is applied to assess the feasibility and spontaneity of the adsorption process. When calculating Gibbs free energy of change, the following must be noted:

- Negative ΔG° value indicates a spontaneous process, where the adsorption process is always feasible.
- Positive ΔG° value indicates the nonfeasibility and nonspontaneity of the adsorption process.
- At equilibrium, the Gibbs free energy change (ΔG°) is zero, and the adsorption process is spontaneous in neither directions.

The equation below can be used to measure the changes in equilibrium constant with variations in temperature

$$\Delta G^o = -\mathrm{RT}\ \ln K_C$$

An example is the use of this equation to evaluate the adsorption of Ni(II) over baker's yeast, which was reported to be spontaneous due to negative ΔG° value and an exothermic process due to negative value of enthalpy change ΔH (Padmavathy, 2008).

10.2 Enthalpy change

The change in enthalpy (ΔH°) is the energy supplied as a heat at a constant pressure when commonly the system does no extra work, and it is determined with a calorimeter by measuring the change in temperature at a constant pressure (Atkins and Paula, 2009). In the adsorption process, enthalpy change provides an insight into the mechanism and nature of the adsorption. Usually, it is calculated from the Van't Hoff equation. The following must be noted when calculating enthalpy change:

- A negative ΔH° value implies an exothermic process of adsorption.
- A positive ΔH° value indicates an endothermic process. For example, the positive value of ΔH° (89.0 kJ mol^{-1}) indicates that the sorption process of malachite green dye over chemically modified rice husk is endothermic (Chowdhury et al., 2011).

The magnitude and sign of standard free energy change (ΔG°) rely on thermodynamic parameters such as standard entropy change (ΔS°) and standard enthalpy change (ΔH°) as shown in the below figure (Fig. 3.7).

10.3 Entropy change

The entropy change (ΔS°) during an adsorption is attained using Van't Hoff equation. Positive ΔS° shows the adsorbent's affinity toward the adsorbate. It indicates increased randomness at the interface of solid/liquid, with some structural changes in the adsorbate and adsorbent.

For example, positive ΔS° (0.108 kJ K^{-1} mol^{-1}) reveals the affinity of the bentonite adsorbent for adsorbing Cu(II) (Gupta, 1998). ΔS° was positive on the kinetics as well as thermodynamics of lead ions adsorption over lateritic nickel ores (Mahapatra et al., 2009). The adsorption of malachite dye by the biosorbent, pleurotuseryngii, was reported to be spontaneous as the value of the entropy change was negative (Wu et al., 2019).

10.4 Isosteric heat of adsorption

Isosteric heat of adsorption (ΔH_X) is considered the ratio of the infinitesimal changes in the adsorbate enthalpy to the infinitesimal changes in the quantity adsorbed under a constant pressure and temperature (Giraldo et al., 2019). Simply put, it is the heat of adsorption formed at a constant amount of the

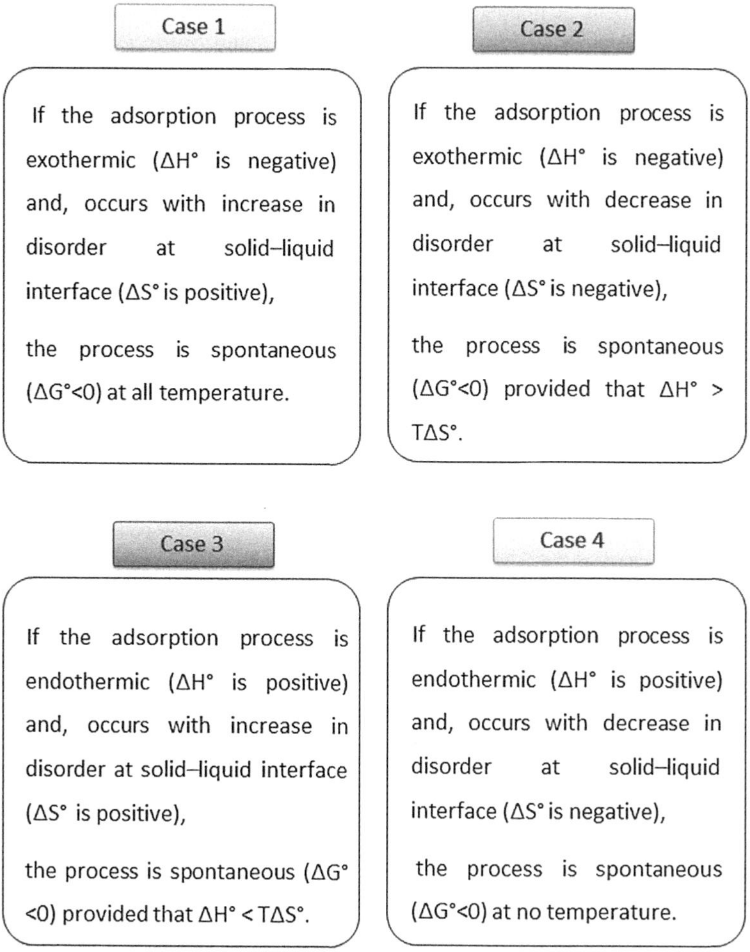
Case 1
If the adsorption process is exothermic (ΔH° is negative) and, occurs with increase in disorder at solid–liquid interface (ΔS° is positive),
the process is spontaneous (ΔG°<0) at all temperature.
Case 2
If the adsorption process is exothermic (ΔH° is negative) and, occurs with decrease in disorder at solid–liquid interface (ΔS° is negative),
the process is spontaneous (ΔG°<0) provided that ΔH° > TΔS°.
Case 3
If the adsorption process is endothermic (ΔH° is positive) and, occurs with increase in disorder at solid–liquid interface (ΔS° is positive),
the process is spontaneous (ΔG° <0) provided that ΔH° < TΔS°.
Case 4
If the adsorption process is endothermic (ΔH° is positive) and, occurs with decrease in disorder at solid–liquid interface (ΔS° is negative),
the process is spontaneous (ΔG°<0) at no temperature.

FIGURE 3.7 Thermodynamics for different cases.

adsorbed adsorbate. In characterizing adsorption processes, ΔH_X is computed by Clausius–Clapeyron equation.

The ΔH_X value calculated was 78–89 kJ mol^{-1} for the adsorption process of malachite green over treated ice husk (Chowdhury et al., 2011). This indicates that isosteric heat of adsorption (ΔH_X) is chemisorption. It is employed as a means to characterize the surface of adsorbents due to the following reasons:

- If the isosteric heat of adsorption of an adsorbent is independent of the quantity of adsorbate it adsorbs, it is homogenous.

- The adsorbent is heterogeneous if its isosteric heat of adsorption varies with the quantity of adsorbate adsorbed (Bae and Snurr, 2010).

10.5 Hopping number

A hopping number (n) denotes the number of hopping done mostly by the adsorbate molecules to find a vacant site onto the surface of the adsorbent during the adsorption process (Higachi et al., 1984). This expression relates the n to surface coverage. The n describes how fast an adsorption process happens. The following is to be noted when dealing with hopping number:

- The small hopping number n indicates a fast adsorption process (Horsfall and Spiff, 2005).
- Acid-treated chrysophyllum albidium biomass, for example, showed faster uptake of Solid and Dissolved Particles (SDP) because of its lower hopping number than salt-treated chrysophyllum albidium biomass, which has a larger hopping number (Menkiti et al., 2014; Ebelegi et al., 2020).

10.6 Adsorption potential

Adsorption potential (A) describes the chemical potential occurring as adsorbate molecules move from the solution onto the adsorbent surface during the adsorption. It calculates the capability of an adsorbent to uptake adsorbate molecules at a constant temperature. The adsorption potential theory is a theory of thermodynamics, which relates to the macroscopic performance of adsorption equilibrium rather than involving the microcosmic mechanism.

10.7 Adsorption density

Adsorption density is defined as a parameter to evaluate the packing of adsorbate molecules onto the adsorbent surface and is typically analyzed at a constant temperature. In an adsorption process, this parameter can be calculated from experimental data using the equation shown in Table 3.4. For example, this parameter was used as the basis for comparing the multiple-site adsorption of Pb, Cu, Zn, and Cd ions onto amorphous iron oxyhydroxides (Benjamin, 1981).

10.8 Sticking probability

Sticking probability (sorption probability, S^*) displays the potential of an adsorbate to stay onto an adsorbent indefinitely. Sticking probability is considered a function of the adsorbate/adsorbent system. When its value lies between zero and unity (i.e., $0 < S^* < 1$), it best serves its purpose. Sticking probability is independent of the system's operating temperatures

TABLE 3.4 Mathematical thermodynamic equations of adsorption experimental data.

Gibbs free energy of change ($\Delta G°$)	$\Delta G° = -RT \ln K_C$	**R = universal gas constant ($Jmol^{-1}\,kg^{-1}\,K^{-1}$)** **$T$ = temperature (K) K_C = equilibrium constant.**
Enthalpy, entropy, and free energy changes	$\ln\left(\frac{q_e}{C_e}\right) = -\frac{\Delta H}{RT} + \frac{\Delta S}{R}$	Plot of $\ln\left(\frac{q_e}{C_e}\right)$ versus $\frac{1}{T}$ Slope $= -\frac{\Delta H}{R}$ Intercept $= \frac{\Delta S}{R}$ $\Delta G = \Delta H - T\Delta S$
	$\ln K_D = \frac{\Delta S°}{R} - \frac{\Delta H°}{RT}$ $K_D = \frac{q_e}{C_e}$	$\Delta H°$, and $\Delta S°$ can be obtained from the linear plot, K_D is the equilibrium constant at standard conditions ($L\,mg^{-1}$). While $\Delta H°$, and $\Delta S°$ can be obtained from the linear plot of $\ln K_D$ and $1/T$
Entropy change ($\Delta S°$)	$\Delta G° = \Delta H° - T\Delta S°$	
Enthalpy change ($\Delta H°$)	$\log K_C = \frac{\Delta S°}{2.303R} - \frac{\Delta H}{2.303RT}$ K_C (Distribution coefficient) $= \frac{C_a}{C_e}$ Van't Hoff equation	R = universal gas constant ($J\,mol^{-1}\,K^{-1}$), T = temperature (K) C_a = amount of adsorbate adsorbed at equilibrium ($mg\,L^{-1}$) C_e = equilibrium concentration of adsorbate in solution ($mg\,L^{-1}$)
Isosteric heat of adsorption (ΔH_X)	$\frac{d(\ln C_e)}{dT} = -\frac{\Delta Hx}{RT^2}$	Clausius–Clapeyron equation

Continued

TABLE 3.4 Mathematical thermodynamic equations of adsorption experimental data.—cont'd

Gibbs free energy of change ($\Delta G°$)	$\Delta G° = -\mathbf{RT} \ln K_C$	**R = universal gas constant (Jmol^{-1} kg^{-1} K^{-1})** **T = temperature (K) K_C = equilibrium constant.**
Hopping number (n)	$n = \frac{1}{(1-\theta)\theta}$	
Adsorption potential (A)	$A = -RT \ln \frac{C_o}{C_e}$	C_o and C_e are initial and final concentration of the adsorbate solution (mol mg^{-1}), R = universal gas constant (KJ mol^{-1} K^{-1}), and T = absolute temperature at which the sorption occurred
Adsorption density (ρ)	$\rho = Z_r C_e \exp - \left[\frac{\Delta G^0}{RT}\right]$	ρ = adsorption density Z = valency of adsorbed ion R = effective radius of adsorbed ion C_e = equilibrium concentration of adsorbate solution (mol mg^{-1}) R = universal gas constant (KJ mol^{-1} K^{-1}) T = absolute temperature (K)
Sticking probability (S^*)	$\ln(1-\theta) = \ln S^* + \frac{E_a}{RT}$	
Activation energy (E_a)	$\ln k_2 = -\frac{E_a}{RT} + \text{constant}$ $\ln k_2 = -\frac{E_a}{RT} + \ln k_o$	Arrhenius equation Plot of $\ln k_2$ versus $\frac{1}{T}$ Slope $= -\frac{E_a}{R}$

ΔH: is the change in enthalpy, is used in order to evaluate if the energy absorbed or evolved during the adsorption process. ΔG: the Gibbs's free energy, is used in order to obtain if the process is spontaneity. ΔS: the entropy, is used in order to obtain the randomness of the process. K_D: is the equilibrium constant at standard conditions (L mg^{-1}).

TABLE 3.5 Types of adsorption processes based on the Gibbs free energy and enthalpy values.

Adsorption process	Thermodynamic characteristics
Spontaneous	(ΔS°) negative
Nonspontaneous	(ΔS°) positive
Exothermic	(ΔH°) negative
Endothermic	(ΔH°) positive

(Ebelegi et al., 2020). Its value can be calculated from experimental data using the equation shown in Tables 3.4 and 3.5. An example is a report on how acid treatment improved the sticking probability of Ni(II) on pumpkin wastes (Horsfall and Spiff, 2005).

10.9 Activation energy

The activation energy (E_a) is the minimum energy needed to cause a chemical reaction. E_a of adsorption processes is predicted by the Arrhenius Equation. When you plot ln k_2 versus $1/T$, you get a straight line from which you may calculate the activation energy (E_a) from the slope of the linear plot.

$$\ln k_2 = -\frac{E_a}{RT} + \ln k_o$$

When dealing with activation energy, the following must be noted:

- Negative E_a, value suggests a low temperature, which points to processes where the sorption process is termed to be exothermic (Yakout and Elsherif, 2010).
- The presence of an energy barrier in the adsorption process is indicated by a positive E_a value, indicating that the system requires energy (increasing temperature) to drive the adsorption process. Endothermic adsorption is the name given to this type of sorption.

11. Conclusions

The development of adsorption processes is highly necessary. Thus, it is important to investigate and understand the interplay between adsorption, micropore diffusion, and reactions of such processes. These developments in literature have been stimulated by a dramatic increase in the adsorption research that has led to major discoveries ranging from new microporous adsorbent materials to new theoretical approaches. One of the most important aspects to comprehend before deciding on a material's suitability as an adsorbent is its adsorption kinetics. In an adsorption process, linear or nonlinear analysis of kinetics is employed. Adsorption reaction models, for the most part, define the rate at which adsorption takes place without providing exact information on the adsorption mechanism. On the other hand, adsorption diffusion models define the key stages that envision adsorption mechanisms.

The effect of several kinetic and thermodynamic parameters on adsorption processes is investigated. Thermodynamic parameters are necessary for determining the impact of heat on adsorption procedures, as well as for defining the spontaneity of adsorption processes. To choose the optimal model, the goodness-of-fit index (coefficient of correlation or sum of squares) is used. The distribution of error function is affected by whether the adsorption kinetics are linear or nonlinear.

References

Al-Musawi, T.J., Brouers, F., Zarrabi, M., 2016. Kinetic modeling of antibiotic adsorption onto different nanomaterials using the Brouers-Sotolongo fractal equation. Environ. Sci. Pollut. Control Ser. 24, 4048–4057.

Atkins, P.W., Paula, J.D., 2009. Physical Chemistry, eighth ed. Oxford University Press, Oxford.

Badruzzman, M., Westerhoff, P., Knappe, D.R.U., 2004. Intra particle diffusion and adsorption of arsenate onto granular ferric hydroxide (GFH). Water Res. 38, 4002–4012.

Bae, Y.-S., Snurr, R.Q., 2010. Optimal isosteric heat of adsorption for hydrogen storage and deliver using metal-organic frameworks. Microporous Mesoporous Mater. 132, 300–303.

Bansal, P., Hall, M., Realff, M.J., Lee, J.H., Bommarius, A.S., 2009. Modeling cellulase kinetics on lignocellulosic substrates. Biotechnol. Adv. 27 (6), 833–848.

Benjamin, M.M., 1981. Multiple-site adsorption of Cd, Cu, Zn and Pb on amorphous iron oxy-hydroxide. J. Colloid Interface Sci. 79, 209–221.

Blanchard, G., Maunaye, M., Martin, G., 1984. Removal of heavy metals from waters by means of natural zeolites. Water Res. 18 (12), 1501–1507.

Boschi, C., Maldonado, H., Ly, M., Guibal, E., 2011. Cd(II) biosorption using Lessonia kelps. J. Colloid Interface Sci. 357, 487–496.

Boyd, G.E., Adamson, A.W., Myers, L.S., 1947. The exchange adsorption of ions from aqueous solutions by organic zeolites. II. Kinetics. J. Am. Chem. Soc. 69, 2836–2848.

Brouers, F., 2014. The fractal (BSF) kinetics equation and its approximations. J. Mod. Phys. 5, 1594–1601.

Caroni, A.L.P.F., de Lima, C.R.M., Pereira, M.R., Fonseca, J.L.C., 2009. The kinetics of adsorption of tetracycline on chitosan particles. J. Colloid Interface Sci. 340, 182–191.

Castillejos, E., Rodríguez-Ramos, I., Soria Sánchez, M., Muñoz, V., Guerrero-Ruiz, A., 2011. Phenol adsorption from water solutions over microporous and mesoporous carbon surfaces: a real time kinetic study. Adsorption 17, 483–488.

Chowdhury, S., Mishra, R., Saha, P., Kushwaha, P., 2011. Adsorption thermodynamics, kinetics and isosteric heat of adsorption of malachite green onto chemically modified rice husk. Desalination 265, 159–168.

Crank, J., 1979. The Mathematics of Diffusion. Clarendon Press, Oxford, UK.

Crini, G., Peindy, H., Gimbert, F., Robert, C., 2007. Removal of C.I. Basic green 4 (malachite green) from aqueous solutions by adsorption using cyclodextrin-based adsorbent: kinetic and equilibrium studies. Separ. Purif. Technol. 53, 97–110.

Delgado-González, L., 2021. Granulated apatite filters for phosphorous retention in treatment wetlands: experience from full-scale applications. J. Water Proc. Eng. 40, 10192.

Do, D.D., Wang, K., 1998. Dual diffusion and finite mass exchange model for adsorption kinetics inactivated carbon. AlChE J. 44, 68–82.

Ebelegi, A.N., Ayawei, N., Wankasi, D., 2020. Interpretation of adsorption thermodynamics and kinetics. Open J. Phys. Chem. 10 (03), 166–182.

El Bardiji, N., Ziate, K., Naji, A., Saidi, M., 2020. Fractal-like kinetics of adsorption applied to the solid/solution interface. ACS Omega 5, 5105–5115.

El Boundati, Y., Ziat, K., Naji, A., Saidi, M., 2019. Generalized fractal-like adsorption kinetic models: application to adsorption of copper on Argan nut shell. J. Mol. Liq. 276, 15–26.

El-Khaiary, M.I., Malash, G.F., 2011. Common data analysis errors in batch adsorption studies. Hydrometallurgy 105, 314–320.

Elkady, M.F., Ibrahim, A.M., El-Latif, M.M.A., 2011. Assessment of the adsorption kinetics, equilibrium and thermodynamic for the potential removal of reactive red dye using eggshell biocomposite beads. Desalination 278, 412–423.

Elovich, S.Y., Larinov, O.G., 1962. Theory of adsorption from solutions of non-electrolytes on solid (I) equation adsorption from solutions and the analysis of its simplest form, (II) verification of the equation of adsorption isotherm from solutions. Izv. Akad. Nauk. SSSR, Otd. Khim. Nauk. 2, 209–216.

Erdoğan, B.C., Ulku, S., 2012. Cr (VI) sorption by using clinoptilolite and bacteria loaded clinoptilolite rich mineral. Microporous Mesoporous Mater. 152, 253–261.

Erosa, M.S.D., Medina, T.I.S., Mendoza, R.N., Rodriguez, M.A., Guibal, E., 2001. Cadmium sorption on chitosan sorbents: kinetic and equilibrium studies. Hydrometallurgy 61, 157–167.

Fashi, F., Ghaemi, A., Moradi, P., 2018. Piperazine-modified activated alumina as a novel promising candidate for CO_2 capture: experimental and modeling. Greenh. Gas Sci. Technol. 9, 37–51.

Frusawa, T., Smith, J.M., 1973. Fluid-particle and intraparticle mass transport rates in slurries. Ind. Eng. Chem. Fundam. 12, 197–203.

Gaspard, S., Altenor, S., Passe-Coutrin, N., Ouensanga, A., Brouers, F., 2006. Parameters for a new kinetic equation to evaluate activated carbon efficiency for water treatment. Water Res. 40, 3467–3477.

Gaulke, M., Guschin, V., Knapp, S., Pappert, S., Eckl, W., 2016. A unified kinetic model for adsorption and desorption: applied to water on zeolite. Microporous Mesoporous Mater. 233, 39–43.

George, W.K., Emik, S., Öngen, A., Kurtulus Özcan, H., Aydın, S., 2018. Modelling of Adsorption Kinetic Processes—Errors, Theory and Application, Advanced Sorption Process Applications. Serpil Edebali, IntechOpen. https://doi.org/10.5772/intechopen.80495.

Giraldo, L., Rodriguez-Estupiñán, P., Moreno-Piraján, J.C., 2019. Isosteric heat: comparative study between Clausius–Clapeyron, CSK and adsorption calorimetry methods. Processes 7, 203.

Guo, X., Wang, J.L., 2019. The phenomenological mass transfer kinetics model for Sr^{2+} sorption onto spheroids primary microplastics. Environ. Pollut. 250, 737–745.

Guo, X., Wang, J.L., 2019a. A general kinetic model for adsorption: theoretical analysis and modeling. J. Mol. Liq. 288. Article 111100.

Guo, X., Wang, J.L., 2019c. Sorption of antibiotics onto aged microplastics in freshwater and seawater. Mar. Pollut. Bull. 149. Article 110511.

Guo, X., Liu, Y., Wang, J.L., 2019. Sorption of sulfamethazine onto different types of microplastics: a combined experimental and molecular dynamics simulation study. Mar. Pollut. Bull. 145, 547–554.

Guo, X., Chen, C., Wang, J.L., 2019b. Sorption of sulfamethoxazole onto six types of microplastics. Chemosphere 228, 300–308.

Gupta, V.K., 1998. Equilibrium uptake, sorption dynamics, process development and column operations for the removal of copper and nickel from aqueous solution and wastewater using activated slag, a low-cost adsorbent. Ind. Eng. Chem. Res. 37, 192–202.

Hai, N.T., You, S.-J., Hosseini-Bandegharaei, A., Chao, H.-P., 2017. Mistakes and inconsistencies regarding adsorption of contaminants from aqueous solutions: a critical review. Water Res. 120, 88–116.

Hamissa, B.,A.M., Brouers, F., Ncibi, M.C., Seffen, M., 2014. Kinetic modeling on methylene blue sorption onto *Agave americana* fibers: fractal kinetics and regeneration studies. Separ. Sci. Technol. 48, 2834–2842.

Higachi, K., Ito, H., Oishi, I., 1984. Principles of Adsorption and Adsorption Processes. Wiley, New York, pp. 71–73.

Hines, A.L., Maddox, R.N., 1985. Mass Transfer: Fundamentals and Applications. Prentice-Hall.

Ho, Y.S., McKay, G., 1998. A comparison of chemisorption kinetic models applied to pollutant removal on various sorbents. Process Saf. Environ. Protect. 76 (4), 332–340.

Ho, Y.S., Mckay, G., 1999. The sorption of lead (II) ions on peat. Water Res. 33, 578–584.

Horsfall, M., Spiff, A.I., 2005. Effect of 2-mercapto ethanoic acid treatment of fluted pumpkin waste (*Telfairia occidentalis* Hook. F.) on the sorption Ni^{2+} ions from aqueous solution. J. Sci. Ind. Res. 64, 613–620.

Hu, J.Q., Yang, S.Z., Guo, L., Xu, X., Yao, T., Xie, F., 2017. Microscopic investigation on the adsorption of lubrication oil on microplastics. J. Mol. Liq. 227, 351–355.

Hu, Q., Wang, Q., Feng, C., Zhang, Z., Lei, Z., Shimizu, K., 2018. Insights into mathematical characteristics of adsorption models and physical meaning of corresponding parameters. J. Mol. Liq. 254, 20–25.

Hussein, Z., Kumar, R., Meghavatu, D., 2018. Kinetics and thermodynamics of adsorption process using a spent-FCC catalyst. Int. J. Eng. Technol. 7, 84–287.

Juan, R.-S., Chen, M.-L., 1997. Application of the Elovich equation to the kinetics off metal sorption with solvent-impregnated resins. Ind. Eng. Chem. Res. 36, 813–820.

Lagergren, S., 1898. About the theory of so-called adsorption of soluble substances. K. Sven. Vetenskapsakad. Handl. 24, 1–39.

Largitte, L., Pasquier, R., 2016a. A review of the kinetics adsorption models and their application to the adsorption of lead by an activated carbon. Chem. Eng. Res. Des. 109, 495–504.

Largitte, L., Pasquier, R., 2016b. New models for kinetics and equilibrium homogeneous adsorption. Chem. Eng. Res. Des. 112, 289–297.

Lazaridi, N.K., Asouhidou, D.D., 2003. Kinetics of sorptive removal of chromium(IV) from aqueous solutions by calcined Mg-Al-CO_3 hydrotalcite. Water Res. 37, 2875–2882.

Leyva-Ramos, R., Geankoplis, C.J., 1985. Model simulation and analysis of surface diffusion of liquids in porous solids. Chem. Eng. Sci. 40, 799–807.

Lima, É.C., Adebayo, M.A., Machado, F.M., 2015. Kinetic and Equilibrium Models of Adsorption, Carbon Nanomaterials as Adsorbents for Environmental and Biological Applications. Springer, pp. 33–69.

Liu, S., Liu, Y., Jiang, L., Zeng, G., Li, Y., Zeng, Z., Wang, X., Ning, Q., 2019. Removal of 17b-Estradiol from water by adsorption onto montmorillonite-carbon hybrids derived from pyrolysis carbonization of carboxymethyl cellulose. J. Environ. Manag. 236, 25–33.

Mahapatra, M., Khatun, S., Anand, S., 2009. Kinetics and thermodynamics of lead (II) adsorption onto lateritic nickel ores of Indian origin. Chem. Eng. J. 155, 184–190.

Malash, G.F., El-Khaiary, M.I., 2010. Piecewise linear regression: a statistical method for the analysis of experimental adsorption data by the intraparticle-diffusion models. Chem. Eng. J. 163, 256–263.

Mathews, A.P., Weber, W.J.J., 1976. Effects of external mass transfer and intraparticle diffusion on adsorption rates in slurry reactors. AICHE Symp. Ser. 166, 91–107.

Mathews, A.P., Weber, W.J.J., 1977. Effects of external mass transfer and intraparticle diffusion on adsorption rates in slurry reactors. AICHE Symp. Ser. 73, 91–98.

Mclintock, I., 1967. The Elovich equation in chemisorption kinetics. Nature 216, 1204–1205.

Meng, F.W., 2005. Study on a Mathematical Model in Predicting Breakthrough Curves of Fixed-Bed Adsorption onto Resin Adsorbent. MS Thesis, Nanjing University, Nanjing, pp. 28–36.

Menkiti, M.C., Aneke, M.C., Ejikeme, P.M., Onukwuli, O.D., Menkiti, N.U., 2014. Adsorptive treatment of brewery effluent using activated chrysophyllum albidum seed shell carbon. SpringerPlus 3. Article No. 213.

Moussavi, G., Alahabadi, A., Yaghmaeian, K., Eskandari, M., 2013. Preparation, characterization and adsorption potential of the NH4Cl-induced activated carbon for the removal of amoxicillin antibiotic from water. Chem. Eng. J. 17, 119–128.

Muhtaseb, A.H., Ibrahim, K.A., Albadarin, A.B., Ali-khashman, O., Walker, G.M., Ahmad, M.N.M., 2011. Remediation of phenol-contaminated water by adsorption using poly(methyl methacrylate) (PMMA). Chem. Eng. J. 168, 691–699.

Ngah, W.S.W., Kamari, A., Koay, Y., 2004. Equilibrium and kinetics studies of adsorption of copper(II) on chitosan and chitosan/PVA beads. Int. J. Biol. Macromol. 34, 155–161.

Ocampo-Pérez, R., Abdel Daiem, M.M., Rivera-Utrilla, J., Méndez-Díaz, J., Sánchez-Polo, M., 2012. Modeling adsorption rate of organic micropollutants present in landfill leachates onto granular activated carbon. J. Colloid Interface Sci. 385, 174–182.

Ofomaja, A.E., 2010. Intraparticle diffusion process for lead (II) biosorption onto mansonia wood sawdust. Bioresour. Technol. 101, 5868–5876.

Özer, A., 2007. Removal of Pb (II) ions from aqueous solutions by sulphuric acid-treated wheat bran. J. Hazard Mater. 14, 753–761.

Özer, A., Akkaya, G., Turabik, M., 2005. The biosorption of Acid Red 337 and Acid Blue 324 on Enteromorpha prolifera: the application of nonlinear regression analysis to dye biosorption. Chem. Eng. J. 112, 181–190.

Padmavathy, V., 2008. Biosorption of nickel(II) ions by baker's yeast: kinetic, thermodynamic and desorption studies. Bioresour. Technol. 99, 3100–3109.

Podder, M.S., Majumder, C.B., 2017. Biosorption of as(III) and as(V) on the surface of TW/ $MnFe_2O_4$ composite from wastewater: kinetics, mechanistic and thermodynamics. Appl. Water Sci. 7, 2689–2715.

Prasad, R.K., Srivastava, S.N., 2009. Sorption of distillery spent wash onto fly ash: kinetics and mass transfer studies. Chem. Eng. J. 146, 90–97.

Qiu, H., Lu, L.V., Pan, B.C., Zhang, Q.J., Zhang, W.M., Zhang, Q.X., 2009. Critical review in adsorption kinetic models. J. Zhejiang Univ. Sci. 10, 716–724.

Reichenberg, D., 1953. Properties of ion-exchange resins in relation to their structure. III. Kinetics of exchange. J. Am. Chem. Soc. 75, 589–597.

Ritchie, A.G., 1977. Alternative to the Elovich equation for the kinetics of adsorption of gases on solids. J. Chem. Soc., Faraday Trans. 1: Phys. Chem. Condens. Phases 73, 1650–1653.

Roy, D., Wang, G.-T., Adrian, D.D., 1993. A simplified solution technique for carbon adsorption model. Water Res. 27, 1033–1040.

Ruthven, D.M., 1984a. Principles of Adsorption and Adsorption Processes. Wiley, London.

Ruthven, D.M., 1984b. Principles of Adsorption and Adsorption Processes. Wiley, New York.

Sabarinathan, G., Karuppasamy, P., Vijayakumar, C.T., Arumuganathan, T., 2019. Development of methylene blue removal methodology by adsorption using molecular polyoxometalate: kinetics, thermodynamics and mechanistic study. Microchem. J. 146, 315–326.

Saha, P., Cowdhury, S., 2011. Insights into Adsorption Thermodynamics.

Schwaab, M., Steffani, E., Barbosa-Coutinho, E., Severo Júnior, J.B., 2017. Critical analysis of adsorption/diffusion modelling as a function of time square root. Chem. Eng. Sci. 173, 179–186.

Sharma, P., Das, M.R., 2012. Removal of a cationic dye from aqueous solution using graphene oxide nanosheets: investigation of adsorption parameters. J. Chem. Eng. Data 58, 151–158.

Simonin, J.P., 2016. On the comparison of pseudo-first order and pseudo-second order rate laws in the modeling of adsorption kinetics. Chem. Eng. J. 300, 254–263.

Souza, R.P., Dotto, G.L., Salau, N.P.G., 2017. Detailed numerical solution of pore volume and surface diffusion model in adsorption systems. Chem. Eng. Res. Des. 122, 298–307.

Thaligari, S.K., Srivastava, V.C., Prasad, B., 2016. Adsorptive desulfurization by zinc-impregnated activated carbon: characterization, kinetics, isotherms, and thermodynamic modeling. Clean Technol. Environ. Policy 18, 1021–1030.

Traegner, U.K., Suidan, M.T., 1989. Evaluation of surface and film diffusion coefficients for carbon adsorption. Water Res. 23, 267–273.

Tran, H.N., You, S.-J., Chao, H.-P., 2017. Fast and efficient adsorption of methylene green 5 on activated carbon prepared from new chemical activation method. J. Environ. Manag. 188, 322–336.

Tseng, R.L., Wu, P.H., Wu, F.C., Juang, R.S., 2014. A convenient method to determine kinetic parameters of adsorption processes by nonlinear regression of pseudo-nth-order equation. Chem. Eng. J. 237, 153–161.

Turner, A., Holmes, L.A., 2015. Adsorption of trace metals by microplastic pellets in fresh water. Environ. Chem. 12, 600–610.

Velinov, N., Najdanovic, S., Vucic, M.R., Mitrovic, J., Kostic, M., Bojic, D., Bojic, A., 2019. Biosorption of loperamide by cellulosic-Al_2O_3 hybrid: optimization, kinetic, isothermal and thermodynamic studies. Cellul. Chem. Technol. 53, 175–189.

Viegas, R.M.C., Campinas, M., Costa, H., et al., 2014. How do the HSDM and Boyd's model compare for estimating intraparticle diffusion coefficients in adsorption processes. Adsorption 20, 737–746.

Walter, W.J., 1984. Evolution of a technology. J. Environ. Eng. 110 (5), 899–917.

Wang, Y., Gao, B.-Y., Yue, W.-W., Yue, Q.-Y., 2007. Adsorption kinetics of nirates from aqueous solutions onto modified wheat residue. Colloids Surf. A Physicochem. Eng. Asp. 308, 1–5.

Wang, B., Gao, B., Wan, Y., 2018. Comparative study of calcium alginate, ball-milled biochar, and their composites on aqueous methylene blue adsorption. Environ. Sci. Pollut. Res. 26, 1–7.

Weber, W.J., Morris, J.C., 1962. Advances in water pollution research: removal of biologically resistant pollutant from wastewater by adsorption. Proceedings of 1st International Conference on Water Pollution Symposium, vol. 2, pp. 231–266.

Weber, W.J., Morris, J.C., 1963. Kinetics of adsorption on carbon from solution. ASCE Sanit. Eng. Div. J. 1, 1–2.

Wu, J., Xia, A., Chen, C., Feng, L., Su, X., Wang, X., 2019. Adsorption thermodynamics and dynamics of three typical dyes onto bio-adsorbent spent substrate of *Pleurotus eryngii*. Int. J. Environ. Res. Publ. Health 16, 2–11.

Xia, Y., Yang, T., Zhu, N., Li, D., Chen, Z., Lang, Q., Liu, Z., Jiao, W., 2019. Enhanced adsorption of Pb (II) onto modified hydrochar: modeling and mechanism analysis. Bioresour. Technol. 288. Article 121593.

Xu, P., Ge, W., Chai, C., Zhang, Y., Jiang, T., Xia, B., 2019. Sorption of polybrominated diphenyl ethers by microplastics. Mar. Pollut. Bull. 145, 260−269.

Yakout, S.M., Elsherif, E., 2010. Batch kinetics isotherm and thermodynamic studies of studies of adsorption of strontium from aqueous solution onto low cost rice-straw based carbons. Carbon-Sci. Technol. 3, 144−153.

Zhang, Q., Crittenden, J., Hristovski, K., Hand, D., Westerhoff, P., 2009. User-oriented batch reactor solutions to the homogeneous surface diffusion model for different activated carbon dosages. Water Res. 43, 1859−1866.

Zhuang, S.T., Liu, Y., Wang, J.L., 2020. Covalent organic frameworks as efficient adsorbent for sulfamerazine removal from aqueous solution. J. Hazard Mater. 383. Article 121126.

Chapter 4

Isotherm models of adsorption processes on adsorbents and nanoadsorbents

1. Isotherm adsorption models

Adsorption isotherm represents the relationship between the adsorbate in the surrounding phase and adsorbate adsorbed on the surface of the adsorbent at equilibrium and constant temperature. This graph is of immense importance to research dealing with adsorption techniques. Adsorption isotherm is fundamentally essential to describe how solutes interact with adsorbents and to optimize the use of adsorbents. In general, isotherm data are correlated using a variety of isotherm models, with the best-fitting model being used to examine adsorption behavior. The adsorption isotherms of distinct pairs are different. Freundlich and Langmuir isotherms are the two primary methods utilized to predict the capacity of adsorption of a certain material. However, adsorption mechanisms can be studied by several ways including the following:

(i) By modeling the adsorption equilibrium results;
(ii) By characterizing the used adsorbent before and after adsorption process;
(iii) By employing the molecular dynamics; and
(iv) By applying the density functional theory (DFT).

Isotherm models of adsorption afford information about the mechanisms involved in adsorption processes, which is important for designing adsorption systems. Furthermore, these models may deliver information on the maximum capacity of adsorption, that is significant for evaluating the performance of adsorbents (Wang and Guo, 2020a,b).

The models of isotherms can be categorized in the following manner based on their physical meaning and their theoretical derivation (Fig. 4.1):

(i) Empirical models: those adsorption isotherm models include linear, Sips, Freundlich, and Toth models that lack a specific physical meaning.

Surface Science of Adsorbents and Nanoadsorbents. https://doi.org/10.1016/B978-0-12-849876-7.00009-9

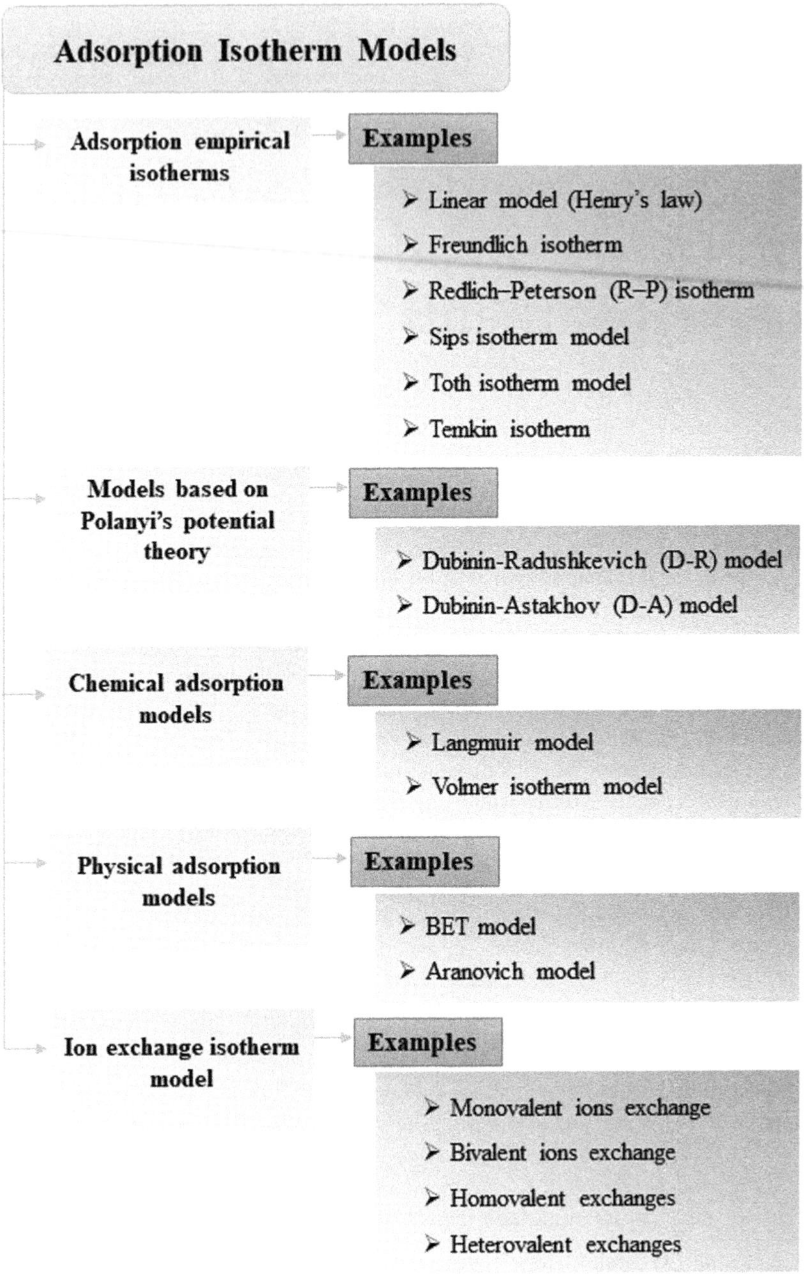
Adsorption Isotherm Models
Adsorption empirical isotherms
Examples
➢ Linear model (Henry's law)
➢ Freundlich isotherm
➢ Redlich–Peterson (R–P) isotherm
➢ Sips isotherm model
➢ Toth isotherm model
➢ Temkin isotherm
Models based on Polanyi's potential theory
Examples
➢ Dubinin-Radushkevich (D-R) model
➢ Dubinin-Astakhov (D-A) model
Chemical adsorption models
Examples
➢ Langmuir model
➢ Volmer isotherm model
Physical adsorption models
Examples
➢ BET model
➢ Aranovich model
Ion exchange isotherm model
Examples
➢ Monovalent ions exchange
➢ Bivalent ions exchange
➢ Homovalent exchanges
➢ Heterovalent exchanges

FIGURE 4.1 General classifications of some of the adsorption isotherms.

(ii) Isotherm models based on Polanyi's potential theory: those models, i.e., the Dubinin–Radushkevich (D–R) and Dubinin–Astakhov (D–A), are considered semiempirical models and commonly applied to model the adsorption of porous adsorbents.

(iii) Chemical adsorption models: the chemical isotherm models are used to describe monolayer adsorption. The chemical adsorption models relate to the formation of chemical bonds.

(iv) Physical adsorption models: the physical isotherms are used to describe multilayer adsorption. The physical adsorption relates to the van der Waals force.

(v) Ion exchange models: much like the literal meaning, ion exchange isotherms can be used to model the adsorption process based on ion exchange.

The chemical, physical, and ion exchange models are considered as theoretical models with rigorous deduction and specific physical meanings (Fig. 4.2).

Several isotherm models are used to investigate the mechanisms of adsorption of different systems. Examples include the linear model, Freundlich model (Freundlich, 1906), Sips model (Sips, 1948), Langmuir model (Langmuir, 1918), Brunauer, Emmett, and Teller (BET) model (Brunauer et al., 1938), and Temkin model (Temkin and Pyzhev, 1940). Empirical models require further investigation using their derivations to obtain clear physical meanings.

The commonly used models are discussed in the following sections.

2. Adsorption empirical isotherms

This class of isotherms includes several models. Some of them are discussed in the following sections.

2.1 Linear model

The linear isotherm model, or Henry's law, is the relation between the equilibrium concentration of adsorbate (C_e) and adsorption capacity at equilibrium (q_e in mg adsorbate/g adsorbent). The model describes the partition of adsorbates between the liquid and solid phases. Generally, the mechanisms of the partition processes include electrostatic interaction, van der Waals interaction, and hydrophobic interaction (Guo et al., 2019a,b).

The adsorption mechanisms as per the linear model are proposed in Fig. 4.3. The linear model describes the condition when the coverage ratio of the sorption sites is low. Consequently, the linear model can support the

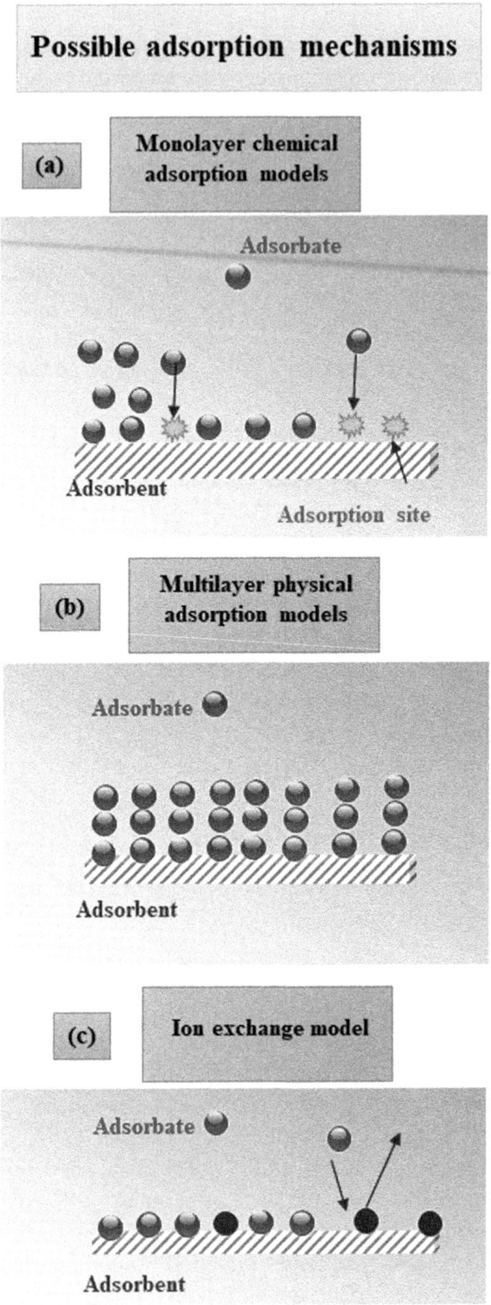
Possible adsorption mechanisms
(a)
Monolayer chemical adsorption models
Adsorbate
Adsorbent
Adsorption site
(b)
Multilayer physical adsorption models
Adsorbate
Adsorbent
(c)
Ion exchange model
Adsorbate
Adsorbent

FIGURE 4.2 Probable main adsorption mechanisms of interaction.

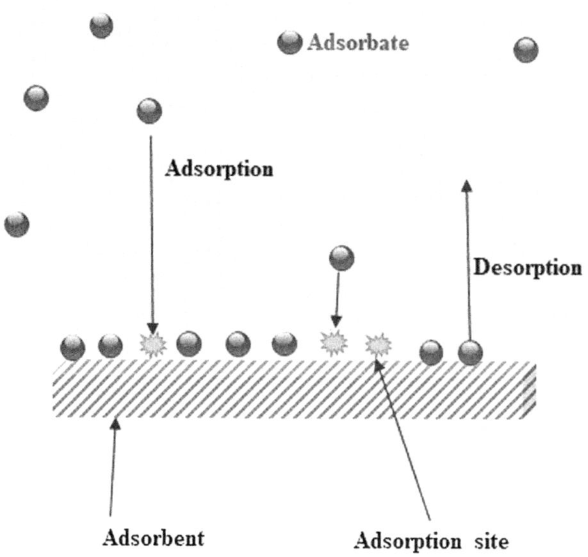
Adsorbate
Adsorption
Desorption
Adsorbent
Adsorption site

FIGURE 4.3 The linear model adsorption mechanisms.

monolayer sorption at low initial adsorbate concentrations of the adsorbate. Langmuir model approximates to Henry's law when the pressure is somehow low in the adsorption of adsorbate molecules on an adsorbent (Khan et al., 2019) (Fig. 4.4; Tables 4.1 and 4.2).

2.2 Freundlich adsorption isotherm

For a given temperature, the Freundlich equation describes how the quantity of gas adsorbed by a unit mass of solid adsorbent changes as the system's pressure changes. This curve is termed as adsorption isotherm (Fig. 4.5). The Freundlich adsorption isotherm is a mathematical relationship between the amount of gas adsorbed on a solid surface and the gas pressure. The isothermal fluctuation of adsorption of an amount of gas adsorbed by unit mass of solid adsorbent with gas pressure was described by Herbert Freundlich (Freundlich, 1909). It is applied to describe nonlinear phenomenon of adsorption (Freundlich, 1906). The relationship can be expressed by the following equation:

$$\frac{X}{m} = K_f . P^{1/n} (n > 1)$$

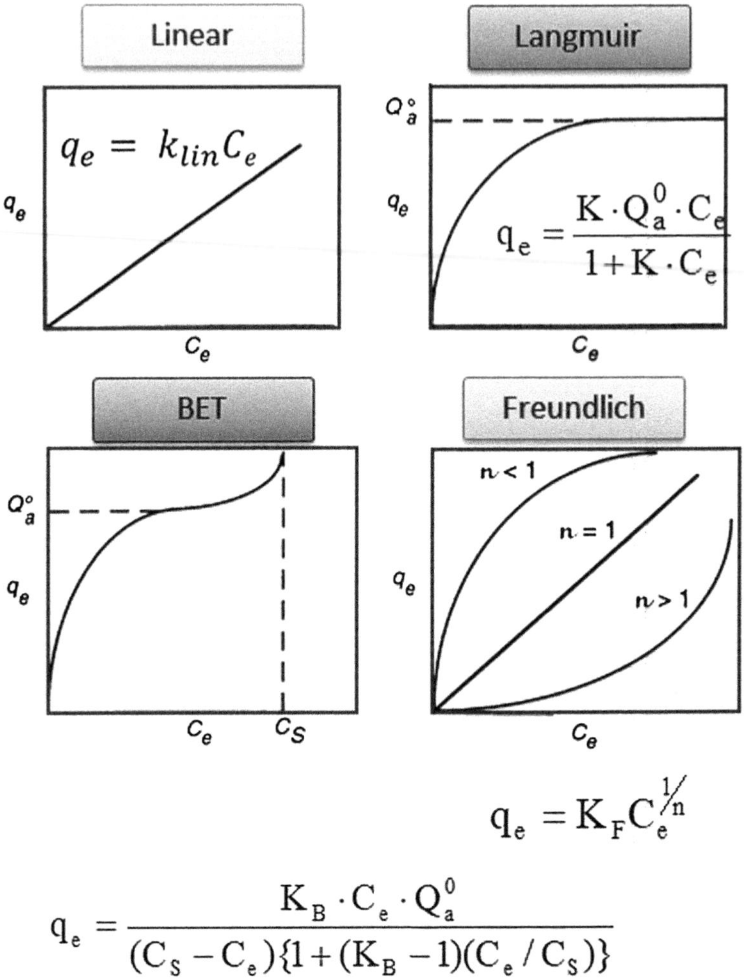
Linear
$q_e = k_{lin}C_e$
q_e
C_e
Langmuir
Q_a^o
q_e
$q_e = \frac{K \cdot Q_a^0 \cdot C_e}{1+K \cdot C_e}$
C_e
BET
Q_a^o
q_e
C_e
C_S
Freundlich
n < 1
n = 1
n > 1
q_e
C_e
$q_e = K_F C_e^{1/n}$
$q_e = \frac{K_B \cdot C_e \cdot Q_a^0}{(C_S - C_e)\{1+(K_B - 1)(C_e / C_S)\}}$

FIGURE 4.4 Comparison of different linear and Langmuir models.

where x refers to the mass of the gas adsorbed on mass (m) of the adsorbent at a pressure P. While k and n are constants depending on the nature of the adsorbent and adsorbate molecules at a certain temperature. The connection is usually depicted as a curve, with the mass of the gas adsorbed per gram of adsorbent plotted against pressure. Physical adsorption decreases with increasing temperature at a fixed pressure, as seen by these charts. At high pressure, these curves always appear to be approaching saturation (Ho and McKay, 1998, 1999).

TABLE 4.1 Mathematical equations for some of the models used in equilibrium isotherm models of adsorption experimental data.

	Linear form	Plot	Calculated coefficient
Linear model	$q_e = KC_e$	q_e versus C_e	K $(\mathrm{L \cdot g^{-1}})$ is the partition coefficient
Langmuir	$\frac{C_e}{q_e} = \frac{C_e}{Q_m} + \frac{1}{Q_m b}$ $\frac{C_e}{q_e} = \frac{1}{K_L q_m} + \frac{C_e}{q_m}$	C_e / q_e versus C_e	$Q_m = 1/\text{slope}$; $b = \text{slope/intercept}$ Langmuir parameter (K_L) and initial concentration (C_o) are used to obtain valuable dimensionless factor (R_L) calculated as $R_L = \frac{1}{1+K_L C_o}$
Freundlich	$\log q_e = \log k_f + \frac{1}{n} \log C_e$ $\ln q_e = \ln K_F + \frac{1}{n} \ln C_e$	$\log q_e$ versus $\log C_e$	$n = 1/\text{slope}$; $k_F = 10^{\text{intercept}}$
Tempkin	$q_e = \frac{RT}{b_T} \ln K_T + \frac{RT}{b_T} \ln C_e$	q_e versus $\ln C_e$	$b = (RT/\text{slope})$; $A = K_T \exp(\text{intercept/slope})$
Dubinin–Radushkevich	$\ln q_e = \ln q_D - B_D \left[RT \ln \left(1 + \frac{1}{C_e} \right) \right]^2$		D–R model is used to obtain the average value of the adsorption free energy (E) as $E = \frac{1}{\sqrt{2 B_D}}$
Redlich–Peterson	$\ln \frac{C_e}{q_e} = \beta \ln C_e - \ln A$		
Toth	$q_e = \frac{K_T C_e}{(\alpha_T + C_e^z)^{1/z}}$		

TABLE 4.2 Examples of some nonlinear equations of the adsorption isotherm models.

Model	Nonlinear form
Langmuir	$q_e = \frac{q_m K_L C_e}{1+K_L C_e}$
Freundlich	$q_e = K_F C_e^{\frac{1}{n}}$
Temkin	$q_e = \frac{RT}{b_T} \ln(K_T C_e)$
Dubinin–Radushkevich	$q_e = q_D \exp^{-B_D \left[RT \ln\left(1+\frac{1}{C_e}\right)\right]^2}$
Redlich–Peterson	$q_e = \frac{AC_e}{1+BC_e^{\beta}}$

where: C_e: is the equilibrium concentration of adsorbate, q_e: is the adsorption capacity at equilibrium (mg adsorbate/g adsorbent), K_L: is the affinity of the adsorption sites (1/unit of concentration), Langmuir parameter (K_L), R_L: the separation factor, which is a valuable dimensionless equilibrium parameter, q_m: is the maximum adsorption capacity (mg adsorbate/g adsorbent), K_F: is the constant of the Freundlich isotherm model (mg adsorbate/g adsorbent), $1/n$: is a constant revealing the adsorption process strength, R: is 0.008314 kJ mol^{-1} K^{-1}, T: is the temperature of the solution (K), b_T: is the constant of the Temkin isotherm model (kJ mol^{-1}), K_T: is the constant of the equilibrium-binding (L g^{-1}), q_D (mg g^{-1}) and B_D (mol^2 kJ^{-1}): are the R–D isotherm model constants, A (L g^{-1}) and B (L $mg^{-1})^{\beta}$: are constants of the Redlich–Peterson isotherm, β: is the exponent of the Redlich–Peterson isotherm model. It is ranging between zero and one. It indicates the adsorbent surface heterogeneity, K_T (mg g^{-1}) is the constant, a_T (mg^z L^{-z}) is the Toth constant, and z is a component which is used to describe the degree of heterogeneity of the adsorption. It is temperature independent, while the value of a_T increases with increasing temperature.

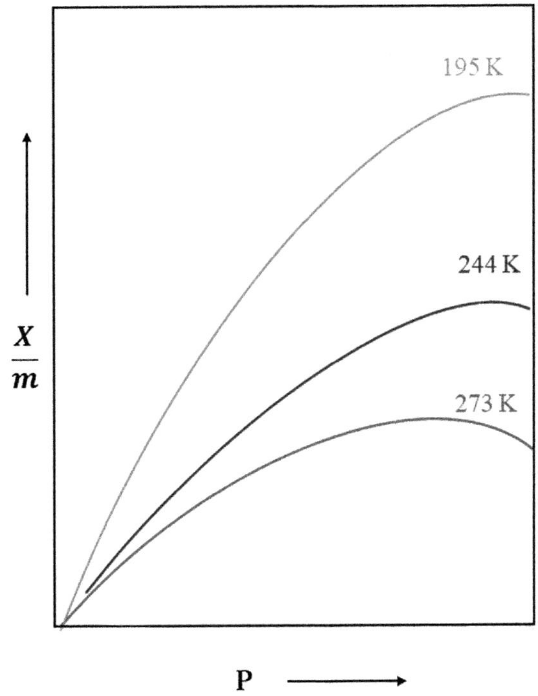

195 K
244 K
273 K
X/m
P

FIGURE 4.5 General Freundlich adsorption isotherm curve.

Moreover, the similar relationship is applicable for the concentrations of a solute adsorbed on the solid surface and concentrations of the solute in the liquid phase, adjusted in the following manner:

$$\frac{X}{m} = K_f.C^{1/n}(n > 1)$$

Both nonlinear and linear equations of the Freundlich model are used. Freundlich equation approximately describes the behavior of adsorption from solution by taking into account the concentration of the solution as opposed to the pressure (Tseng et al., 2014; Elovich and Larinov, 1962; Mclintock, 1967).

Taking log on both sides of the equation gives the following results:

$$\log\frac{x}{m} = \log k_f + \frac{1}{n}\log C_e$$

Or

$$\log q_e = \log k_f + \frac{1}{n}\log C_e$$

The validity of Freundlich isotherm can be verified by plotting log x/m on y-axis (ordinate) and log P on x-axis (abscissa) (Fig. 4.6). If the result is a straight line, the Freundlich isotherm is valid, otherwise it is not. Commonly, the slope of the straight line provides the value of $1/n$ while the intercept on the y-axis provides the value of log k. Thus, Freundlich isotherm roughly describes the behavior of adsorption. The factor 1 can have values between zero and one. Thus, above equation holds good over a limited range of pressure.

To model multilayer adsorption on heterogamous surfaces, the Freundlich isotherm was used (Zaheer et al., 2019). This model represents the adsorption scenario in which the equilibrium coverage fraction is around 50%. As a result, the Freundlich model may reflect both chemical and physical adsorption with a coverage proportion of roughly 50%.

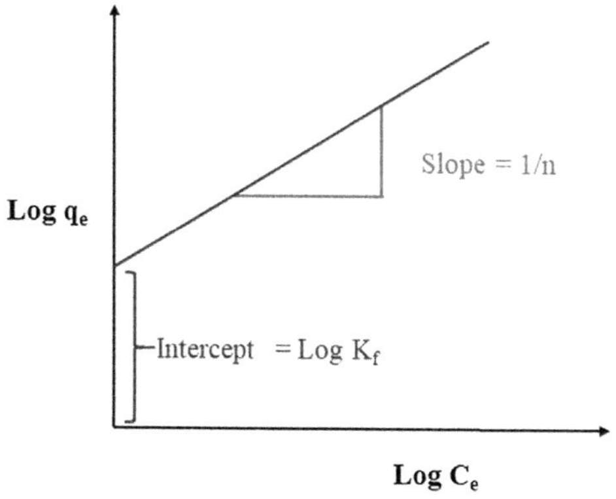
Log q_e
Slope = 1/n
Intercept = Log K_f
Log C_e

FIGURE 4.6 Straight line plot of Freundlich adsorption isotherm equation.

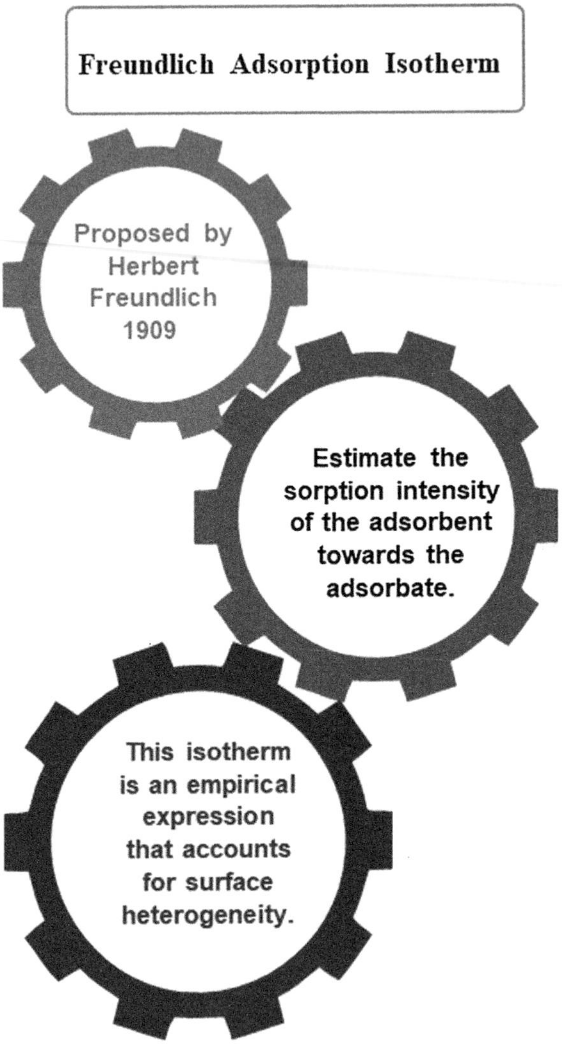
Freundlich Adsorption Isotherm
Proposed by Herbert Freundlich 1909
Estimate the sorption intensity of the adsorbent towards the adsorbate.
This isotherm is an empirical expression that accounts for surface heterogeneity.

FIGURE 4.7 Highlights on the key terms of the Freundlich isotherm model.

Fig. 4.7 displays some highlights on the Freundlich isotherm model with assumptions in Fig. 4.8 and limitations in Fig. 4.9.

2.3 Redlich–Peterson Isotherm

The Redlich–Peterson (R–P) isotherm is considered a three-parameter empirical adsorption hybrid model which incorporates elements from the

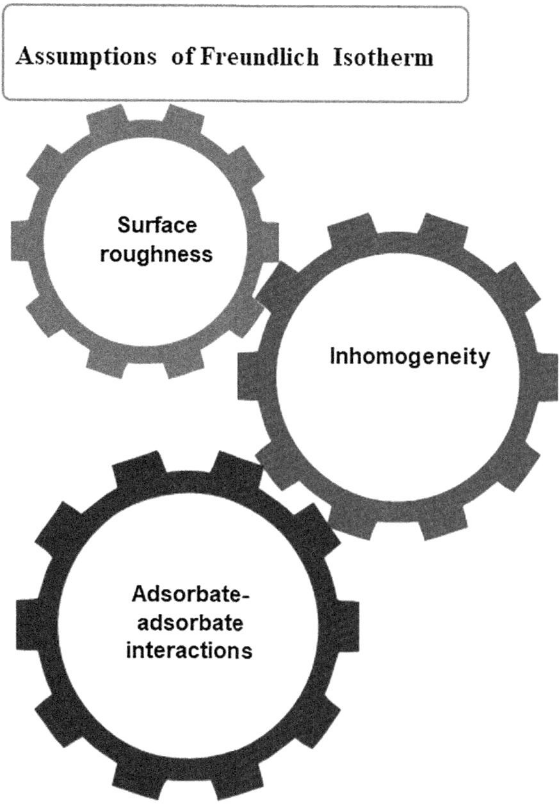
Assumptions of Freundlich Isotherm
Surface roughness
Inhomogeneity
Adsorbate-adsorbate interactions

FIGURE 4.8 Highlights on assumptions of Freundlich isotherm.

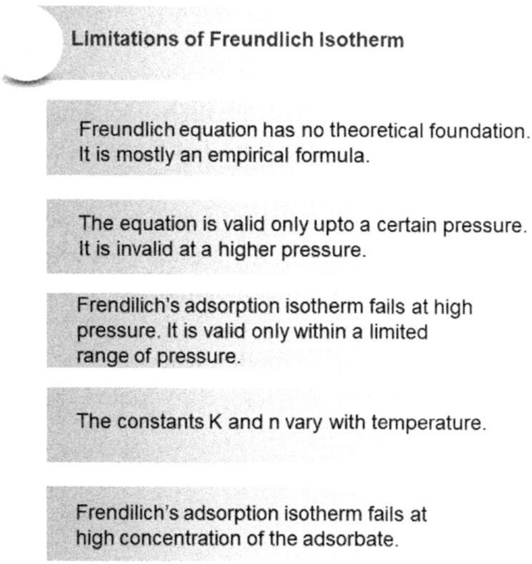
Limitations of Freundlich Isotherm
Freundlich equation has no theoretical foundation. It is mostly an empirical formula.
The equation is valid only upto a certain pressure. It is invalid at a higher pressure.
Frendilich's adsorption isotherm fails at high pressure. It is valid only within a limited range of pressure.
The constants K and n vary with temperature.
Frendilich's adsorption isotherm fails at high concentration of the adsorbate.

FIGURE 4.9 Highlights on limitations of Freundlich isotherm.

FIGURE 4.10 General plot of Redlich–Peterson adsorption isotherm.

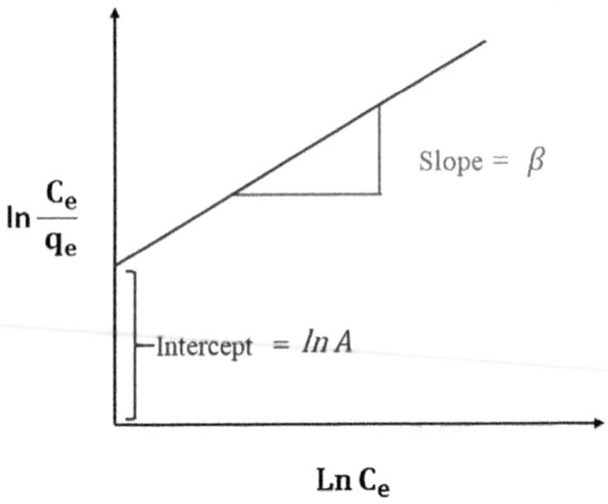

Slope = β
ln Ce/qe
Intercept = ln A
Ln Ce

Langmuir and Freundlich isotherms. Generally, the adsorption mechanism is unique in nature. It does not follow ideal monolayer adsorption characteristics. It has been commonly used in heterogeneous and homogeneous adsorption processes (Redlich and Peterson, 1959).

Redlich–Peterson adsorption capacity constant is achieved via trial and error to attain the maximum linear regression value of the isotherm plot (Fig. 4.10).

2.4 Sips model

The Sips isotherm is a hybrid form of Langmuir and Freundlich models, and it is deduced to predict the adsorption in heterogeneous systems and circumvent the limitation of the rising adsorbate concentration associated with the Freundlich isotherm (Sips, 1948). It is the most applicable three-parameter isotherm model for monolayer adsorption (Ebadi et al., 2015). Furthermore, in some cases, it can be used to describe the heterogeneous or homogeneous processes.

Sips isotherm becomes Freundlich isotherm at low adsorbate concentrations. Sips isotherm predicts monolayer adsorption at high concentrations, similar to the Langmuir isotherm. Consequently, Sips isotherm is used to describe only monolayer adsorption systems. Multilayer adsorption might occur in a high pressure system. Hence, Sips model is used to represent the adsorption at the pressure range of interest, and the *initial concentration* parameter shall not be considered as the ultimate saturation adsorption (Fig. 4.11).

In brief, Sips model becomes the Langmuir model when $n_s = 1$ and becomes the Freundlich model at low C_0. Nevertheless, Sips model does not satisfy Henry's law at low C_0.

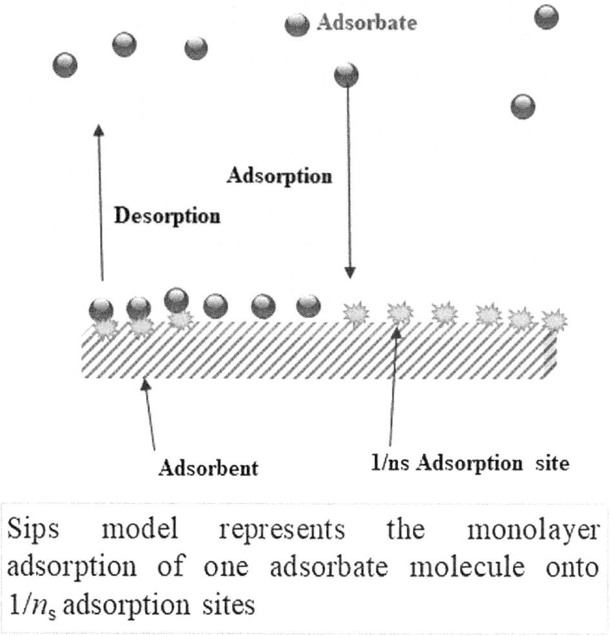
Adsorbate
Adsorption
Desorption
Adsorbent
1/ns Adsorption site
Sips model represents the monolayer adsorption of one adsorbate molecule onto $1/n_s$ adsorption sites

FIGURE 4.11 Adsorption mechanism of Sips isotherm model.

2.5 Toth isotherm model

It is important to widen the applications of the Langmuir model in heterogeneous processes. Thus, Toth developed an equation that assumes the adsorption energies of most adsorption sites as smaller than mean energy (Toth, 1971). When the component describing the degree of heterogeneity of the adsorption process, z, equates to one, Toth model becomes Langmuir isotherm. Generally, larger deviation of z from the unity reveals the adsorption is more of heterogeneous nature.

2.6 Temkin isotherm model

Temkin isotherm model assumes adsorption heat of molecules to decrease linearly with an increase in the coverage of the adsorbent surface. In this model, adsorption is characterized by a uniform distribution of binding energies up to a maximum binding energy. The Temkin model presumes that adsorption is a multilayer process (Temkin and Pyzhev, 1940). Interestingly, the expression of statistical mechanical for the Temkin isotherm was reported by Yang (1993, 2003). The derived equation is substituted into the

FIGURE 4.12 General plot of Temkin adsorption isotherm.

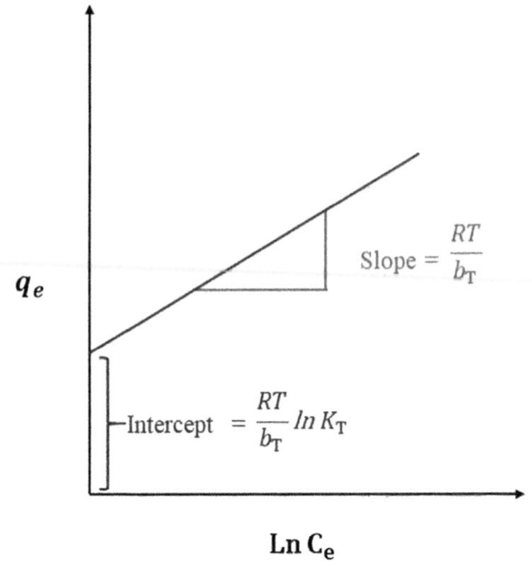

q_e
Slope = $\frac{RT}{b_T}$
Intercept = $\frac{RT}{b_T} \ln K_T$
Ln C_e

Clapeyron–Clausius equation, thus confirming that differential heat of adsorption decreases linearly with an increase in coverage (Figs. 4.12 and 4.13).

3. Adsorption models based on Polanyi's potential theory

This class of isotherms includes several models. As suggested by Wang and Guo (2020a,b), Polanyi's potential theory describes that the adsorption comprises an "adsorption space," where molecules lose potential energies. These energies are temperature independent and increase in the spaces closing to the adsorbent. Mostly, the highest potential energy is attained in the cracks or pores inside the used adsorbent (Polanyi, 1932; Schenz and Manes, 1975). Some of the models are discussed in the following sections.

3.1 Dubinin–Radushkevich model

The D–R equation was developed as an adaptation of the Polanyi potential theory of adsorption. This equation for micropore solids was devised by Dubinin and his colleagues. It was assumed as an empirical isotherm to describe the adsorption of gases onto solids (Dubinin and Radushkevich, 1947). It assumes the distribution of pores in adsorbents to follow Gaussian energy distribution (Polanyi, 1932; Dąbrowski, 2001) (Figs. 4.14 and 4.15).

The D–R isotherm is used to define the adsorption mechanisms; it helps distinguish between chemisorption and physisorption. The D–R isotherm does not reduce to Henry's law at low pressures, which is an essential for thermodynamic consistence.

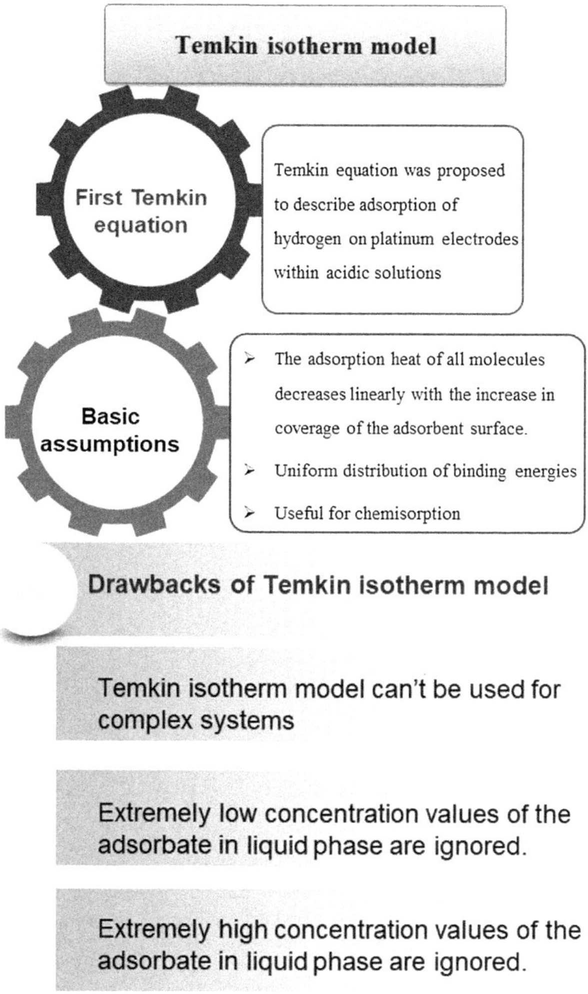
Temkin isotherm model
First Temkin equation
Temkin equation was proposed to describe adsorption of hydrogen on platinum electrodes within acidic solutions
Basic assumptions
The adsorption heat of all molecules decreases linearly with the increase in coverage of the adsorbent surface.
Uniform distribution of binding energies
Useful for chemisorption
Drawbacks of Temkin isotherm model
Temkin isotherm model can't be used for complex systems
Extremely low concentration values of the adsorbate in liquid phase are ignored.
Extremely high concentration values of the adsorbate in liquid phase are ignored.

FIGURE 4.13 Highlights on Temkin isotherm model and its limitations.

3.2 Dubinin–Astakhov model

The D–A model is a more generalized version of the D–R model (Dubinin and Astakhov, 1971). It was developed based on the Polanyi's potential theory as a semiempirical model and was derived based on the statistical mechanical principles (Chen and Yang, 1994). It is applied to investigate the micropore structure of an adsorbent. For the adsorption of adsorbates on adsorbents with mesopores and micropores, the common isotherm model was deduced as the D–A model when the equilibrium coverage fraction was higher than C_e/C_s.

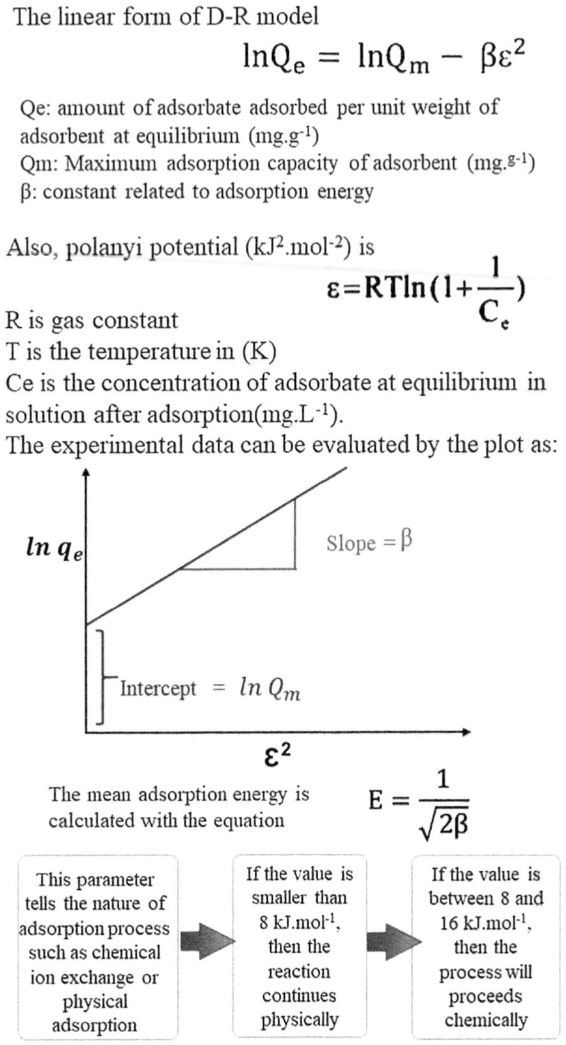
The linear form of D-R model
lnQe = lnQm − βε2
Qe: amount of adsorbate adsorbed per unit weight of adsorbent at equilibrium (mg.g-1)
Qm: Maximum adsorption capacity of adsorbent (mg.g-1)
β: constant related to adsorption energy
Also, polanyi potential (kJ2.mol-2) is
ε=RTln(1+1/Ce)
R is gas constant
T is the temperature in (K)
Ce is the concentration of adsorbate at equilibrium in solution after adsorption(mg.L-1).
The experimental data can be evaluated by the plot as:
ln qe
Slope = β
Intercept = ln Qm
ε2
The mean adsorption energy is calculated with the equation
E = 1/√2β
This parameter tells the nature of adsorption process such as chemical ion exchange or physical adsorption
If the value is smaller than 8 kJ.mol-1, then the reaction continues physically
If the value is between 8 and 16 kJ.mol-1, then the process will proceeds chemically

FIGURE 4.14 Highlights on Dubinin–Radushkevich (D–R) model.

For example, the D–A model was used as a common model for the acetylene adsorption onto MOFs (Cheng and Hu, 2016). This model is used to describe the adsorption in microporous homogeneous systems.

4. Chemical adsorption models

This class of models considers the monolayer adsorption processes where the adsorbate molecules are adsorbed onto the adsorption sites of adsorbents. Both

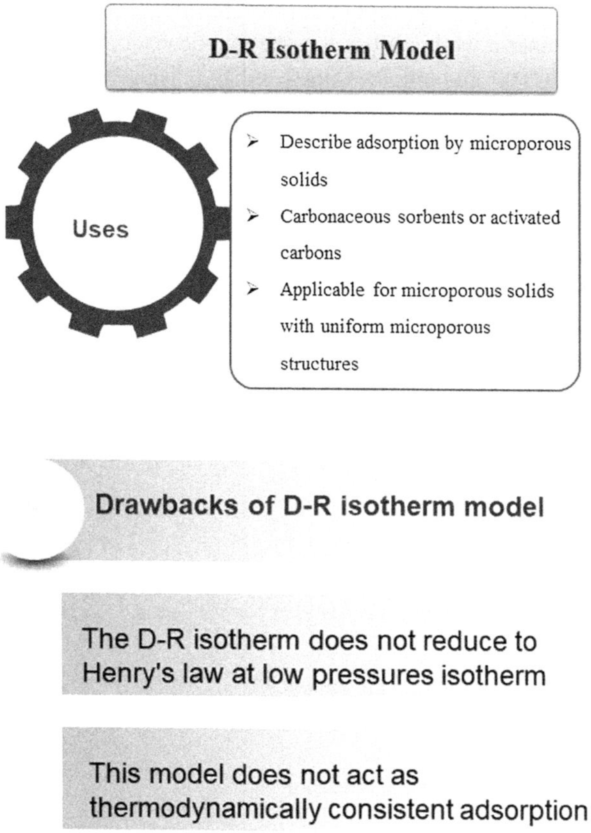
D-R Isotherm Model
Uses
Describe adsorption by microporous solids
Carbonaceous sorbents or activated carbons
Applicable for microporous solids with uniform microporous structures
Drawbacks of D-R isotherm model
The D-R isotherm does not reduce to Henry's law at low pressures isotherm
This model does not act as thermodynamically consistent adsorption

FIGURE 4.15 Uses and drawbacks of Dubinin–Radushkevich (D–R) isotherm model.

Langmuir and Volmer isotherm models are theoretical models with reasonable derivations and with specific physical meanings.

4.1 Langmuir model

Langmuir model explains adsorption by assuming an adsorbate to behave as an ideal gas at isothermal conditions. According to the model, adsorption and desorption are reversible processes (Langmuir, 1916). The Langmuir adsorption isotherms predict adsorption at low adsorption densities and a maximum surface coverage at higher solute metal concentrations. Langmuir adsorption is applicable for monolayer adsorption onto a homogeneous surface when no interaction occurs between adsorbed species.

Furthermore, the Langmuir model can be used to represent monolayer sorption onto the adsorbent's surface and in the pores within the adsorbent.

This can prove that the equilibrium results are sufficiently characterized by the Langmuir isotherm while diffusion is the rate-determining step. However, this model has certain limitations (Souza et al., 2017; Schwaab et al., 2017) (Figs. 4.16 and 4.17).

4.2 Volmer isotherm model

The Volmer model is an adsorption model with a monolayer distribution. It assumes that adsorbate molecules migrate over adsorbent surfaces. The interactions between adsorbates are, however, minimal in this model (Volmer, 1925).

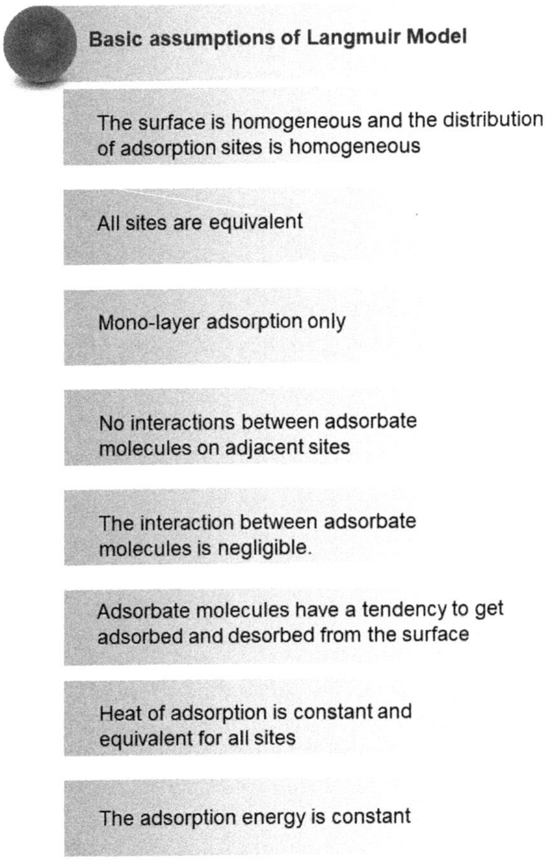
Basic assumptions of Langmuir Model
The surface is homogeneous and the distribution of adsorption sites is homogeneous
All sites are equivalent
Mono-layer adsorption only
No interactions between adsorbate molecules on adjacent sites
The interaction between adsorbate molecules is negligible.
Adsorbate molecules have a tendency to get adsorbed and desorbed from the surface
Heat of adsorption is constant and equivalent for all sites
The adsorption energy is constant

FIGURE 4.16 Some basic assumptions of Langmuir model.

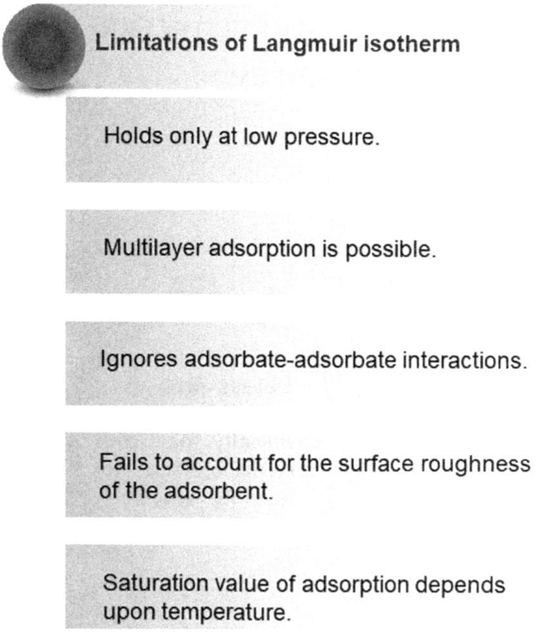
Limitations of Langmuir isotherm
Holds only at low pressure.
Multilayer adsorption is possible.
Ignores adsorbate-adsorbate interactions.
Fails to account for the surface roughness of the adsorbent.
Saturation value of adsorption depends upon temperature.

FIGURE 4.17 Some limitations of Langmuir isotherm.

5. Physical adsorption models

Multilayer adsorption processes are simulated using theoretical physical adsorption isotherm models. The main driving factor of physical adsorption is the van der Waals force. Some models are discussed below.

5.1 BET adsorption isotherm

This theory of multilayer adsorption was proposed by Brunauer, Emmett, and Teller (Brunauer et al., 1938). To depict gas adsorption to multimolecular layers, the BET model was suggested. It's a theoretical multilayer physical sorption model that's been used to calculate porous materials' specific areas as well as pore size distribution (Duong, 1998) (Fig. 4.18).

The following are the assumptions of the isotherm of BET:

- BET theory assumes physisorption to result in the formation of multilayer adsorption. Thus, claiming that adsorption is a multilayer homogeneous process.
- The theory assumes that the solid surface has uniform sites of adsorption, and adsorption at one site does not affect adsorption at neighboring sites.
- The adsorption energy in the first layer differs from other layers.

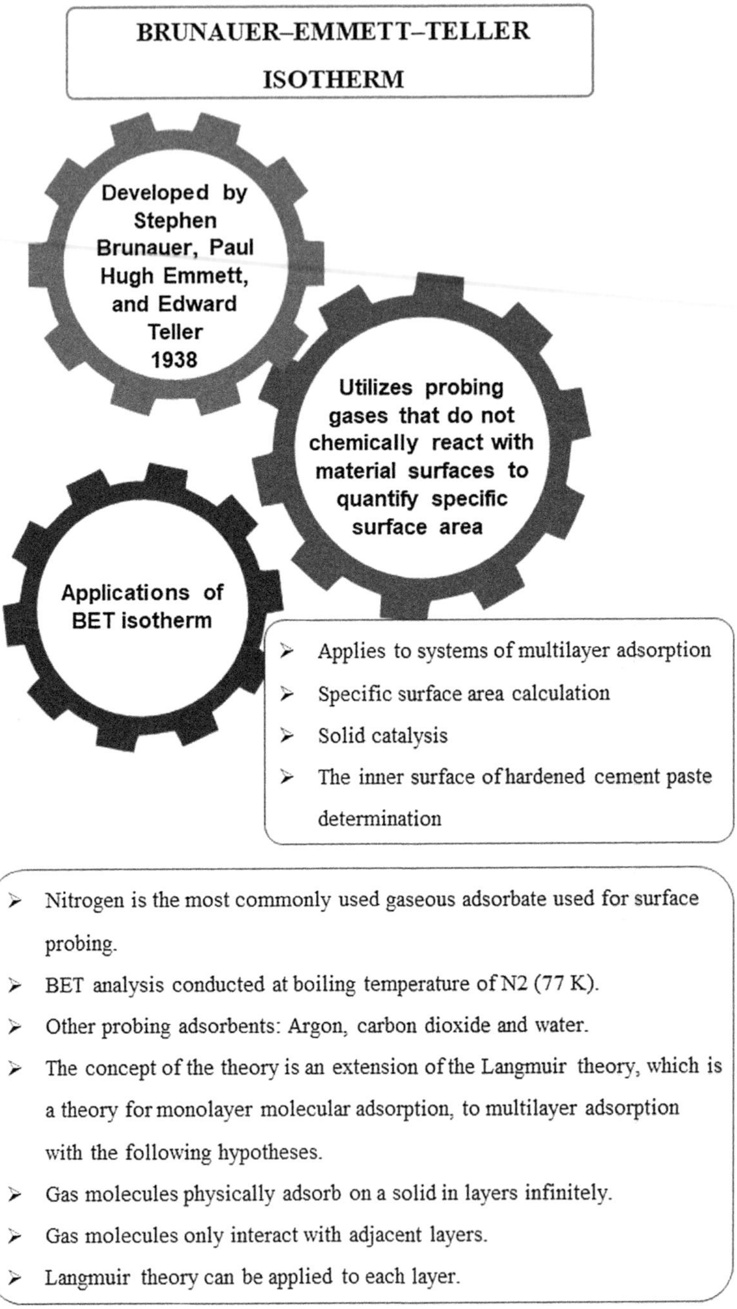
BRUNAUER–EMMETT–TELLER ISOTHERM
Developed by Stephen Brunauer, Paul Hugh Emmett, and Edward Teller 1938
Utilizes probing gases that do not chemically react with material surfaces to quantify specific surface area
Applications of BET isotherm
➢ Applies to systems of multilayer adsorption
➢ Specific surface area calculation
➢ Solid catalysis
➢ The inner surface of hardened cement paste determination
➢ Nitrogen is the most commonly used gaseous adsorbate used for surface probing.
➢ BET analysis conducted at boiling temperature of N2 (77 K).
➢ Other probing adsorbents: Argon, carbon dioxide and water.
➢ The concept of the theory is an extension of the Langmuir theory, which is a theory for monolayer molecular adsorption, to multilayer adsorption with the following hypotheses.
➢ Gas molecules physically adsorb on a solid in layers infinitely.
➢ Gas molecules only interact with adjacent layers.
➢ Langmuir theory can be applied to each layer.

FIGURE 4.18 Highlights of the BET isotherm.

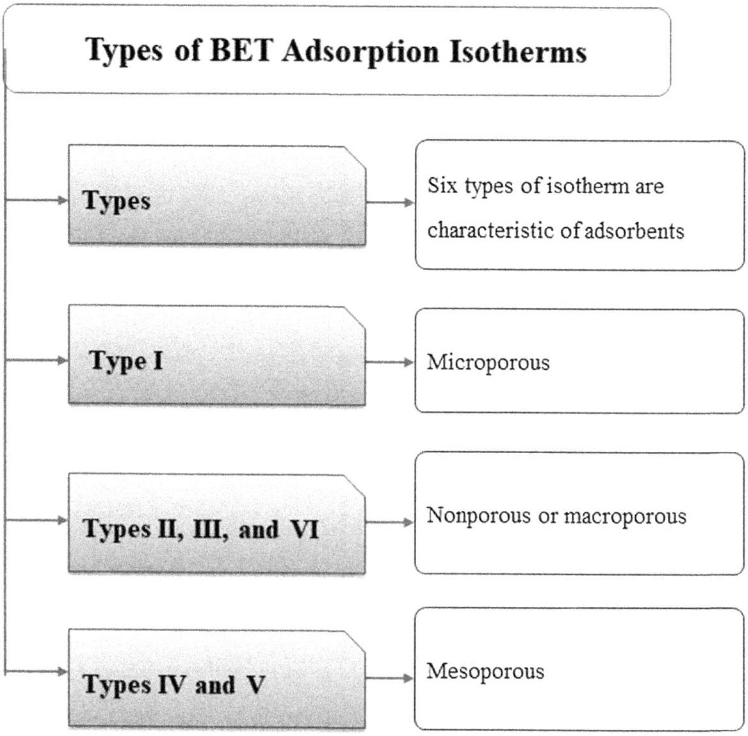
Types of BET Adsorption Isotherms
Types
Six types of isotherm are characteristic of adsorbents
Type I
Microporous
Types II, III, and VI
Nonporous or macroporous
Types IV and V
Mesoporous

FIGURE 4.19 Types of BET isotherms.

- ➢ The adsorption rate equals the desorption rate for each layer.
- ➢ After the formation of the monolayer, the adsorption process can continue with the formation of multilayers involving the second layer, third, and so on.

The greatest amount of adsorbate that may be absorbed by the adsorbent at a given pressure is determined by an adsorption isotherm. The experimental data of different adsorbent–adsorbate combinations must be correlated with different isotherm models to normalize the data. The International Union of Pure and Applied Chemistry (IUPAC) divides adsorption pairs into eight groups based on the type of adsorption isotherms (Thommes et al., 2015; Maeda et al., 2002; Lowell et al., 2004; Hanaor et al., 2014; Galarneau et al., 2018; Rouquerol et al., 2007) (Figs. 4.19, 4.20, and 4.21):

- ➢ Type-I(a) isotherm: this type is used to describe narrow microporous adsorbent that has a pore size ≤ 1 nm.

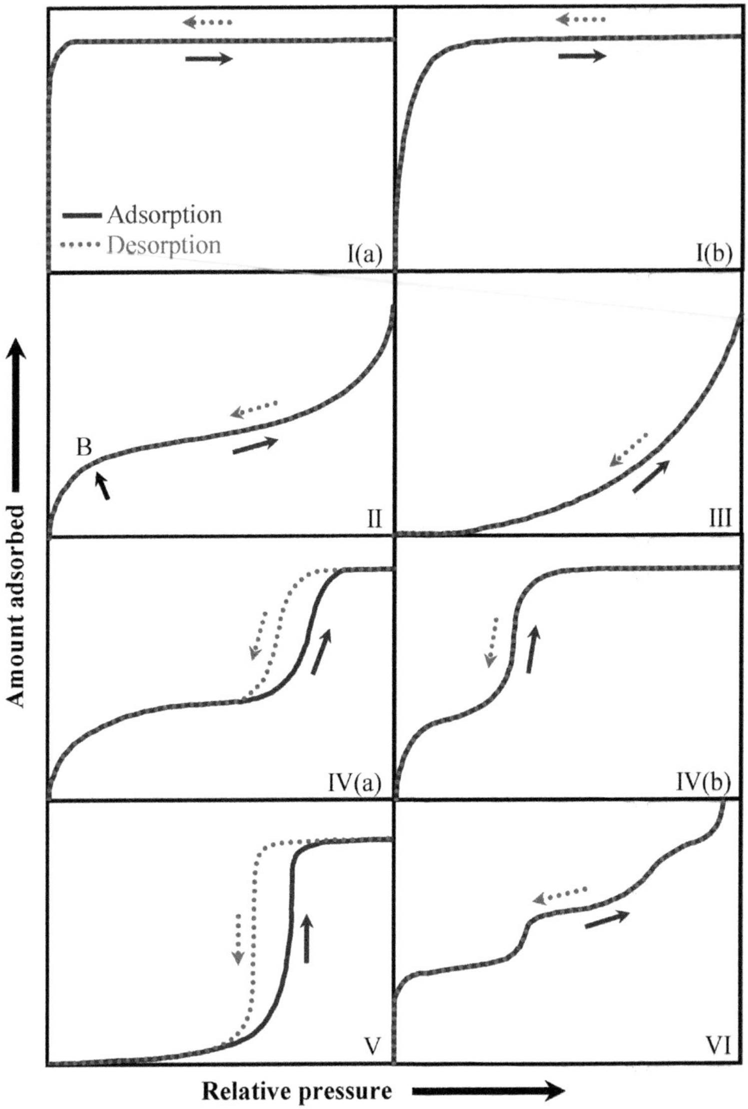
Adsorption
Desorption
I(a)
I(b)
B
II
III
IV(a)
IV(b)
V
VI
Amount adsorbed
Relative pressure

FIGURE 4.20 Classification of Physisorption isotherms (2015 IUPAC) (Rahman et al., 2019; Thommes et al., 2015). B symbolizes monolayer formation.

- The Type-I(b) isotherm: this type is used to describe monolayer adsorption. The adsorption rises steadily with pressure until it reaches a halt at saturation pressure.
- Type-II isotherm: this type is used to describe multilayer adsorption. It is approximately analogous to the Type-I(b) shape, however, the difference

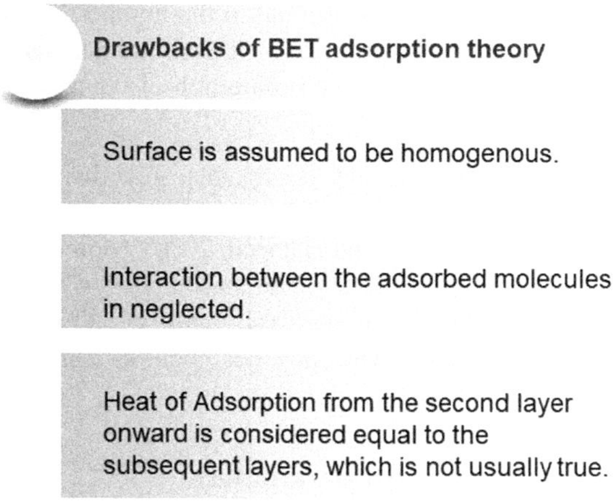
Drawbacks of BET adsorption theory
Surface is assumed to be homogenous.
Interaction between the adsorbed molecules in neglected.
Heat of Adsorption from the second layer onward is considered equal to the subsequent layers, which is not usually true.

FIGURE 4.21 Some drawbacks of BET adsorption theory.

being the absence of plateau in Type-II. Even when the pressure ratio is close to unity, the adsorption continues to rise.

- ≫ Type-III adsorption isotherm: the shape of this type of isotherm is convex (Muttakin et al., 2018). The adsorption is minimal at low pressures, but it increases dramatically at high pressures.
- ≫ Type-IV isotherm: this isotherm is divided into two:
  - Type-IV(a) isotherm: it is with hysteresis, and pore width greater than 4 nm.
  - Type-IV(b) isotherm: it is without hysteresis, and pore width smaller than 4 nm. It is observed for adsorbents that have cylindrical and conical mesoporous with a smaller width, which is completely reversible closed at the tapered end.
- ≫ Type-V isotherm: this type is notable by its characteristic S-shaped isotherm. This type demonstrates a hysteresis loop.
- ≫ Type-VI isotherm: in this type, the adsorption occurs in steps.

5.2 Aranovich model

The Aranovich isotherm is a theoretically corrected polymolecular adsorption isotherm with two parameters. It is applied to represent adsorption over a larger range of adsorbate concentrations (Aranovich, 1992). This method can be used to properly compute the surface areas of porous adsorbents.

The following are the assumptions of the model:

- ≫ Only the "nearest neighbors" interact on the adsorbents' surfaces, which are flat and uniform.

- ≻ The desorption energy is proportional to the number of layers.
- ≻ The model can handle the problem of lateral interactions not being taken into consideration, as well as the prohibition of voids in the adsorbate.

Note:

The linear regression method is suitable, practical, and appropriate. Nevertheless, the adsorption models' linearization can affect the dependent and independent variables (Foo and Hameed, 2010). Moreover, in some cases, this may introduce propagated errors with inaccurate and biased estimation of the model parameters. Nonlinear regression can provide more accurate calculations of model parameters. This, however, requires convenient methods for solving the nonlinear isotherms.

6. Classification based on parameters

There are many other adsorption isotherm models (Ayawei et al., 2017; Foo and Hameed, 2010). Some of these can be classified based on the number of parameters.

6.1 One-parameter isotherm

This is the simplest adsorption isotherm where the amount of surface adsorbate is proportional to the partial pressure of the adsorptive gas or concentration of adsorbate in liquid solution. The most commonly used isotherm in this class is

- ≻ Henry's Isotherm

6.2 Two-parameter isotherm

There are several isotherm models under this category. Examples include the following:

- ≻ Hill-Deboer Model
- ≻ Fowler–Guggenheim Model
- ≻ Langmuir Isotherm
- ≻ Freundlich Isotherm
- ≻ Dubinin–Radushkevich Isotherm
- ≻ Temkin Isotherm
- ≻ Flory–Huggins Isotherm
- ≻ Hill Isotherm
- ≻ Halsey Isotherm
- ≻ Harkin–Jura Isotherm
- ≻ Jovanovic Isotherm
- ≻ Elovich Isotherm
- ≻ Kiselev Isotherm

6.3 Three-parameter isotherms

There are several isotherm models under this category. Examples include the following:

- ➢ Redlich–Peterson Isotherm
- ➢ Sips Isotherm
- ➢ Toth Isotherm
- ➢ Koble–Carrigan Isotherm
- ➢ Kahn Isotherm
- ➢ Radke-Prausniiz Isotherm
- ➢ Langmuir–Freundlich Isotherm
- ➢ Jossens Isotherm

6.4 Four-parameter isotherms

There are several isotherm models under this category. Examples include the following:

- ➢ Fritz–Schlunder Isotherm
- ➢ Baudu Isotherm
- ➢ Weber–van Vliet Isotherm
- ➢ Marczewski–Jaroniec Isotherm

6.5 Five-parameter isotherms

Fritz and Schlunder derived an empirical equation that can fit a wide range of experimental results due to the large number of coefficients in the isotherm (Yaneva et al., 2013). The model is called

- ➢ Fritz and Schlunder five-parameter empirical isotherm model

7. Applications of adsorption isotherms

There are several possible applications of adsorption isotherms. The following points introduce short key highlights on applications of adsorption isotherms:

- ➢ To compute the capacity and percentage removal of adsorbates from a certain media or environment.
- ➢ To acquire the greatest adsorbent absorption and affinity between adsorbent and adsorbate, Langmuir parameters can be applied.
- ➢ Freundlich parameters can be used to obtain adsorption capacity of adsorbents.
- ➢ To calculate the adsorption capacity of adsorbents using Freundlich parameters.

- To calculate the specific surface area and pore size distribution from BET isotherms using BET curves.
- D–R parameters are used for adsorption mechanism.
- Temkin parameters can be used for understanding adsorbent–adsorbate interactions.
- To understand the spontaneity of a system.
- To have more information on the exothermicity of a system.

8. Conclusions

The relationship between the amounts of adsorbate (x) adsorbed on the surface of the adsorbent (m) and pressure (P) at constant temperature is known as the adsorption isotherm. Langmuir isotherms, Freundlich isotherms, BET isotherms, D–R isotherms, and Temkin isotherms are examples of adsorption isotherms. The Langmuir adsorption model assumes that the adsorbent surface has a set number of accessible sites with the same energy. Monolayer adsorption occurs because the surface is homogeneous.

The Freundlich isotherm is an empirical equation that accounts for surface heterogeneity caused by multilayer adsorption as well as the exponential distribution of adsorbent active sites and their energies toward the adsorbate. At greater pressures, the Freundlich adsorption isotherm failed.

The BET theory is used to determine specific surface area in multilayer adsorption systems, and it typically uses probing gases that do not chemically react with adsorbent surfaces. The theory concept is an extension of the Langmuir theory for monolayer molecular adsorption used for multilayer adsorption. The assumption that the surface is homogeneous is not necessarily and always the case. The interaction between the adsorbed molecules on the side has been overlooked.

References

Aranovich, G.L., 1992. The theory of polymolecular adsorption. Langmuir 8, 736–739.

Ayawei, N., Ebelegi, A., Wankasi, D., 2017. Modelling and interpretation of adsorption isotherms. J. Chem. 2017. Article ID 3039817.

Brunauer, S., Emmet, P.H., Teller, E., 1938. Adsorption of gases in multimolecular layers. J. Am. Chem. Soc. 60, 309–319.

Chen, S.G., Yang, R.T., 1994. Theoretical basis for the potential theory adsorption isotherms. The Dubinin-Radushkevich and Dubinin-Astakhov equations. Langmuir 10, 4244–4249.

Cheng, P., Hu, Y.H., 2016. Dubinin-Astakhov model for acetylene adsorption on metal-organic frameworks. Appl. Surf. Sci. 377, 349–354.

Dąbrowski, A., 2001. Adsorption–from theory to practice. Adv. Colloid Interface Sci. 93, 135–224.

Dubinin, M.M., Astakhov, V.A., 1971. Development of the concepts of volume filling of micropores in the adsorption of gases and vapors by microporous adsorbents communication 1. Carbon adsorbents. Bull. Acad. Sci. USSR, Div. Chem. Sci. 20, 3–7.

Dubinin, M.M., Radushkevich, L.V., 1947. The equation of the characteristic curve of the activated charcoal. Proc. Acad. Sci. USSR Phys. Chem. Sect. 55, 331–337.

Duong, D.D., 1998. Adsorption Analysis: Equilibria and Kinetics. Imperial College Press, London.

Ebadi, R., Saadi, Z., Fazaeli, R., Fard, N.E., 2015. Monolayer and multilayer adsorption isotherm models for sorption from aqueous media. Kor. J. Chem. Eng. 32, 787–799.

Elovich, S.Y., Larinov, O.G., 1962. Theory of adsorption from solutions of non-electrolytes on solid (I) equation adsorption from solutions and the analysis of its simplest form, (II) verification of the equation of adsorption isotherm from solutions. Izv. Akad. Nauk. SSSR, Otd. Khim. Nauk. 2, 209–216.

Foo, K.Y., Hameed, B.H., 2010. Insights into the modeling of adsorption isotherm systems. Chem. Eng. J. 156, 2–10.

Freundlich, H.M.F., 1906. Über die adsorption in lösungen. Z. Phys. Chem. 57, 385–470.

Freundlich, H., 1909. Kapillarchemie, eine Darstellung der Chemie der Kolloide und verwandter Gebiete. Akademische Verlagsgesellschaft.

Galarneau, A., Mehlhorn, D., Guenneau, F., Coasne, B., Villemot, F., Minoux, D., Aquino, C., Dath, J.-P., 2018-10-31. Specific surface area determination for microporous/mesoporous materials: the case of mesoporous FAU-Y zeolites. Langmuir. Am. Chem. Soc. 34 (47), 14134–14142.

Guo, X., Liu, Y., Wang, J.L., 2019a. Sorption of sulfamethazine onto different types of microplastics: a combined experimental and molecular dynamics simulation study. Mar. Pollut. Bull. 145, 547–554.

Guo, X., Chen, C., Wang, J.L., 2019b. Sorption of sulfamethoxazole onto six types of microplastics. Chemosphere 228, 300–308.

Hanaor, D.A.H., Ghadiri, M., Chrzanowski, W., Gan, Y., 2014. Scalable surface area characterization by electrokinetic analysis of complex anion adsorption. Langmuir 30 (50), 15143–15152.

Ho, Y.S., McKay, G., 1998. A comparison of chemisorption kinetic models applied to pollutant removal on various sorbents. Process Saf. Environ. Protect. 76 (4), 332–340.

Ho, Y.S., Mckay, G., 1999. The sorption of lead (II) ions on peat. Water Res. 33, 578–584.

Khan, A., Szulejko, J.E., Samaddar, P., Kim, K.H., Eom, W., Ambade, S.B., Han, T.H., 2019. The effect of diverse metal oxides in graphene composites on the adsorption isotherm of gaseous benzene. Environ. Res. 172, 367–374.

Langmuir, I., 1916. The constitution and fundamental properties of solids and liquids. J. Am. Chem. Soc. 38, 2221–2295.

Langmuir, I., 1918. The adsorption of gases on plane surfaces of glass, mica and platinum. J. Am. Chem. Soc. 40, 1361–1403.

Lowell, S., Shields, E., Martin, T., Matthias, T., 2004. Characterization of Porous Solids and Powders: Surface Area, Pore Size and Density, first ed. Springer, Dordrecht, The Netherlands.

Maeda, N., Chen, N., Tirrell, M., Israelachvili, J.N., 2002. Adhesion and friction mechanisms of polymer-on-polymer surfaces. Science 297 (5580), 379–382.

Mclintock, I., 1967. The Elovich equation in chemisorption kinetics. Nature 216, 1204–1205.

Muttakin, M., Mitra, S., Thu, K., Ito, K., Saha, B.B., 2018. Theoretical framework to evaluate minimum desorption temperature for IUPAC classified adsorption isotherms. Int. J. Heat Mass Tran. 122, 795–805.

Polanyi, M., 1932. Section III.—Theories of the adsorption of gases. A general survey and some additional remarks. Introductory paper to section III. Trans. Faraday Soc. 28, 316–333.

Rahman, M.M., Muttakin, M., Pal, A., Shafiullah, A.Z., Saha, B.B., 2019. A statistical approach to determine optimal models for IUPAC-classified adsorption isotherms. Energies 12, 4565.

Redlich, O., Peterson, D.L., 1959. A useful adsorption isotherm. J. Phys. Chem. 63, 1024–1026.

Rouquerol, J., Llewellyn, P., Rouquerol, F., 2007. Is the bet equation applicable to microporous adsorbents? Stud. Surf. Sci. Catal. 160, 49–56. Elsevier.

Schenz, T.W., Manes, M., 1975. Application of the Polanyi adsorption potential theory to adsorption from solution on activated carbon. VI. Adsorption of some binary organic liquid mixtures. J. Chem. Phys. 79, 604–609.

Schwaab, M., Steffani, E., Barbosa-Coutinho, E., Severo Júnior, J.B., 2017. Critical analysis of adsorption/diffusion modelling as a function of time square root. Chem. Eng. Sci. 173, 179–186.

Sips, R., 1948. On the structure of a catalyst surface. J. Chem. Phys. 16, 490–495.

Souza, R.P., Dotto, G.L., Salau, N.P.G., 2017. Detailed numerical solution of pore volume and surface diffusion model in adsorption systems. Chem. Eng. Res. Des. 122, 298–307.

Temkin, M.J., Pyzhev, V., 1940. Kinetics of ammonia synthesis on promoted iron catalyst. Acta Phys. Chim. USSR 12, 327–356.

Thommes, M., Kaneko, K., Neimark, A.V., Olivier, J.P., Rodriguez-Reinoso, F., Rouquerol, J., Sing, K.S.W., 2015. Physisorption of gases, with special reference to the evaluation of surface area and pore size distribution (IUPAC technical report). Pure Appl. Chem. 87, 1051–1069.

Toth, J., 1971. State equations of the solid gas interface layer. Acta Chem. Acad. Hung. 69, 311–317.

Tseng, R.L., Wu, P.H., Wu, F.C., Juang, R.S., 2014. A convenient method to determine kinetic parameters of adsorption processes by nonlinear regression of pseudo-nth-order equation. Chem. Eng. J. 237, 153–161.

Volmer, M., 1925. Thermodynamische folgerungen aus der zustandsgleichung Fur adsorbierte stoffe. Z. Phys. Chem. 115, 253–261.

Wang, J., Guo, X., 2020a. Adsorption kinetic models: physical meanings, applications, and solving methods. J. Hazard Mater. 390, 122156.

Wang, J., Guo, X., 2020b. Adsorption isotherm models: classification, physical meaning, application and solving method. Chemosphere 258, 127279.

Yaneva, Z.L., Koumanova, B.K., Georgieva, N.V., 2013. Linear regression and nonlinear regression methods for equilibrium modelling of p-nitrophenol biosorption by *Rhizopus oryzae*: comparison of error analysic criteria. J. Chem. 2013, 10. Article ID 517631.

Yang, C., 1993. Statistical mechanical aspects of adsorption systems obeying the Temkin isotherm. J. Phys. Chem. 97, 7097–7101.

Yang, R.T., 2003. Adsorbents: Fundamentals and Applications. Wiley-Interscience, Hoboken, NJ, ISBN 0471297410.

Zaheer, Z., Aisha, A.A., Aazam, E.S., 2019. Adsorption of methyl red on biogenic Ag@Fe nanocomposite adsorbent: isotherms, kinetics and mechanisms. J. Mol. Liq. 283, 287–298.

Chapter 5

Development and synthesis of nanoparticles and nanoadsorbents

1. Introduction

Nanomaterials that can be used as adsorbents are different in size, shape, and structure. They can be spherical, conical, hollow, rod, coiled, plane, cylindrical, and asymmetrical. Nanomaterials can be crystalline or amorphous. Generally, nanomaterials are classified into nanoclays, nanoparticles, and nanoemulsions (Fig. 5.1). Several nanoparticles are highly effective, efficient, economically viable, and reusable as nanoadsorbents.

Nanoparticles are present in nanocomposites or nanostructures forms. Nanostructures are made from basic units or blocks with a small dimensionality, i.e., zero, one, two, and three dimensions (Fig. 5.2). In zero-dimensional nanoparticles, the moment of electrons is cramped in all three dimensions, e.g., quantum dots. If electrons move only in x-direction, they are one-dimensional nanoparticles, e.g., quantum wires. While in two-dimensional thin films, free electrons can move freely in x, y. In three-dimensional nanostructured materials, free electrons can move freely in x, y, z directions. Table 5.1 lists the description of various terms associated with nanomaterials.

The characteristics and unique properties of nanoparticles make them perfect for numerous uses in science and industrial engineering (Fig. 5.3). There are many nanoadsorbents used for several uses including wastewater treatment, gas separation, membrane, oil separation, and oil upgrading. They have been found to be highly effective compared to bulk materials.

2. Classification of nanoadsorbents

Nanomaterials that can be used as adsorbents can be classified based on their synthesis and nature as well as their role in adsorption applications, which is

Surface Science of Adsorbents and Nanoadsorbents. https://doi.org/10.1016/B978-0-12-849876-7.00004-X

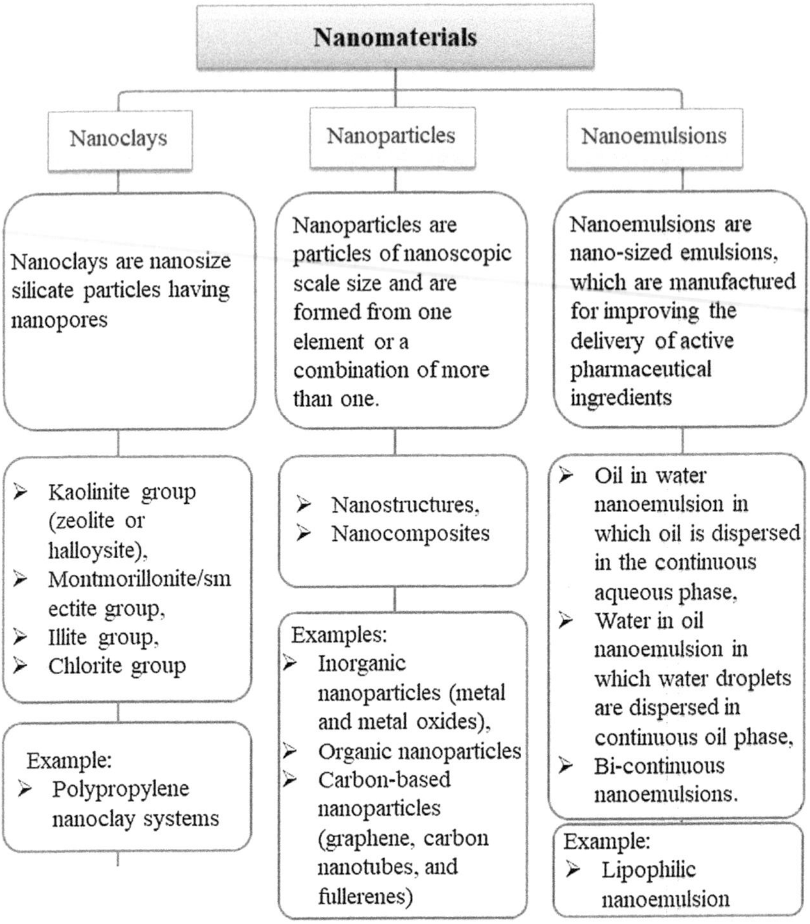
Nanomaterials
Nanoclays
Nanoparticles
Nanoemulsions
Nanoclays are nanosize silicate particles having nanopores
Nanoparticles are particles of nanoscopic scale size and are formed from one element or a combination of more than one.
Nanoemulsions are nano-sized emulsions, which are manufactured for improving the delivery of active pharmaceutical ingredients
Kaolinite group (zeolite or halloysite),
Montmorillonite/smectite group,
Illite group,
Chlorite group
Nanostructures,
Nanocomposites
Oil in water nanoemulsion in which oil is dispersed in the continuous aqueous phase,
Water in oil nanoemulsion in which water droplets are dispersed in continuous oil phase,
Bi-continuous nanoemulsions.
Example:
Polypropylene nanoclay systems
Examples:
Inorganic nanoparticles (metal and metal oxides),
Organic nanoparticles
Carbon-based nanoparticles (graphene, carbon nanotubes, and fullerenes)
Example:
Lipophilic nanoemulsion

FIGURE 5.1 Classification of nanomaterials into nanoclay, nanoparticles and nanoemulsions.

dependent on their innate surface property and additional external functionalization (Fig. 5.4). This class includes:

(i) nanoparticles like metallic nanoparticles (silver NPs), metallic oxide NPs (aluminum trioxide or titanium dioxide), nanostructured mixed oxides (nanostructured binary iron–titanium mixed oxide particles), and magnetic NPs (iron di- and trioxides),

(ii) carbonaceous nanomaterials (CNMs) that are based on sorbent properties, which include carbon nanotubes (CNTs), carbon nanoparticles (CNPs), and carbon nanosheets (CNSs),

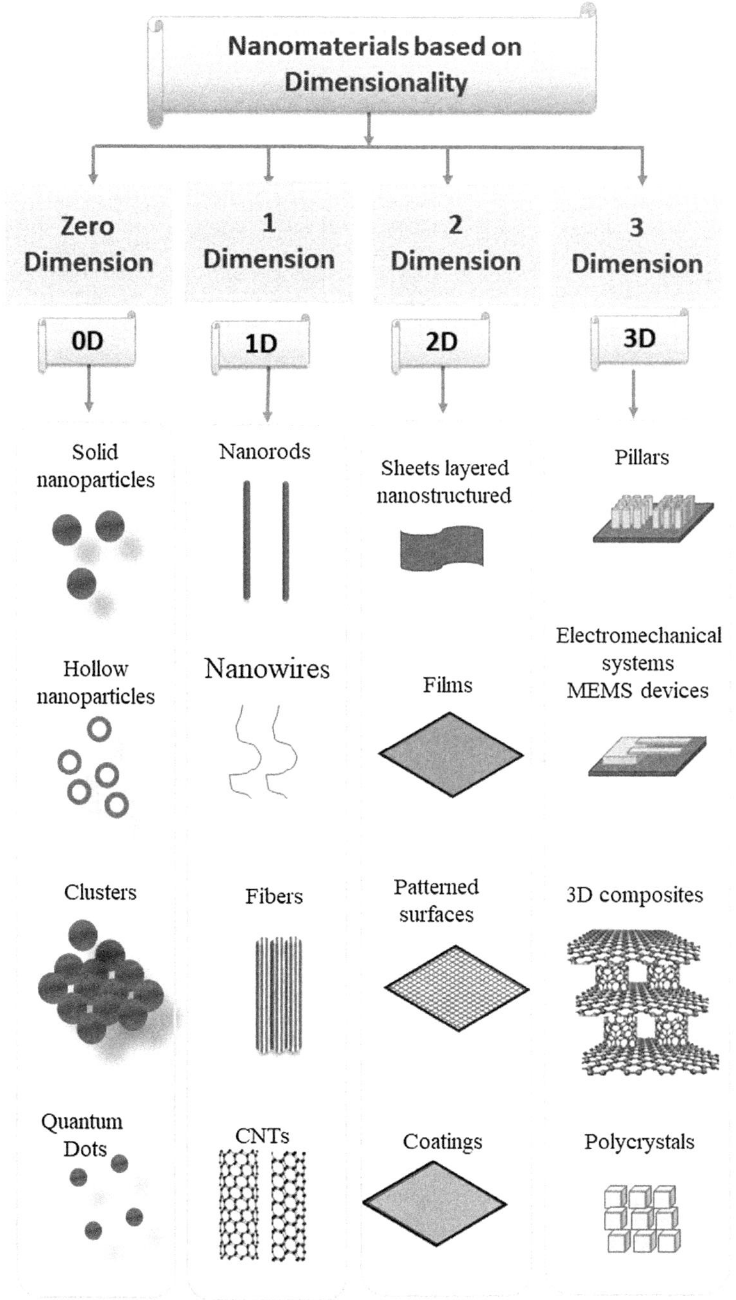
Nanomaterials based on Dimensionality
Zero Dimension
1 Dimension
2 Dimension
3 Dimension
0D
1D
2D
3D
Solid nanoparticles
Nanorods
Sheets layered nanostructured
Pillars
Hollow nanoparticles
Nanowires
Films
Electromechanical systems MEMS devices
Clusters
Fibers
Patterned surfaces
3D composites
Quantum Dots
CNTs
Coatings
Polycrystals

FIGURE 5.2 Classification of nanomaterials based on dimensionality.

TABLE 5.1 A description of numerous terminologies used under science-related topics.

Terminology	Definition
Science	Science is described as the methodical pursuit and application of knowledge and understanding of the natural and social worlds using evidence-based methods.
Chemistry	Chemistry is the scientific study of the properties and behavior of matter. Chemistry is the study of substances, that is, elements and compounds.
Nanoscience	Nanoscience is defined as the study of fundamental relationships between physical properties and material dimensions on the nanometer scale (a scale covering 1–100 nm). Nanoscience is the study of phenomena and manipulation of materials at atomic, molecular, and macromolecular scales, where properties differ significantly from those at a larger scale. It is the study of fundamental relationships between physical properties and material dimensions on the nanometer scale (a scale covering 1–100 nm).
Nanotechnology	Nanotechnology is a discipline of science and engineering devoted to creating, producing, and using structures, devices, and systems at the nanoscale, or with one or more dimensions of the order of 100 nm (one billionth of a millimeter) or less. Nanotechnology is the technology at the nanoscale level in which materials or devices are developed via controlling matter at the nanoscale length to stimulate the distinctive properties of the nanomaterial. Nanotechnology extends to atom by atom arrangements. At this scale, we enter the regime of quantum technology.
Picotechnology	Picotechnology refers to technology at the picoscale level in which materials, devices, or systems are possibly developed via controlling matter at the picoscale length to stimulate the unique properties of the material at the pico-level. It is intended to parallel the term nanotechnology. It is a hypothetical future level of technological manipulation of matter, on the scale of trillionths of a meter or picoscale (10^{-12}).
Femtotechnology	Femtotechnology is a hypothetical term used in reference to the structuring of matter on the scale of a femtometer (10^{-15} m).

TABLE 5.1 A description of numerous terminologies used under science-related topics.—cont'd

Terminology	Definition
Engineering	The application of scientific concepts to the design and construction of machines, structures, and other products is known as engineering.
Nanomanufacturing	Nanomanufacturing is the process of producing items at the nanoscale level using either bottom-up or top-down methods.
Engineered nanomaterials	Engineered nanomaterials are chemical compounds or materials with particle sizes in at least one dimension ranging from 1 to 100 nm. Engineered nanoparticles are well known for their many functional advantages derived from their unique physical and chemical features.

(iii) silicon nanomaterials (SiNMs), including silicon nanotubes (SiNTs), silicon nanoparticles (SiNPs), and silicon nanosheets (SiNSs),
(iv) nanofibers (NFs) and nanoclays,
(v) polymer-based nanomaterials (PNMs),
(iv) xerogels and aerogels.

3. Classification of nanoadsorbents

Classification of nanoadsorbents based on material production and role in the sorption process is given in Fig. 5.5. Nanoparticles can also be classified into organic, inorganic (metallic), and carbon and mixed oxide nanostructures and magnetic-based nanoparticles. The most significant and distinguishing qualities of any nanoadsorbent are its enormous surface area and porosity. A wide range of materials such as activated carbon, zeolites, pillared clays, metal-organic frameworks, polymers, and others are accessible (Kim et al., 2004; Kumari et al., 2019). Table 5.2 lists the description of terminologies used under nanotechnology-related topics.

3.1 Organic nanoparticles

Organic nanoparticles are materials with two or more dimensions with a size in the range of 1–100 nm. They are made up of synthetic and natural organic molecules that self-assemble. The nanoparticles' shape, size, and chemical composition play a major role in determining the specific properties and activity of the nanomaterials. Micelles, dendrimers, ferritin, and liposomes are examples of well-known organic nanoparticles.

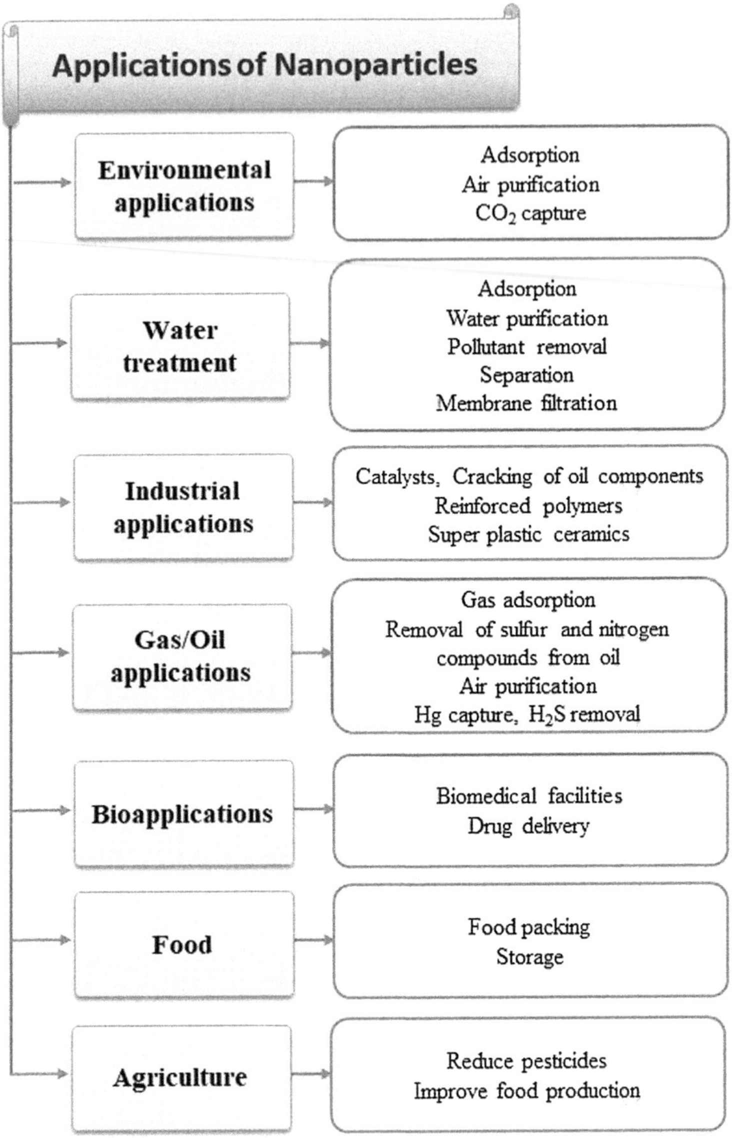
Applications of Nanoparticles
Environmental applications
Adsorption
Air purification
CO_2 capture
Water treatment
Adsorption
Water purification
Pollutant removal
Separation
Membrane filtration
Industrial applications
Catalysts, Cracking of oil components
Reinforced polymers
Super plastic ceramics
Gas/Oil applications
Gas adsorption
Removal of sulfur and nitrogen compounds from oil
Air purification
Hg capture, H_2S removal
Bioapplications
Biomedical facilities
Drug delivery
Food
Food packing
Storage
Agriculture
Reduce pesticides
Improve food production

FIGURE 5.3 Examples of the applications of nanoparticles and nanoadsorbents.

3.2 Inorganic nanoparticles

Inorganic nanoparticles usually are manufactured from inorganic salts precipitation. Metal and metal oxide particles are the most typical types of

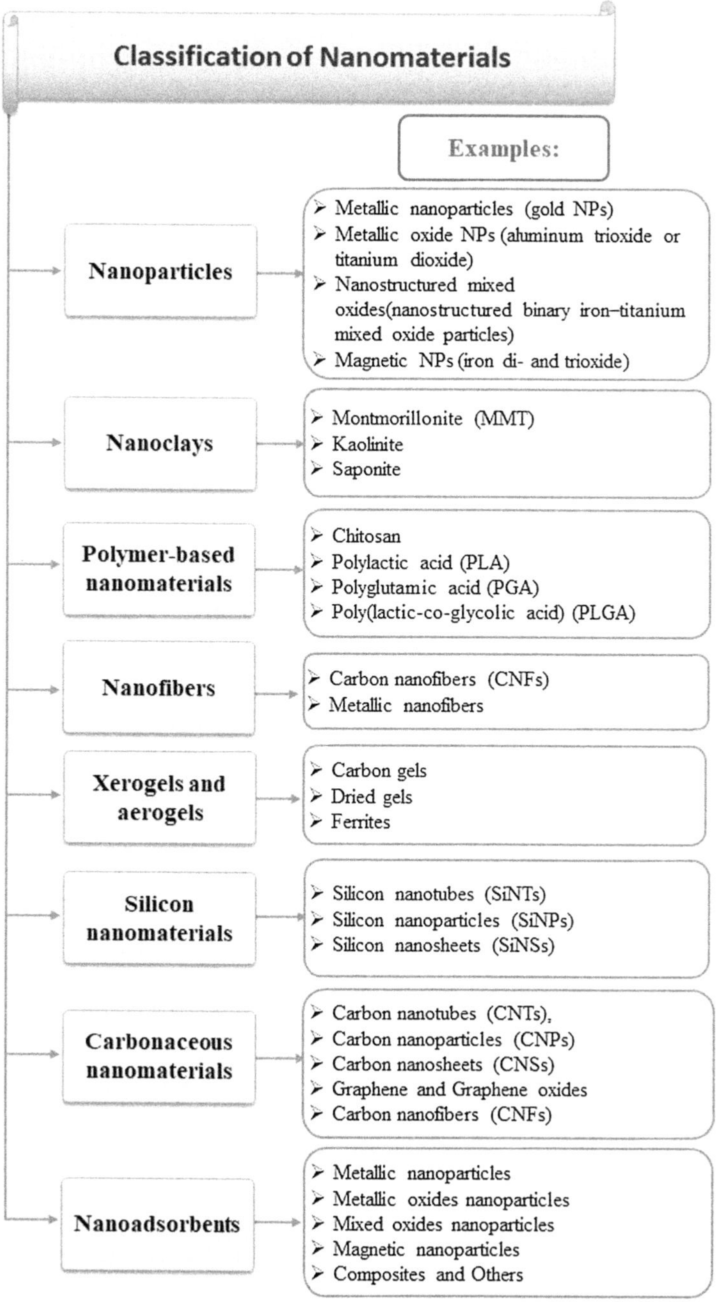

Classification of Nanomaterials
Examples:
Nanoparticles
➢ Metallic nanoparticles (gold NPs)
➢ Metallic oxide NPs (aluminum trioxide or titanium dioxide)
➢ Nanostructured mixed oxides(nanostructured binary iron–titanium mixed oxide particles)
➢ Magnetic NPs (iron di- and trioxide)
Nanoclays
➢ Montmorillonite (MMT)
➢ Kaolinite
➢ Saponite
Polymer-based nanomaterials
➢ Chitosan
➢ Polylactic acid (PLA)
➢ Polyglutamic acid (PGA)
➢ Poly(lactic-co-glycolic acid) (PLGA)
Nanofibers
➢ Carbon nanofibers (CNFs)
➢ Metallic nanofibers
Xerogels and aerogels
➢ Carbon gels
➢ Dried gels
➢ Ferrites
Silicon nanomaterials
➢ Silicon nanotubes (SiNTs)
➢ Silicon nanoparticles (SiNPs)
➢ Silicon nanosheets (SiNSs)
Carbonaceous nanomaterials
➢ Carbon nanotubes (CNTs),
➢ Carbon nanoparticles (CNPs)
➢ Carbon nanosheets (CNSs)
➢ Graphene and Graphene oxides
➢ Carbon nanofibers (CNFs)
Nanoadsorbents
➢ Metallic nanoparticles
➢ Metallic oxides nanoparticles
➢ Mixed oxides nanoparticles
➢ Magnetic nanoparticles
➢ Composites and Others

FIGURE 5.4 Categorization of nanomaterials based on structure and nature.

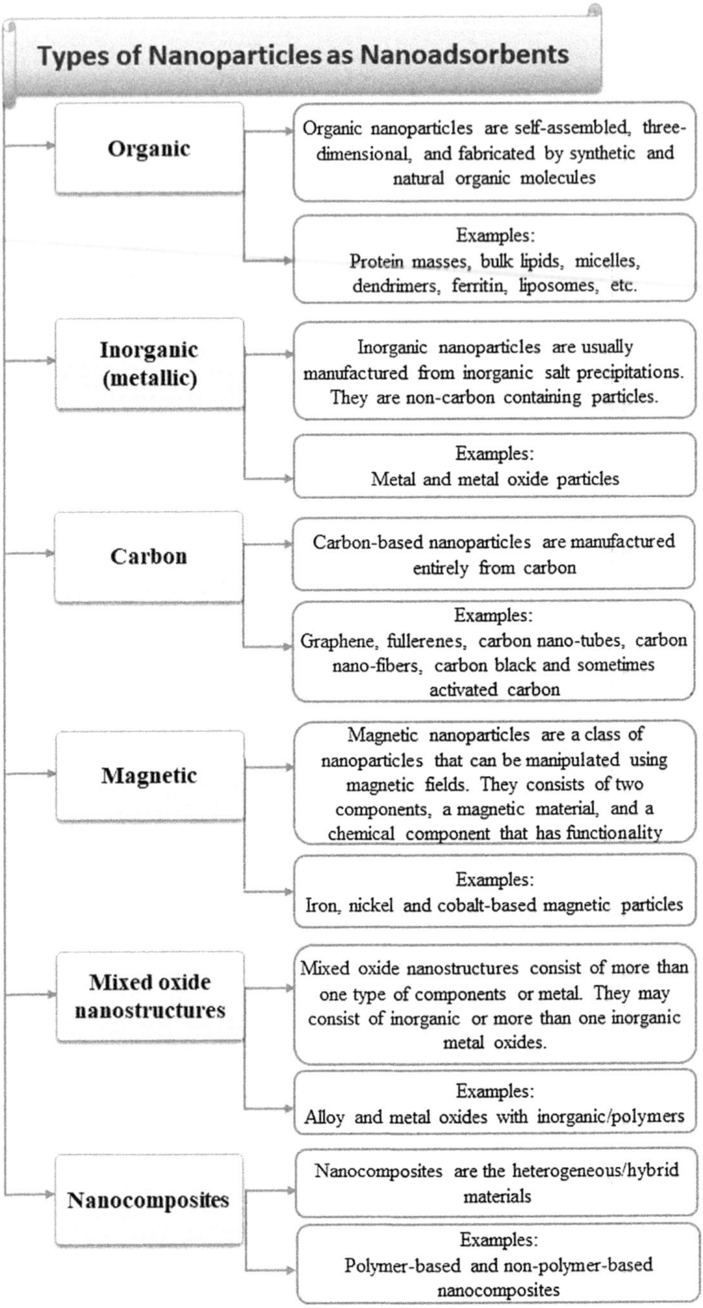
Types of Nanoparticles as Nanoadsorbents
Organic
Organic nanoparticles are self-assembled, three-dimensional, and fabricated by synthetic and natural organic molecules
Examples:
Protein masses, bulk lipids, micelles, dendrimers, ferritin, liposomes, etc.
Inorganic (metallic)
Inorganic nanoparticles are usually manufactured from inorganic salt precipitations. They are non-carbon containing particles.
Examples:
Metal and metal oxide particles
Carbon
Carbon-based nanoparticles are manufactured entirely from carbon
Examples:
Graphene, fullerenes, carbon nano-tubes, carbon nano-fibers, carbon black and sometimes activated carbon
Magnetic
Magnetic nanoparticles are a class of nanoparticles that can be manipulated using magnetic fields. They consists of two components, a magnetic material, and a chemical component that has functionality
Examples:
Iron, nickel and cobalt-based magnetic particles
Mixed oxide nanostructures
Mixed oxide nanostructures consist of more than one type of components or metal. They may consist of inorganic or more than one inorganic metal oxides.
Examples:
Alloy and metal oxides with inorganic/polymers
Nanocomposites
Nanocomposites are the heterogeneous/hybrid materials
Examples:
Polymer-based and non-polymer-based nanocomposites

FIGURE 5.5 Classification of nanoadsorbents based on material production and role in sorption process.

TABLE 5.2 A description of terminologies used under nanotechnology-related topics.

Terminology	Definition
Nanostructured materials	Nanostructured material refers to materials that have structural elements, molecules, crystallites, or clusters with dimensions in the 1–100 nm range.
Nanomaterials	A nanomaterial is a nanomaterial that has at least one of its dimensions in the nanoscale range of 1–100 nm.
Nanoparticles	A nanoparticle is a small particle of matter that ranges between 1 and 100 nm (nm) in at least one of its diameters.
Nano-object	A nano-object is a distinct material with one dimension, two dimensions, or three external dimensions in the 1–100 nm range.
Nanocomposites	The nanocomposite is a multiphase solid material where one of the phases has one, two, or three dimensions of less than 100 nm or structures having nanoscale repeat distances between the different phases that make up the material. Nanocomposites are heterogeneous/hybrid materials that are produced by the mixtures of polymers with inorganic solids (clays to oxides) at the nanometric scale.
Aspect ratio	The aspect ratio of a nano-object can be described as the ratio of the length of the major axis to the width of the minor axis of a material.
Nanospheres	Nanospheres are nanoparticles that have an aspect ratio of 1.
Nanorods	The term nanorods can be used when the shortest and longest axes have various lengths. Nanorods normally have a width in the range of 1–100 nm and an aspect ratio >1.
Nanofibers	A nanofiber is traditionally defined as a cylindrical structure with an outer diameter below 1000 nm and an aspect ratio—the ratio between length and width—greater than 50. It is a nano-object with one or two dimensions in the nanoscale range.
Nanowires	Nanowires are analogues to nanorods but with a higher aspect ratio.
Nanotubes	Nanotubes are made by “winding” single sheets of graphite with honeycomb structures into very long, thin tubes that have a stable, strong, and flexible structure.
Nanotubes	Hollow nanofibers are called nanotubes.

noncarbon containing particles. Among the most prominent metal oxides nanoparticles are titania, silica, alumina, zirconium, iron oxides, and zinc oxide nanoparticles, modified with various functionalities. Inorganic nanoparticles are used as adsorbents for several applications including wastewater treatment, gas separation, and CO_2 capture.

3.3 Carbon-based nanoparticles

Carbon-based nanoparticles are made exclusively from carbon, e.g., graphene, fullerenes, carbon nanotubes (CNTs), carbon nanofibers, carbon black, and activated carbon. Carbon-based adsorbents like porous carbon, graphene, and fullerenes exhibit high thermal stability and high adsorption capability (Rao et al., 2007). CNTs are an allotrope of carbon and have a cylindrical-shaped structural form rolled up in a tubelike structure (Iijima, 1991). CNTs can be in two forms:

(i) Single-Walled Carbon Nanotubes (SWCNTs), which have a single graphene sheet rolled up as a tube;

(ii) Multi-Walled Carbon Nanotubes (MWCNTs), which have multiple graphene sheets rolled up as a tubelike structure.

CNTs are among the best nanosorbents that are very efficient toward the adsorption of inorganic and organic pollutants in aqueous media and for the adsorption of gases (Gupta and Saleh, 2013).

3.4 Magnetic-based nanoparticles

Magnetic-based nanoparticles are materials with two or more dimensions with size in the range of 1–100 nm. They are manipulated using magnetic fields. They comprise of two components: a magnetic material, often iron, nickel, and cobalt, and a chemical component that has functionality. Different qualities are required for each conceivable application of magnetic nanoparticles. Therefore, the synthesis route should be selected carefully to allow preparing magnetic-based nanoparticles with the required properties.

3.5 Mixed oxide nanostructures

In chemistry, mixed oxide nanostructures are oxides that contain cations of more than one chemical element or cations of a single element in several states of oxidation. The term mixed oxide nanostructure is commonly applied to solid ionic compounds that contain the oxide anion O_2^- and two or more element cations. Metals in various oxidation states can be combined in varied ratios to generate a wide range of materials. These new materials, which vary in physical, chemical, and morphological properties, can be used in a variety of scientific and technological domains.

Because of the existence of significant densities of edge/corner sites and defects, mixed metal oxide nanostructures have distinctive surface morphologies and have a more reactive surface. According to morphological research, nanocrystals are more polyhedral and so have more flaws. Such flaws could be of the Frankel or Schottky type (vacancies), or they could appear as a typical edge, corner, or crystal plane shapes. Surface gas reactions can approach the stoichiometric range when a ceramic material with a large surface area shows roughly 35% of the ceramic moieties on the surface (Khaleel and Richards, 2001).

An example of this class is alumina. When compared to other commercially available solid-phase materials, the synthesis of alumina nanoparticles is very straightforward and inexpensive. Controlled hydrolysis of aluminum alkoxides is the most common process for making fine alumina (Al_2O_3) nanoparticles, such as aerogels and xerogels. Simple aluminum salts could be beneficial for pyrolysis of an ethanolic solution of appropriate metal alkoxides to make alumina nanoparticles or aerogels like $(TiO_2)_x(Al_2O_3)_{1-x}$. Hydrothermal synthesis and vaporization of volatile precursors, followed by thermal degradation in a reactor, are two other common techniques. The latter is known as the aerosol method (Gubin et al., 2005).

3.6 Nanocomposites

Metal oxide nanoparticles are often characterized by poor mechanical strength. Thus, they are typically impregnated or supported by other large-sized porous supports. Examples of supports are nanomaterials such as zeolites and carbon nanostructures to form nanocomposites that have comparatively better usability and applications (Saravanan et al., 2016). The nanocomposite of carbon nanotube/tungsten oxide (MWCNT/WO3) was synthesized and its catalytic activity for rhodamine B removal from wastewater under sunlight was studied (Saleh, 2022a,b).

Nanocomposites are heterogeneous/hybrid materials made at the nanometric scale by combining polymers with inorganic solids (clays to oxides). The process of in situ growth and polymerization of biopolymer and the inorganic matrix is commonly used to create nanocomposites. Types of nanocomposites include:

- polymer based such as polymer with ceramic, inorganic/polymer, inorganic with organic hybrid, and polymer with silicate layers.
- nonpolymer-based nanocomposites. Examples are metal with metal oxides, ceramic with metal nanoparticles, and ceramic with other ceramic nanocomposites.

Nanocomposites properties that have indicated substantial improvements:

- Increased chemical resistance
- Permeability of gases, water, and solvents are reduced

- ➢ Mechanical properties include strength, bulk modules, with stands limit, etc.
- ➢ Enhanced stiffness
- ➢ Thermal stability
- ➢ Hinders flame and reduce smoke generations
- ➢ More surface appearance
- ➢ Improved electrical conductivity
- ➢ Enhance optical clarity as compared to conventionally filled polymers

4. Approaches for preparation of adsorbents and nanoadsorbents

There are several types of materials that are used as adsorbents. The main classes of adsorbents and nanoadsorbents can be generally classified as shown in Fig. 5.6.

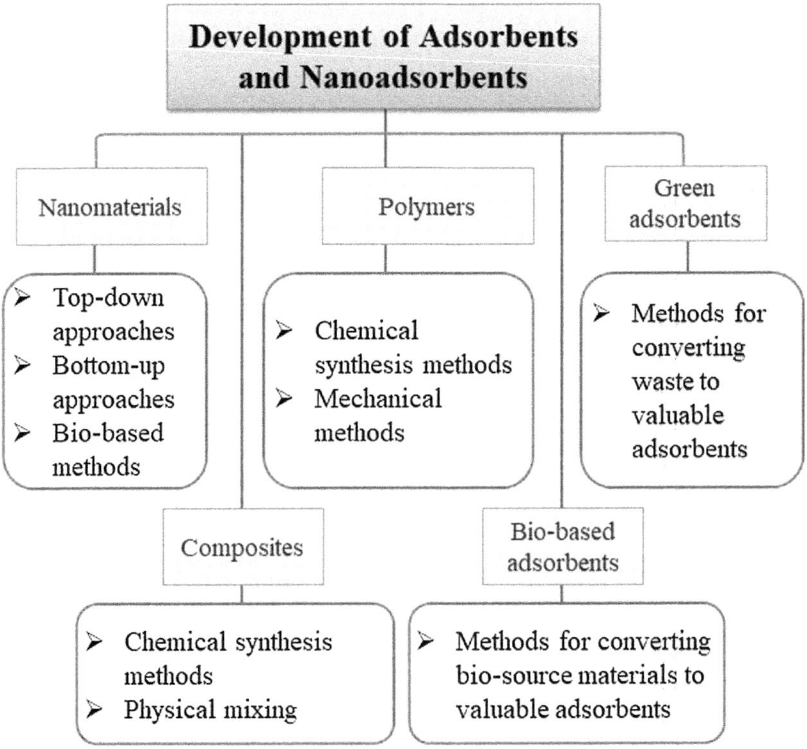
Development of Adsorbents and Nanoadsorbents
Nanomaterials
Polymers
Green adsorbents
➢ Top-down approaches
➢ Bottom-up approaches
➢ Bio-based methods
➢ Chemical synthesis methods
➢ Mechanical methods
➢ Methods for converting waste to valuable adsorbents
Composites
Bio-based adsorbents
➢ Chemical synthesis methods
➢ Physical mixing
➢ Methods for converting bio-source materials to valuable adsorbents

FIGURE 5.6 Classification of some adsorbents with their possible preparation approaches.

The preparation approaches of adsorbents and nanoadsorbents rely mainly on the type, precursors, and the desired properties. Generally, there are several methods and procedures for the preparation of adsorbents and materials in general. These methods can be classified into physical, chemical, and biological or the hybrid of more than one of these methods (Fig. 5.7). Guidelines of the general steps for the preparation of adsorbents and nanoadsorbents are shown in Fig. 5.8. Depending on the type of the material to be prepared, the method and procedure are to be selected.

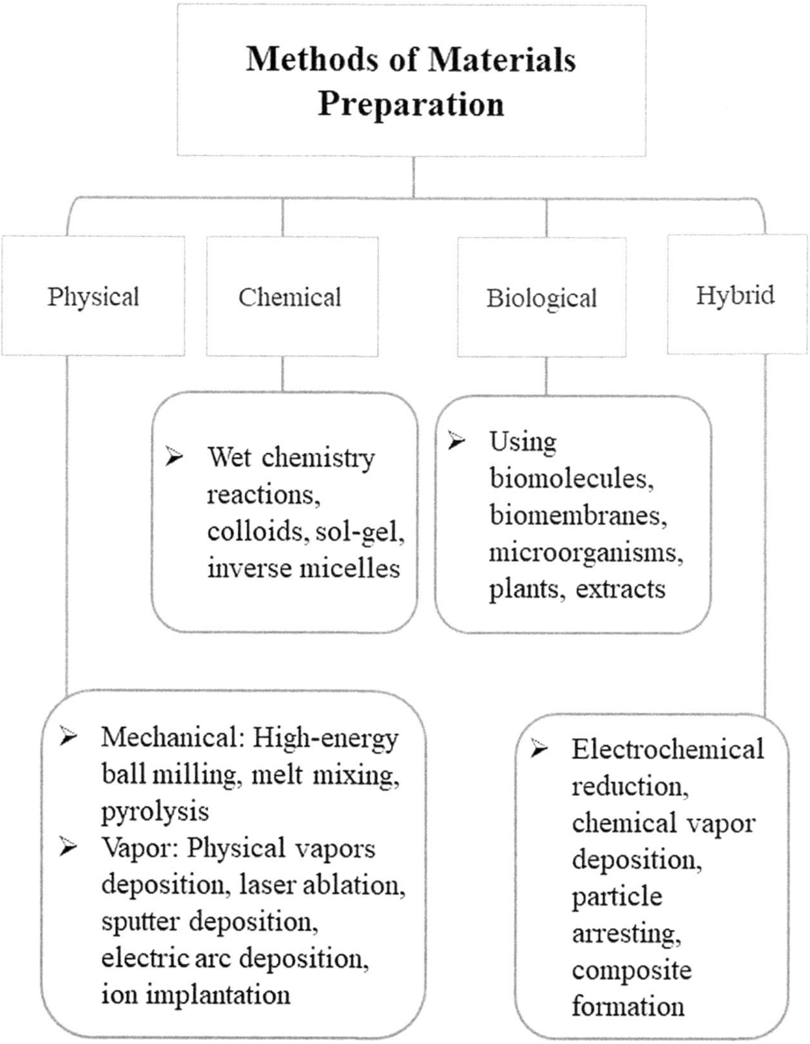
Methods of Materials Preparation
Physical
Chemical
Biological
Hybrid
➢ Wet chemistry reactions, colloids, sol-gel, inverse micelles
➢ Using biomolecules, biomembranes, microorganisms, plants, extracts
➢ Mechanical: High-energy ball milling, melt mixing, pyrolysis
➢ Vapor: Physical vapors deposition, laser ablation, sputter deposition, electric arc deposition, ion implantation
➢ Electrochemical reduction, chemical vapor deposition, particle arresting, composite formation

FIGURE 5.7 Examples of different methods used for the synthesis of materials.

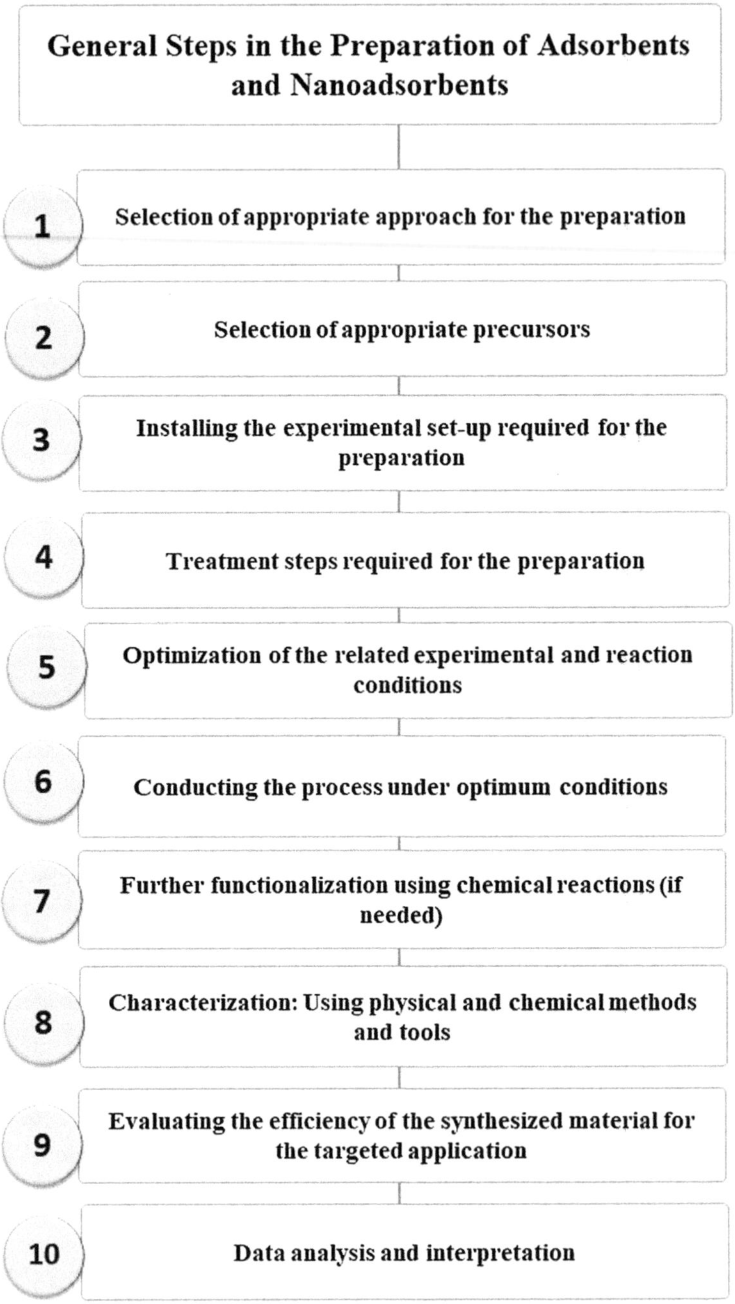
General Steps in the Preparation of Adsorbents and Nanoadsorbents
1
Selection of appropriate approach for the preparation
2
Selection of appropriate precursors
3
Installing the experimental set-up required for the preparation
4
Treatment steps required for the preparation
5
Optimization of the related experimental and reaction conditions
6
Conducting the process under optimum conditions
7
Further functionalization using chemical reactions (if needed)
8
Characterization: Using physical and chemical methods and tools
9
Evaluating the efficiency of the synthesized material for the targeted application
10
Data analysis and interpretation

FIGURE 5.8 Guideline of general steps in the preparation of adsorbents and nanoadsorbents.

5. Preparation of nanomaterials

The synthesis of adsorbents like nanomaterials and nanostructures is an important aspect of nanoscience and nanotechnology. Nanomaterials have sparked increased interest due to their unique optical, magnetic, electrical, mechanical, and chemical properties compared to bulk materials. Nanoadsorbents can have new physical properties and uses, especially when nanomaterials are developed with the necessary size, shape, morphology, crystal structure, and chemical composition.

Nanomaterials have gotten a lot of press since their physical, chemical, electrical, and magnetic properties differ dramatically from their higher-dimensional counterparts and are influenced by their form and size. Several systems have been advanced to synthesize nanomaterials with controlled shape, size, dimensionality, and structure (Saleh, 2021a,b; Liang, 2021; Yan et al., 2015; Liu et al., 2014; Kalhapure et al., 2015). The qualities of materials determine their performance. The properties of materials are determined by the atomic structure, composition, microstructure, defects, and interfaces, all of which are influenced by the synthesis' thermodynamics and kinetics. Methods for nanomaterials' preparation are classified into: (a) Top-down approach and (b) Bottom-up approach (Fig. 5.9).

5.1 Top-down approach

The bulk material is broken down into nanosized structures or particles in a top-down method (Khanna et al., 2019). Top-down synthesis techniques are an evolution of those used to create micron-sized particles. They are naturally simpler and rely on bulk material removal or division or bulk fabrication process downsizing, to produce the desired structure with adequate attributes. Nanomaterials are created by etching crystal planes out of substrates or removing crystal planes that are already there (Fig. 5.10).

The methods used under this approach are high-energy ball milling, mechanical milling, condensation, atomization, sputtering, arc discharge, explosion, coating, lithography, etching, electrospinning, laser ablation, atomic force manipulation, aerosol spray, and other physical methods (Palencia, 2021; Jiang, 2021).

Advantages

- ➢ This method is useful for creating long-range order structures and making macroscopic linkages.
- ➢ It is useful for making macroscopic linkages.

Limitations

- ➢ The disadvantage of this method is that the surface structure is imperfect. Such flaws would have a substantial impact on nanostructures and nanomaterials' physical and chemical properties. Lithography nanowires, for

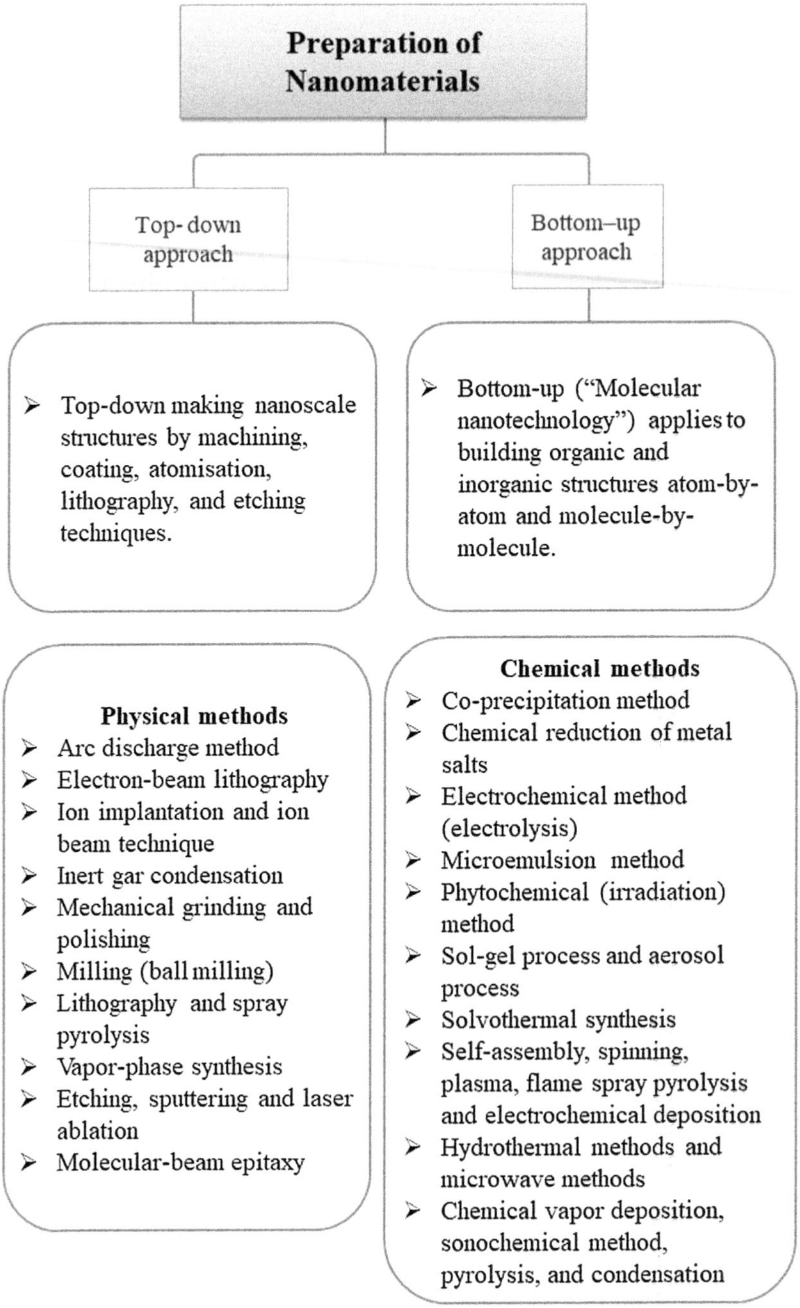
Preparation of Nanomaterials
Top- down approach
Bottom–up approach
➢ Top-down making nanoscale structures by machining, coating, atomisation, lithography, and etching techniques.
➢ Bottom-up ("Molecular nanotechnology") applies to building organic and inorganic structures atom-by-atom and molecule-by-molecule.
Physical methods
➢ Arc discharge method
➢ Electron-beam lithography
➢ Ion implantation and ion beam technique
➢ Inert gar condensation
➢ Mechanical grinding and polishing
➢ Milling (ball milling)
➢ Lithography and spray pyrolysis
➢ Vapor-phase synthesis
➢ Etching, sputtering and laser ablation
➢ Molecular-beam epitaxy
Chemical methods
➢ Co-precipitation method
➢ Chemical reduction of metal salts
➢ Electrochemical method (electrolysis)
➢ Microemulsion method
➢ Phytochemical (irradiation) method
➢ Sol-gel process and aerosol process
➢ Solvothermal synthesis
➢ Self-assembly, spinning, plasma, flame spray pyrolysis and electrochemical deposition
➢ Hydrothermal methods and microwave methods
➢ Chemical vapor deposition, sonochemical method, pyrolysis, and condensation

FIGURE 5.9 Some of the main approaches for the synthesis of nanomaterials.

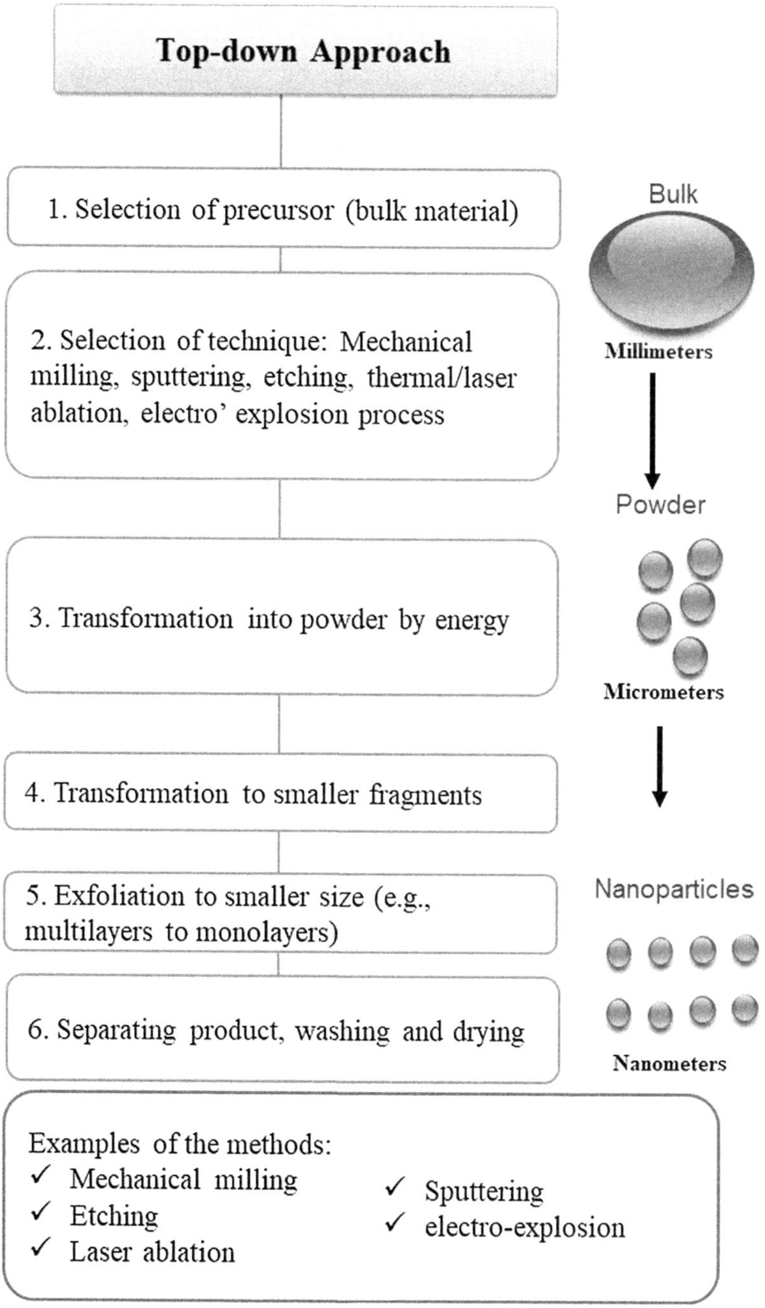
Top-down Approach
1. Selection of precursor (bulk material)
2. Selection of technique: Mechanical milling, sputtering, etching, thermal/laser ablation, electro' explosion process
3. Transformation into powder by energy
4. Transformation to smaller fragments
5. Exfoliation to smaller size (e.g., multilayers to monolayers)
6. Separating product, washing and drying
Bulk
Millimeters
Powder
Micrometers
Nanoparticles
Nanometers
Examples of the methods:
✓ Mechanical milling
✓ Etching
✓ Laser ablation
✓ Sputtering
✓ electro-explosion

FIGURE 5.10 Guideline for the steps of the top-down approach for nanomaterials' preparation.

example, are not smooth and may have a lot of contaminants and structural imperfections on their surface.

- Top-down approach is not suitable for forming evenly shaped nanomaterials.
- With the use of the top-down approach, it is not easy to obtain small size of nanoparticles even with the use of high energy.
- The shortage of the surface structure has a substantial impact on the surface chemistry and properties of nanomaterials.
- Top-down approach may cause considerable crystallographic loss to the shape of the prepared nanoparticles.

5.2 Bottom-up approach

The bottom-up strategy is an alternative that has the potential to produce less waste and hence be more cost-effective (Fig. 5.11). It refers to the atom-by-atom, molecule-by-molecule, or cluster-by-cluster construction of a substance from the bottom-up. Many of these approaches are still under development or are only now being employed for commercial nanopowder manufacture. Some of the well-known bottom-up procedures reported for the manufacture of luminous nanoparticals include oraganometallic chemical route, reverse-micelle route, sol–gel synthesis, colloidal precipitation, hydrothermal synthesis, template-assisted sol–gel, electrodeposition, and so on.

Because of several advantages, including fewer flaws, more homogeneous chemical composition, and better ordering, the bottom-up approach is more often used in nanoparticle production. In a bottom-up approach, the Gibbs free energy, thermodynamic equilibrium, and kinetic approaches are shown to be the major strategies for nanoparticle production. Examples of the methods under bottom-up approach are:

- thermolysis (e.g., pyrolysis and spray pyrolysis), hydrothermal methods,
- oraganometallic chemical route, reverse micelle route, interfacial synthesis, micelles and microemulsions, soft and hard templating methods, reverse micelle methods
- hydrothermal synthesis, chemical vapor deposition (CVD),
- chemical reduction, electrochemical synthesis, electrodeposition methods,
- sol–gel synthesis, template-assisted sol–gel, colloidal precipitation, arrested precipitation, solvothermal synthesis, solvated metal atom dispersion,
- photochemical synthesis, sonochemical routes,
- biological methods, and
- hybrid methods.

Advantages

- The bottom-up approach incorporates the use of several solvents. Because it includes more chemistry than a top-down method, it is good for

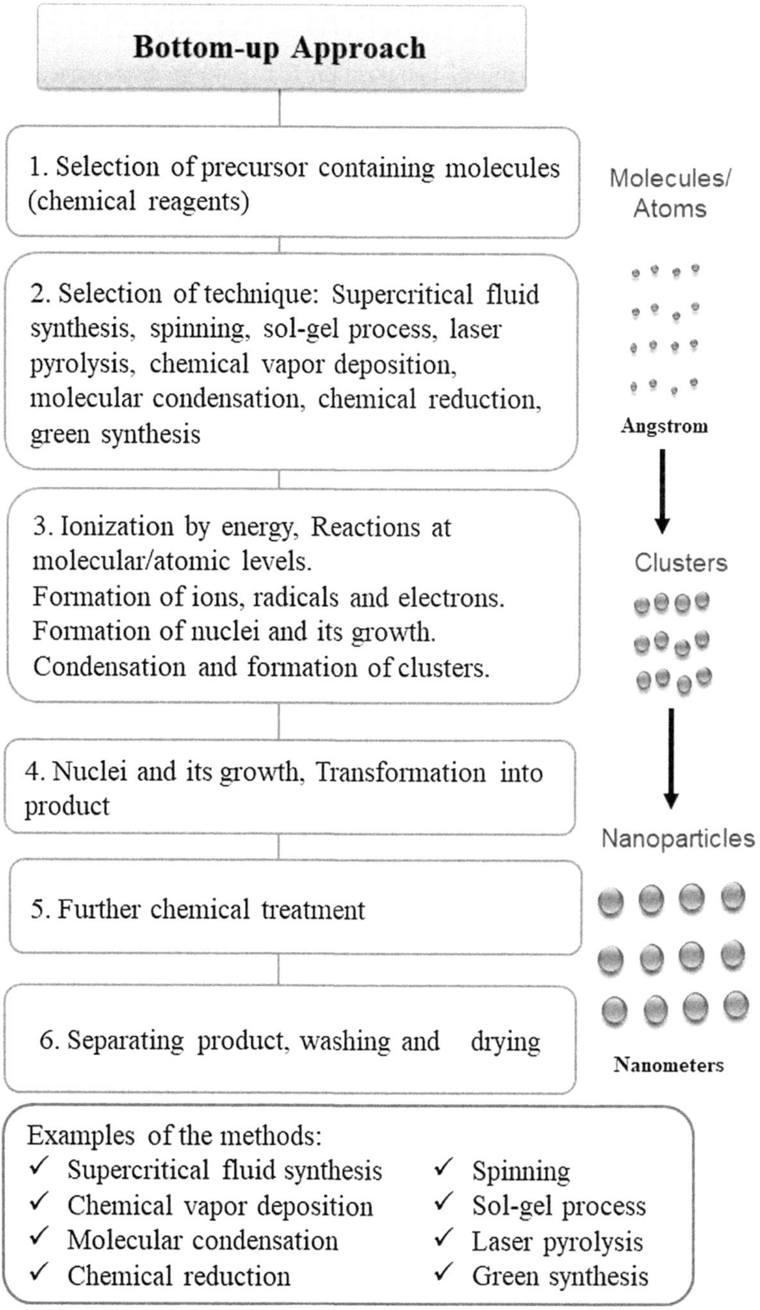
Bottom-up Approach
1. Selection of precursor containing molecules (chemical reagents)
2. Selection of technique: Supercritical fluid synthesis, spinning, sol-gel process, laser pyrolysis, chemical vapor deposition, molecular condensation, chemical reduction, green synthesis
3. Ionization by energy, Reactions at molecular/atomic levels.
Formation of ions, radicals and electrons.
Formation of nuclei and its growth.
Condensation and formation of clusters.
4. Nuclei and its growth, Transformation into product
5. Further chemical treatment
6. Separating product, washing and drying
Molecules/ Atoms
Angstrom
Clusters
Nanoparticles
Nanometers
Examples of the methods:
✓ Supercritical fluid synthesis
✓ Chemical vapor deposition
✓ Molecular condensation
✓ Chemical reduction
✓ Spinning
✓ Sol-gel process
✓ Laser pyrolysis
✓ Green synthesis

FIGURE 5.11 Guideline for the steps of the bottom-up approach for nanomaterials' preparation.

assembly and establishes a short-range order at nanoscale dimensions with fewer flaws.

- It demonstrates a chemical composition that is more homogeneous and well-ordered.
- On substrates, nanoparticles are created by stacking atoms on top of each other, resulting in crystal planes. Those crystal planes are also stacked on top of one another.

Some nanoparticles can be made using either method. Graphene, for example, can be made from graphite powder by grinding (top-down approach) or by CVD, which uses the carbon atoms' layer (bottom-up approach). The bottom-up strategy, on the other hand, is the sole way to create well-dispersed and fine nanoparticles.

Limitations

- It requires very skilled personnel to perform the preparation.
- It requires solvents and chemical reagents.
- It is not as easy as the other approach to scale up the production, though it is doable.

5.2.1 Hydrothermal method

The hydrothermal process is typically carried out in a pressurized container known as an "Autoclave," which allows pressure and temperature to be controlled and maintained. The temperature at the boiling point of water can be raised during nanomaterial manufacturing, allowing the vapor to become saturated. This approach has been widely employed in the manufacture of various nanoparticles (Yang et al., 2001) (Fig. 5.12).

The advantage of hydrothermal technology is that it can be used to regulate reaction pressure, temperature, solvent characteristics, and solution composition, as well as additives to control the size of the particles and their morphology, crystalline phase, and surface chemistry (Carp et al., 2004).

5.2.2 Solvothermal method

The solvothermal approach is similar to the hydrothermal method, with the exception that it does not require water as a solvent (Fig. 5.13). Surprisingly, when organic solvents or compounds with high boiling temperatures are used, this approach is efficient in the nanomaterials preparation. Furthermore, compared to the hydrothermal approach, this process allows for more precise control of material size and shape. With or without the inclusion of surfactants, this process produces nanorods (nanomaterials) with or without the addition of required surfactants.

5.2.3 Thermolysis of metal-containing compounds

The thermolysis of metal-containing compounds can be done in closed or open systems as well as in isothermal and nonisothermal circumstances, with both

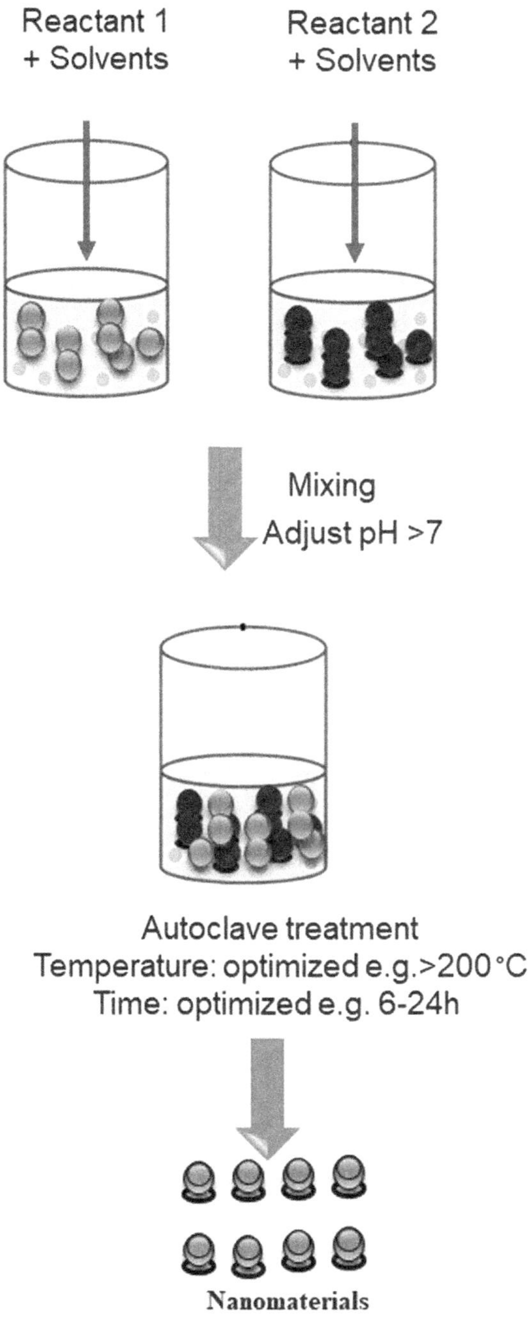
Reactant 1
+ Solvents
Reactant 2
+ Solvents
Mixing
Adjust pH >7
Autoclave treatment
Temperature: optimized e.g.>200°C
Time: optimized e.g. 6-24h
Nanomaterials

FIGURE 5.12 The main steps of the hydrothermal process for the synthesis of nanomaterials.

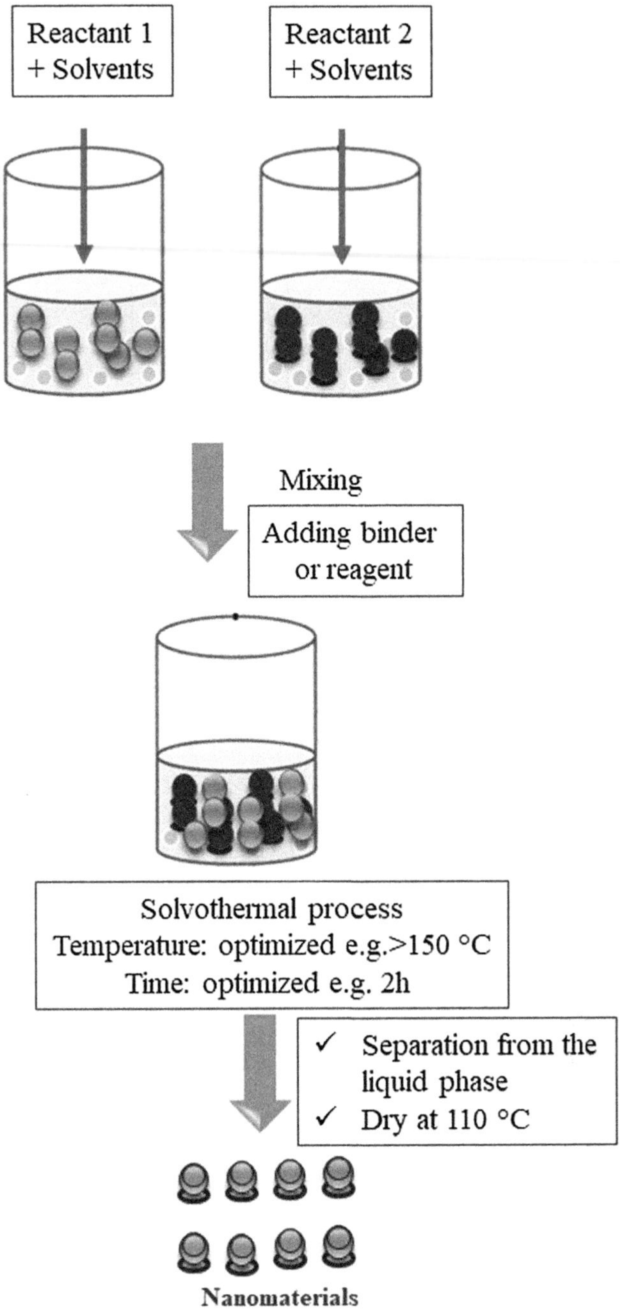
Reactant 1
+ Solvents
Reactant 2
+ Solvents
Mixing
Adding binder
or reagent
Solvothermal process
Temperature: optimized e.g.>150 °C
Time: optimized e.g. 2h
✓ Separation from the liquid phase
✓ Dry at 110 °C
Nanomaterials

FIGURE 5.13 Main steps of the solvothermal process for the synthesis of nanomaterials.

external and internal heating sources. Different ways allow adjusting the degree of conversion and defining the equipment for the experimental research of compound thermolysis, depending on the tasks.

In a one-step CVD synthesis of nanodispersed Fe oxides, $[Fe(OBu)_3]_2$ has been proposed as the MCC (Mathur et al., 2002). When the reaction is performed in a liquid medium in the presence of surfactants or polymers, it is possible to stabilize the resulting amorphous nanoparticles with diameters of up to 10 nm. A good example of $Fe(CO)_5$ thermolysis in two stages has been described. At 100°C, an iron oleate complex is produced from $Fe(CO)_5$ and oleic acid, which then decomposes at 300°C to provide primary small nanoparticles (4−11 nm). They are transformed into crystalline -Fe nanoparticles after being kept at 500°C, as proven by X-ray diffraction. Laser photolysis of volatile MCC (most commonly metal carbonyls) can also be used for this (Hyeon et al., 2001).

5.2.4 Chemical vapor deposition method

High-performance thin nanofilms are made using the chemical vapor deposition (CVD) process. The substrate is treated with volatile precursors that act on the surface of the substrate to form the desired films in this approach. Gas flow through the reaction chamber usually eliminates volatile by-products. The quality of the deposited nanomaterials on the surface is influenced by a number of factors, including temperature, reaction rate, and precursor quantity (Kim et al., 2004). CVD was used to form tin (Sn^{4+})-doped titania nanoparticle films, according to the study (Cao et al., 2004). Another CVD approach was used to make a doped titania nanoparticle, in which titania crystallizes into rutile structures based on the kind and amount of cations present in the chemical processes (Gracia et al., 2004). The advantage of this process is that the nanofilm coating is constant, but it has a number of drawbacks, including the higher temperatures required for chemical reactions and the difficulty of scaling up (Sudarshan, 2003).

Another example is ZnO. Chemical vapor deposition (CVD), electrodeposition, vapor phase transfer, thermal evaporation employing ZnO powders, and other methods of production affect the sizes and forms of ZnO nanoparticles. Nanostructured zinc oxide with large surface areas can also be formed via a matrix-assisted approach, in which metal oxide precursors are homogenously doped on the matrix surface with a large surface area. The matrix is then removed by calcination at high temperatures. A matrix-aided approach was used to make ZnO nanostructures with varying amounts of doped precursors on the activated carbon surface (Park et al., 2010).

5.2.5 Thermal decomposition and pulsed laser ablation

Decomposing metal alkoxides, salts, heat, or electricity can all be used to make doped metals. Furthermore, the properties of nanomaterials are highly influenced by the rate at which the precursor's concentrations in the processes flow and the environment in which they are formed. The thermal breakdown of titanium alkoxide at 1200°C has been reported to produce titania nanoparticles with a diameter of less than 30 nm (Kim et al., 2005). Another study used a pulsed laser ablation technique to create titania nanoparticles with a diameter of 3–8 nm (Liang et al., 2004). The solution combustion process was also used to make doped anatase titania nanoparticles (Nagaveni et al., 2004b). However, the relatively high cost, limited yield, and not easy in maintaining the morphology and structure of the plant are all downsides of this strategy.

5.2.6 Templating method

The templating method is a technique for creating materials with comparable morphologies. The synthesis of nanomaterials via the templating method has recently become quite popular. By combining morphological qualities with reactive deposition, this technology allows for the creation of a large number of novel materials with a consistent and regulated morphology by simply modifying the surface of the nanomaterials. A number of templates have been created in recent years to synthesize various nanomaterials (Iwasaki et al., 2004; Jinsoo et al., 2005). This technology has some drawbacks, such as expensive synthetic methods where templates are removed, typically by calcination, which raises manufacturing costs and increases the risk of contamination (Bavykin et al., 2006).

5.2.7 Combustion method

Combustion synthesis (CS), also known as self-propagating high-temperature synthesis (SHS), is a low-cost process for making a variety of industrially relevant nanostructures. Nowadays, combustion synthesis is a widely used method for the creation of nanomaterials. It is based on fast heating of a solution containing redox groups. This approach produces nanoparticles that are highly crystalline and have a huge surface area (Nagaveni et al., 2004a). To make the crystalline materials, the temperature is raised to around 650°C for 1–2 min throughout the manufacturing (Deganello, 2017).

The classification of CS processes is generally based on the physical nature of the initial reaction medium:

- ➢ Conventional SHS is used as initial reactants in solid state (condensed phase combustion).
- ➢ Solution combustion synthesis is processing where the initial reaction medium is an aqueous solution (Fig. 5.14).
- ➢ Synthesis of nanoparticles in flame is known as gas-phase combustion.

5.2.8 Method of the gas phase

Because it can be done chemically or physically, this approach is useful for making thin films. Nanomaterials are generated in the gas phase as a result of a chemical reaction or decomposition of a precursor (Jones and Chalker, 2003;

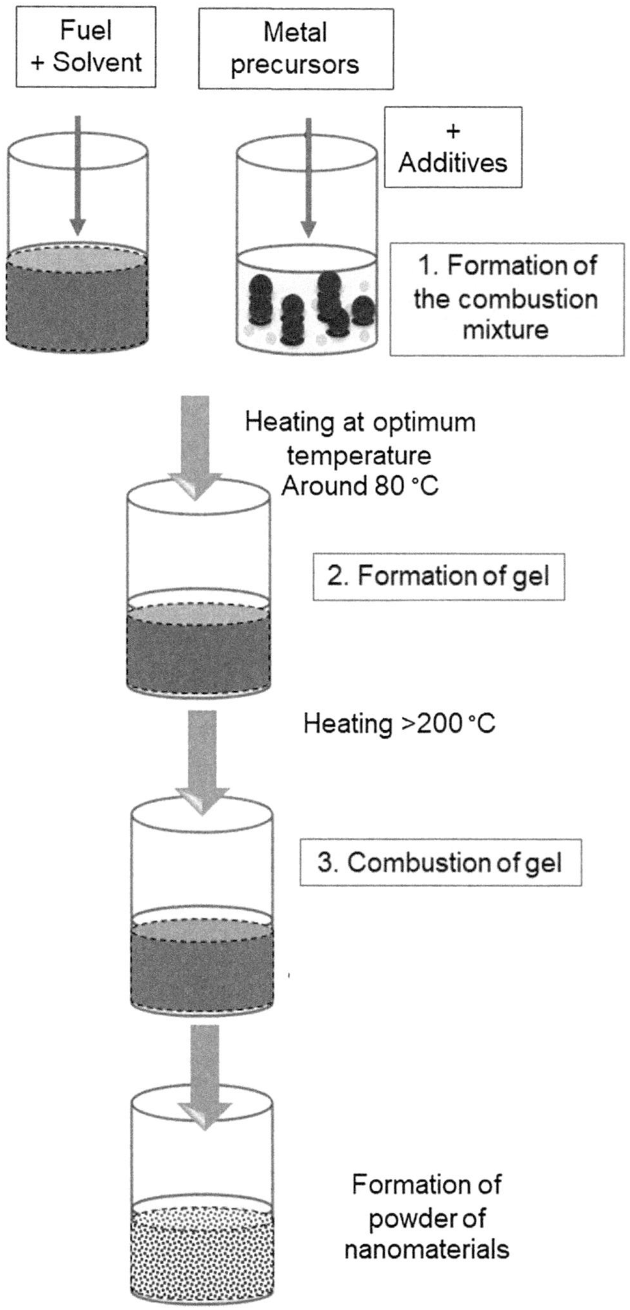
Fuel
+ Solvent
Metal
precursors
+
Additives
1. Formation of
the combustion
mixture
Heating at optimum
temperature
Around 80 °C
2. Formation of gel
Heating >200 °C
3. Combustion of gel
Formation of
powder of
nanomaterials

FIGURE 5.14 Illustration of the major steps involved in the solution-combustion synthesis.

Lee et al., 2017). Physical vapor deposition (PVD) is also a technology that can be utilized to create thin film deposition. Surprisingly, films are produced without the use of chemical transitions utilizing the gas phase approach. Electron beam (E-beam) evaporation is the process of producing titania thin films by heating the titania material with an electron beam and producing electrons. There are numerous advantages of using E-beam evaporation to deposit titania over CVD, including smoothness and improved conductivity (Van de Krol et al., 1997).

5.2.9 Sol–gel method

This approach has a number of benefits, including the ability to impregnate or coprecipitate nanomaterials, which can be utilized to introduce dopants. The sol–gel approach has been utilized to create diverse oxide materials (Fernandez-Garcia et al., 2004), and this process provides for better control of texture creation, chemical reaction, and morphological features of solid materials. The capacity to scale up with high purity nanomaterials is a major advantage of the sol–gel process (Kolen'ko et al., 2005). The hydrolysis and polymerization processes of the precursors produce a colloidal suspension in the sol–gel technology. Inorganic metal salts or metal organic compounds are commonly used as precursors (Pierre, 1998). Furthermore, any factor that affects either or both processes is expected to affect the gel formation qualities, and these parameters are referred to as sol–gel technique factors. Type of solvent, water content, acid or basic content, and different types of precursor, precursor concentration, and temperature are all elements to consider. The structure of the first gel formation is influenced by these parameters. The wet gel can then be matured in a different solvent after this stage. Aging is the period of time between the production of a gel and its drying (Chen and Mao, 2007).

Sol–gel is the most popular method for preparing photocatalysts, such as titania or doped titania. Wet chemical procedures (such as sol–gel) have the benefit of allowing the synthesis of nanometer-sized crystalline titania powder with high purity at a low temperature. Modified sol–gel processes, such as ultrasonic aided sol–gel, aerogel method, which is similar to sol–gel, sol–gel and photoreductive decomposition, precipitation, two-step wet chemical approach, and extremely low temperature precipitation, have also been employed (Bettinelli et al., 2007; Peng et al., 2008).

5.2.9.1 Sol–gel process

The sol–gel process is a wet-chemical technique that is widely used to deposit nanocomposite films. In this process, sol (or solution) containing sources for component materials, such as metal alkoxides and metal chlorides precursors for metal oxides, metallic nanoparticles for metals, tetraethoxysilane for silica matrix, catalyzers, stabilizers, and other additives for porosity generation, were prepared first.

The sol then undergoes hydrolysis and polycondensation reactions to evolve gradually toward the formation of a gel-like network. The basic structure or morphology of the solid phase can range anywhere from discrete colloidal particles to continuous chainlike polymer networks.

The formation of the nanocomposite film from the sol–gel precursor involves either dip coating or spin coating on a substrate, decomposition, and pyrolysis of compounds, removal of water and organics from the resulting network followed by nucleation and growth of the crystallites. The thermal decomposition behavior of the gel precursor plays an important role in crystallites size and in film porosity.

For example,

Basic chemical reactions

hydrolysis(1) $Si(OR)_4 + H_2O \longrightarrow (OR)_3SiOH + ROH$

polycondensation(2)

$(OR)_3SiOH + HOSi(OR)_3 \longrightarrow (OR)_3Si\text{-}O\text{-}Si(OR)_3 + H_2O$ (2)

$$C_2H_5O-\underset{\underset{OC_2H_5}{|}}{\overset{\overset{OC_2H_5}{|}}{Si}}-OC_2H_5 + H_2O \longrightarrow C_2H_5O-\underset{\underset{OC_2H_5}{|}}{\overset{\overset{OC_2H_5}{|}}{Si}}-OH + C_2H_5OH$$

$$C_2H_5O-\underset{\underset{OC_2H_5}{|}}{\overset{\overset{OC_2H_5}{|}}{Si}}-OH + OH-\underset{\underset{OC_2H_5}{|}}{\overset{\overset{OC_2H_5}{|}}{Si}}-OC_2H_5 \longrightarrow C_2H_5O-\underset{\underset{OC_2H_5}{|}}{\overset{\overset{OC_2H_5}{|}}{Si}}-O-\underset{\underset{OC_2H_5}{|}}{\overset{\overset{OC_2H_5}{|}}{Si}}-OH + H_2O$$

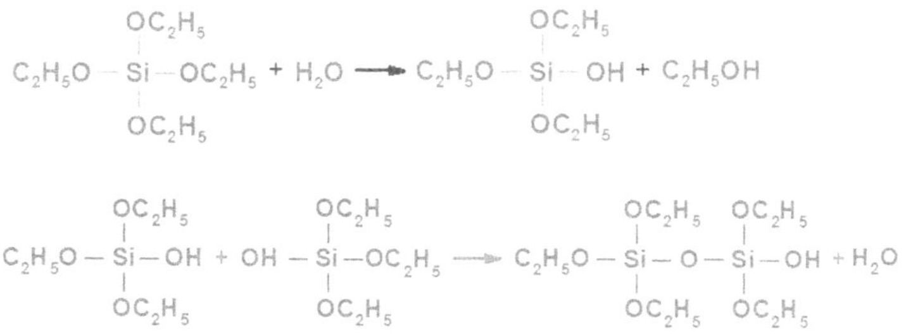

5.2.9.2 Sol–gel advantages

Advantages of sol–gel processes include:

➢ Sol–gel is an excellent technique for preparing high purity multicomponent films. Various types of nanocomposite films have been prepared by the sol–gel process and used as active materials for gas sensors.
➢ Sol–gel synthesis is a very viable alternative method to produce nanocrystalline elemental, alloy, and composite powders in an efficient and cost-effective manner.
➢ Sol–gel process involves the formation of sol, followed by a formation of gel (Fig. 5.15).
➢ Sol is a colloidal suspension of solid particles in a liquid phase.
➢ Gel refers to the interconnected network formed between phases.
➢ Obtaining of materials impossible to produce using some other methods.
➢ Homogeneity in molecular scale due to reaction in solution.
➢ Very high purity of obtained materials. By this method, one can able to get uniform small-sized powder.

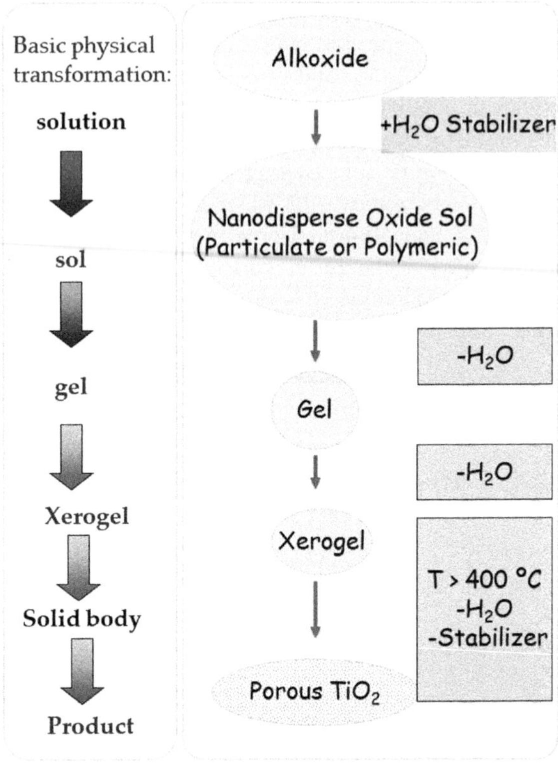
Basic physical transformation:
solution
sol
gel
Xerogel
Solid body
Product
Alkoxide
+H2O Stabilizer
Nanodisperse Oxide Sol (Particulate or Polymeric)
-H2O
Gel
-H2O
Xerogel
T > 400 °C
-H2O
-Stabilizer
Porous TiO2

FIGURE 5.15 Main steps for the sol–gel process used for the preparation of materials.

- Low process temperature in comparison with other methods (i.e., melting of glass).
- Obtaining thin layers on different base materials (glass, metal, plastic). For example, by this method without high temperature, high-density new compositions of glass can be obtained.
- Easy to do coating for films. For example, by this method, films with special porosity can be prepared.
- It can be used for improved adhesion.
- By this method, metal (inorganic)—organic composites can be obtained.

Disadvantages

- This method involves mostly the use of expensive starting compounds or precursors and solvents.
- Difficulties to obtaining large-scale monolits, free from crack.
- Sol–gel reactions are not reversible and impossible to completely inhibit.

➢ A commonly encountered problem is that procedures might result in amorphous or low crystallinity products, which necessitate annealing for crystallization or additional crystallization.

6. Biotechnological approach

Synthesis of nanomaterials with biotechnological instruments is deemed safe and environmentally friendly (Fig. 5.16). It falls within the category of green technology and nanobiotechnology because it involves the use of reducing and stabilizing agents during the preparation process (Parveen et al., 2021; Li et al., 2011; Joerger et al., 2000). Biotechnological approaches are thought to use a bottom-up strategy, in which metal ions are converted to nanomaterials using (Aarthye, 2021; Seifipour et al., 2020; Parashar et al., 2009; Luechinger et al., 2009):

➢ biological and plants extracts,
➢ microorganisms like bacteria, fungi, and actinomycetes, and
➢ algae, biomolecules, microseaweeds, and enzymes.

Advantages:

➢ Biotechnological approach yields better-organized nanoparticles compared.
➢ No toxic materials are used.
➢ It can be combined with the bottom-up approach.

Limitations

➢ It requires biological experience.
➢ It requires biotoxicity tests.
➢ There might be some interference from the biocomponents presented in the extracts.
➢ Reproducibility issues.

7. Microwave-assisted synthesis of nanomaterials

Microwave synthesis is used with microwave irradiation, which has a penetration characteristic that makes it possible to homogeneously heat up the reaction solution (Fig. 5.17). Microwave-assisted synthesis works via the stimulation of the material's dipoles in an external field using microwave electromagnetic radiations and is usually used in conjunction with a well-known synthesis approach (Corradi et al., 2005). It is advantageous because the synthesis method may be adjusted to produce nanomaterials of the desired size and shape. The alignment or orientation of molecules by an external electrical field can result in the generation of internal heat, which reduces the

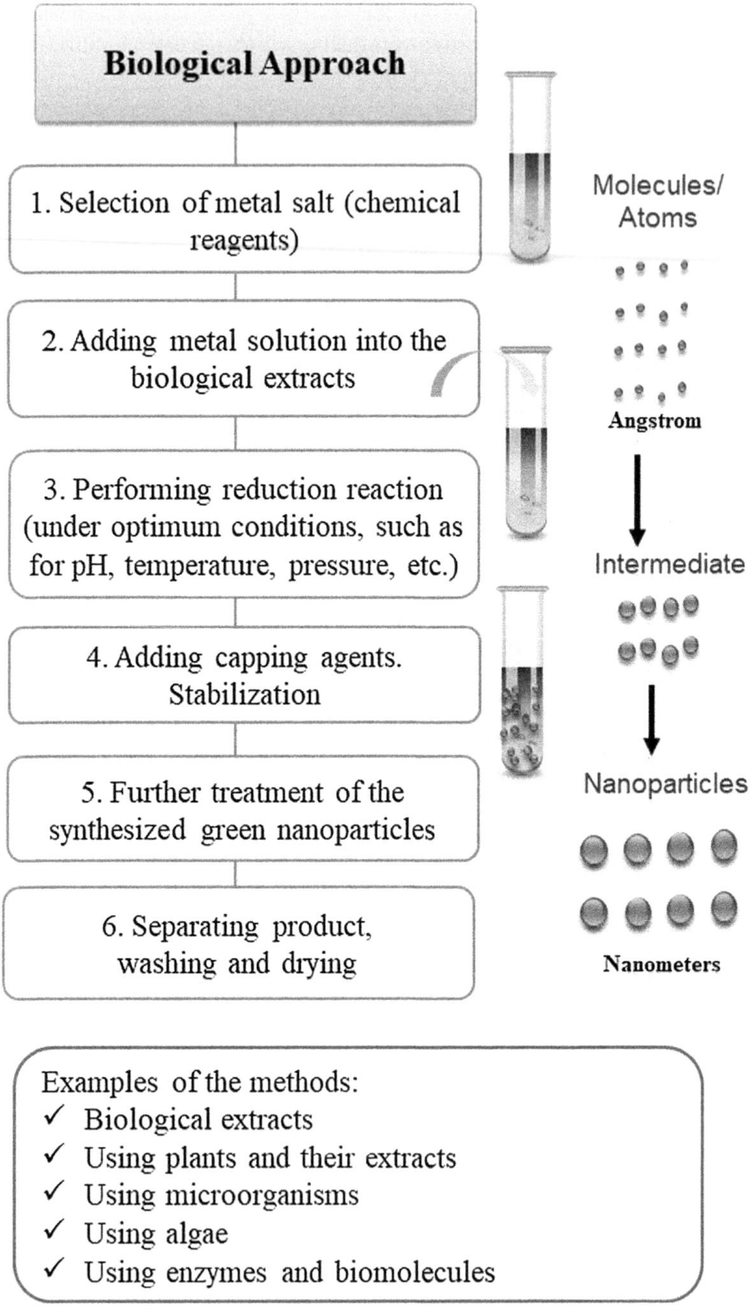
Biological Approach
1. Selection of metal salt (chemical reagents)
2. Adding metal solution into the biological extracts
3. Performing reduction reaction (under optimum conditions, such as for pH, temperature, pressure, etc.)
4. Adding capping agents. Stabilization
5. Further treatment of the synthesized green nanoparticles
6. Separating product, washing and drying
Molecules/ Atoms
Angstrom
Intermediate
Nanoparticles
Nanometers
Examples of the methods:
✓ Biological extracts
✓ Using plants and their extracts
✓ Using microorganisms
✓ Using algae
✓ Using enzymes and biomolecules

FIGURE 5.16 Guideline for the main steps of the biological approach for nanomaterials' preparation.

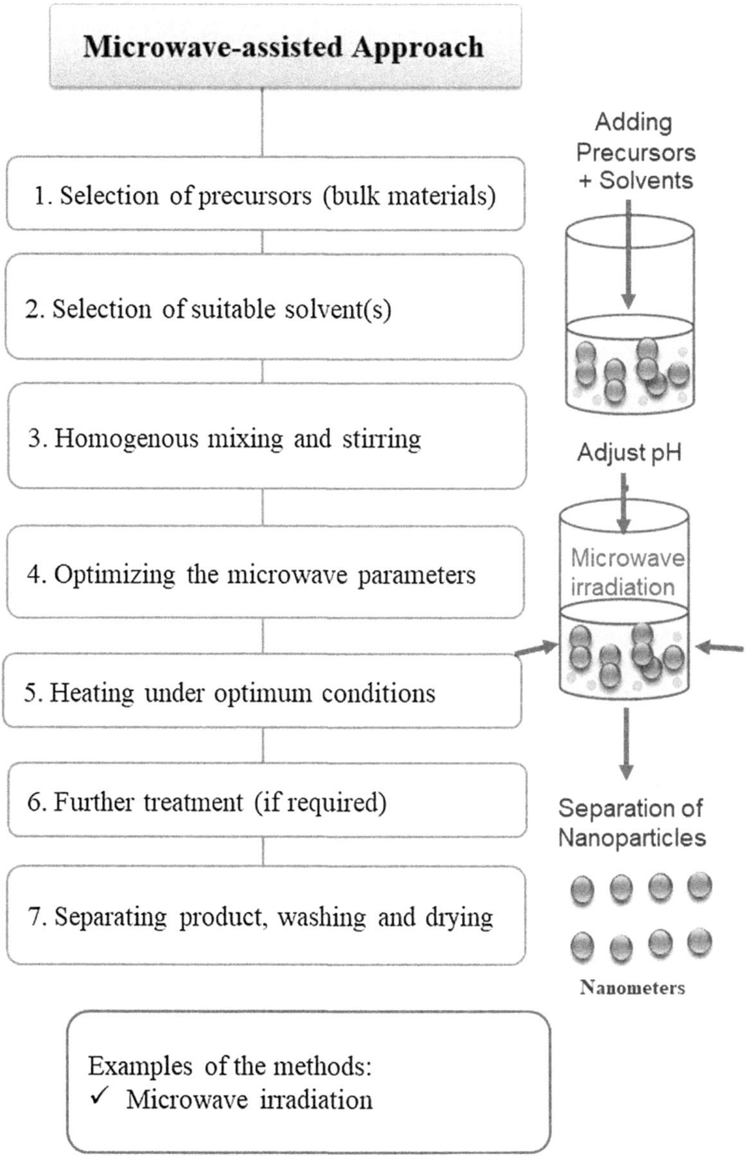
Microwave-assisted Approach
1. Selection of precursors (bulk materials)
2. Selection of suitable solvent(s)
3. Homogenous mixing and stirring
4. Optimizing the microwave parameters
5. Heating under optimum conditions
6. Further treatment (if required)
7. Separating product, washing and drying
Examples of the methods:
✓ Microwave irradiation
Adding Precursors + Solvents
Adjust pH
Microwave irradiation
Separation of Nanoparticles
Nanometers

FIGURE 5.17 Guideline for the main steps of the microwave approach for nanomaterials' preparation.

amount of time and energy required for processing. It is particularly because of microwaves' heating homogeneity. The time reaction is reduced by adopting microwave-assisted synthesis steps. The microwave approach provides a control of the properties of the nanomaterials (Inamuddin and Ali, 2018).

Advantages

- ➢ The microwave-assisted solution pathway is a cost-effective wet-chemical strategy for the synthesis of nanomaterials that has additional benefits such as rapid volumetric heating, high reaction rates, size and shape control via reaction parameter adjustment, and energy efficiency.
- ➢ In microwave synthesis, homogeneous heating of the reactants reduces thermal gradients and offers uniform nucleation and growth conditions, resulting in the creation of nanomaterials with a uniform size distribution.
- ➢ It saves energy resources.
- ➢ It can be combined with the bottom-up approach.

Limitations

- ➢ It is not easy to control the heating.
- ➢ It requires additional safety precautions.
- ➢ It is lack of scalability (batch size typically is limited to a few grams).
- ➢ Its safety and health hazards are related to the use of microwave-heating apparatus.
- ➢ It has limited applicability (to materials that absorb microwaves).
- ➢ It has reproducibility issues.

8. Synthesis of polymers

Polymerization, also known as polymer synthesis, is a chemical reaction that involves the covalent bonding of monomers to generate polymer structures. The number of repeating units in the polymer chain determines the length of the chain, which is referred to as the degree of polymerization (DP). Davankov was the first to synthesize cross-linked polymers (Tsyurupa and Davankov, 2002, 2006). External cross-linkers cross-link monomer-based precursors in the presence of a suitable solvent and catalyst, as shown in Fig. 5.18. The polymer chains are linked in solvents using the proper catalyst and a number of cross-linkers. The reaction then occurs quickly, forming a strong chain link with rigid bridges. Strong chains produced across the network prevent the structure from collapsing after the solvent is removed (Leonidas et al., 2017). Soaked polymers become one-phase materials with a lot of porosity and a low density after drying.

Cross-linkers include, for example, tetrachloromethane, monochlorodimethyl ether, 4,4′-bis-(chloromethyl) biphenyl, and p,p′-bis-chloromethyl-l. There are several types of cross-linked polymers including cross-linked polysulfone and polyarylates, styrene divinylbenzene copolymerization, self-condensation of p-xylylene dichloride, microporous triazine polymer, polyurethane–polylactide core–shell particles, and quaternary ammonium-functionalized β-cyclodextrin polymer (Zeng and Huang, 2020; Akpe et al., 2020).

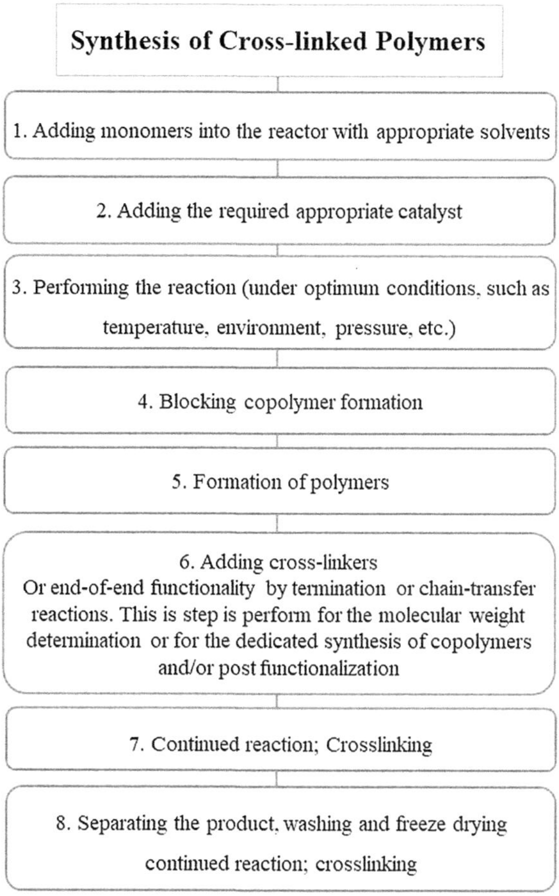
Synthesis of Cross-linked Polymers
1. Adding monomers into the reactor with appropriate solvents
2. Adding the required appropriate catalyst
3. Performing the reaction (under optimum conditions, such as temperature, environment, pressure, etc.)
4. Blocking copolymer formation
5. Formation of polymers
6. Adding cross-linkers
Or end-of-end functionality by termination or chain-transfer reactions. This is step is perform for the molecular weight determination or for the dedicated synthesis of copolymers and/or post functionalization
7. Continued reaction; Crosslinking
8. Separating the product, washing and freeze drying continued reaction; crosslinking

FIGURE 5.18 Protocol for the synthesis of cross-linked polymers.

Some of the methods used to produce cross-linked polymers include Heck coupling, Friedel–Crafts alkylation, Suzuki coupling, Sonogashira coupling, free-radical polymerization, N-alkylation, and Michael addition processes. The resulting polymers can be very porous and stiff, with a large surface area (Khodakarami and Bagheri, 2021). Fig. 5.19 is an example of cross-linked polymer production (McNamee et al., 2013).

The polymers are swollen in appropriate solvents before being cross-linked to produce a functionalized polymer with good porosity and thermal and chemical resilience. Numerous methodologies and metrics, such as weight and volume swelling, as well as inner surface area, can be employed to examine the produced polymers (Tsyurupa and Davankov, 2002).

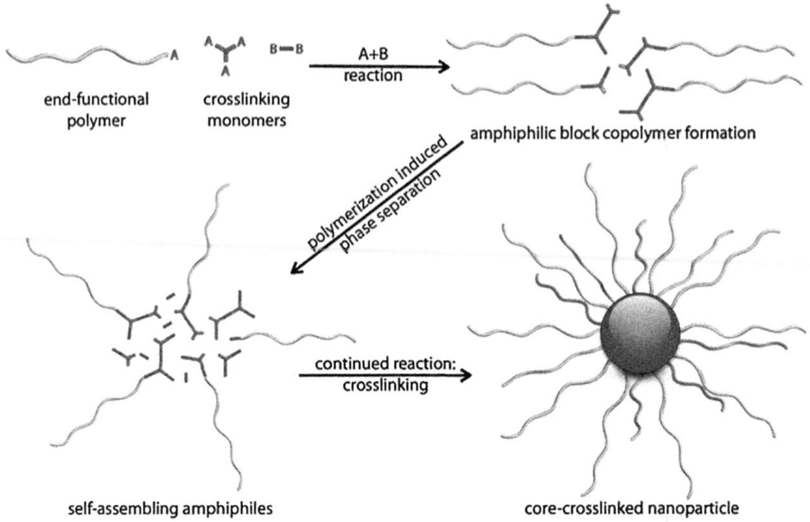
A
A
A
A
B—B
A+B
reaction
end-functional
polymer
crosslinking
monomers
amphiphilic block copolymer formation
polymerization induced
phase separation
continued reaction:
crosslinking
self-assembling amphiphiles
core-crosslinked nanoparticle

FIGURE 5.19 Illustration of the cross-linking of amphiphilic block copolymers.

9. Preparation of nanocomposites

Inorganic nanoclusters, fullerenes, clays, metals, oxides, or semiconductors can be combined with a variety of organic polymers, organic and organometallic chemicals, biological molecules, enzymes, and sol–gel generated polymers to create nanocomposites.

Nanocomposite materials, which are made by combining two or more different building constituents into one material, have unique features that can be explained by their small size, vast surface area, and, of course, the interfacial contact between the phases. Their exceptional potential has been successfully applied to improve the biological potential of a variety of medicines, biomaterials, catalysts, and high-value-added materials (Xin et al., 2012).

There are two ways to prepare polymeric nanocomposites: bottom-up and top-down methods. Fig. 5.20 illustrates the general guidelines for the preparation of polymers (using only polymers or monomers without nanoparticles) or polymeric nanocomposites (using polymers or monomers in presence of nanoparticles).

The basic idea behind nanocomposites is to create a large interface between the nanosized building components and the polymer matrix. The dimensionality of the nanosized heterogeneity or the composition of nanocomposites can be used to classify them. One type is polymeric nanocomposites.

The most common issues with nanocomposites are particle dispersion homogeneity in the polymer matrix and interfacial interactions. Interfacial

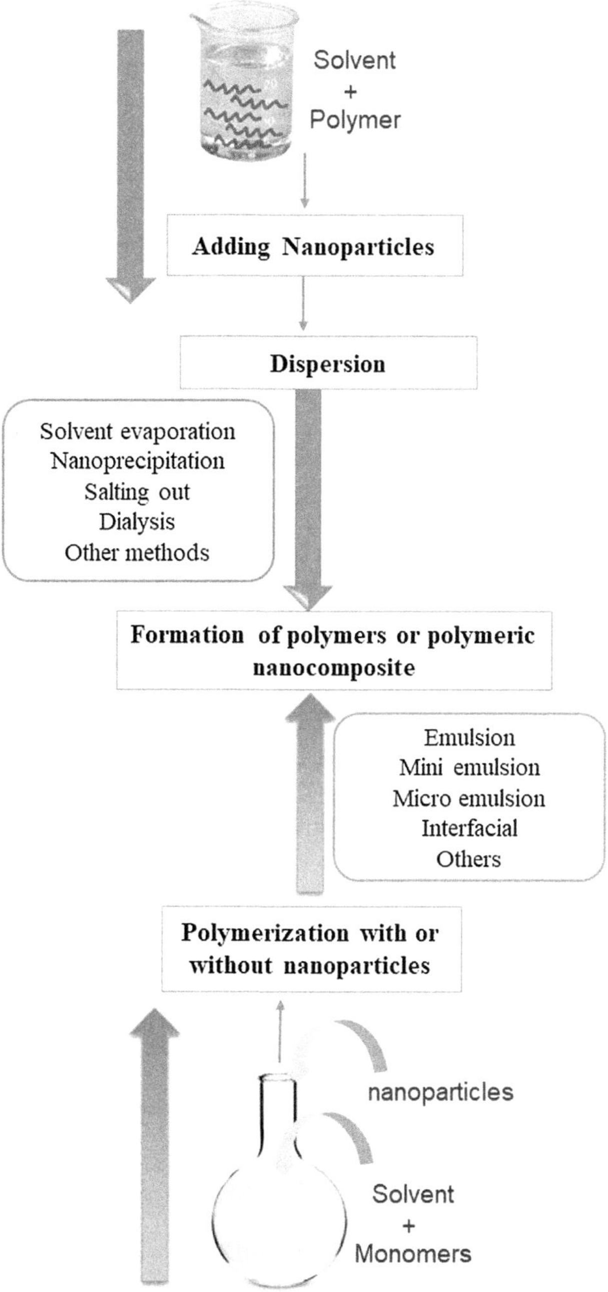
Solvent
+
Polymer
Adding Nanoparticles
Dispersion
Solvent evaporation
Nanoprecipitation
Salting out
Dialysis
Other methods
Formation of polymers or polymeric nanocomposite
Emulsion
Mini emulsion
Micro emulsion
Interfacial
Others
Polymerization with or without nanoparticles
nanoparticles
Solvent
+
Monomers

FIGURE 5.20 Guideline for the steps for the preparation of polymers (using only polymers or monomers without nanoparticles) or polymeric nanocomposites (using polymers or monomers in presence of nanoparticles).

interactions are explored superficially in comparison to the underlying notion of nanocomposites, and the available information is restricted and frequently inconsistent. The interactions of a coated nanoparticle surface, the size and qualities of the available uncoated area, the strength of interfacial adhesion, and the macroscopic properties of nanocomposites are among the challenges in nanocomposites preparation.

10. Conclusions

Some of the processes used to manufacture nanomaterials include chemical vapor deposition, thermal decomposition, hydrothermal synthesis, solvothermal synthesis, pulsed laser ablation, templating method, combustion method, microwave synthesis, gas-phase method, and conventional sol–gel method. The fabrication and process are the key issues in nanoscience and nanotechnology to explore the novel properties and phenomena of nanomaterials to realize their potential applications in science and technology. Many technological approaches/methods have been explored to fabricate nanomaterials. Followings are the key issues or challenges in the fabrication of nanostructured materials using any process or technique:

- ➢ to develop synthesis methods by which the particle size is controlled
- ➢ to develop synthesis methods by which the shape of nanoparticles is controlled
- ➢ to develop synthesis methods by which the structure either crystalline or amorphous is controlled
- ➢ to develop synthesis methods by which particle size distribution (monodispersive: all particles are of the same size) can be obtained

References

Aarthye, P., 2021. Sureshkumar, Green synthesis of nanomaterials: an overview. Mater. Today Proc. https://doi.org/10.1016/j.matpr.2021.04.564.

Akpe, S.G., Ahmed, I., Puthiaraj, P., Yu, K., Ahn, W.-S., 2020. Microporous organic polymers for efficient removal of sulfamethoxazole from aqueous solutions. Microporous Mesoporous Mater. 296, 109979.

Bavykin, D.V., Friedrich, J.M., Walsh, F.C., 2006. Protonated titanates and TiO_2 nanostructured materials: synthesis, properties, and applications. Adv. Mater. 18 (21), 2807–2824.

Bettinelli, M., Dallacasa, V., Falcomer, D., Fornasiero, P., Gombac, V., Montini, T., Romano, L., Speghini, A., 2007. Photocatalytic activity of TiO2 doped with boron and vanadium. J. Hazard Mater. 146 (3), 529–534.

Cao, Y., Yang, W., Zhang, W., Liub, G., Yue, P., 2004. Improved photocatalytic activity of Sn^{4+} doped TiO_2 nanoparticulate films prepared by plasma-enhanced chemical vapor deposition. New J. Chem. 2 (8), 218–222.

Carp, O., Huisman, C.L., Reller, A., 2004. Photoinduced reactivity of titanium dioxide. Prog. Solid State Chem. 32, 33.

Chen, X., Mao, S.S., 2007. Titanium dioxide nanomaterials: synthesis, properties, modifications, and applications. Chem. Rev. 107, 2891.

Corradi, A.B., Bondioli, F., Focher, B., Ferrari, A.M., Grippo, C., Mariani, E., Villa, C., 2005. Conventional and microwave-hydrothermal synthesis of TiO_2 Nanopowders. J. Am. Ceram. Soc. 88, 2639–2641.

Deganello, F., 2017. Nanomaterials for environmental and energy applications prepared by solution combustion based-methodologies: role of the fuel, Mater. Today Off. 4 (4 Part E), 5507–5516.

Fernandez-Garcia, M., Martinez-Arias, A., Hanson, J.C., Rodriguez, J.A., 2004. Nanostructured oxides in chemistry: characterization and properties. Chem. Rev. 104 (9), 4063–4104.

Gracia, F., Holgado, J.P., Caballero, A., Gonzalez-Elipe, A.R., 2004. Structural, optical, and photoelectrochemical properties of Mn^{+}-TiO_2 model thin film photocatalysts. J. Phys. Chem. B 108 (45), 17466–17476.

Gubin, S.P., Yurkov, G.Y., Kataeva, N.A., 2005. Inorg. Mater. 41 (10), 1017–1032. CODEN: INOMAF; ISSN:0020-1685. (Pleiades Publishing, Inc.).

Gupta, V.K., Saleh, T.A., 2013. Sorption of pollutants by porous carbon, carbon nanotubes and fullerene—an overview. Environ. Sci. Pollut. Res. 20, 2828–2843.

Hyeon, T., Lee, S.S., Park, J., Chung, Y., Na, H.B., 2001. Synthesis of highly crystalline and monodisperse maghemite nanocrystallites without a size-selection process. J. Am. Chem. Soc. 123, 798.

Iijima, S., 1991. Helical microtubules of graphitic carbon. Nature 354, 56–58.

Inamuddin, A.M.A., Ali, M., 2018. Applications of Nanocomposite Materials in Drug Delivery. In: A Volume in Woodhead Publishing Series in Biomaterials, Book.

Iwasaki, M., Davis, S.A., Mann, S., 2004. Spongelike macroporous TiO_2 monoliths prepared from starch gel template. J. Sol-Gel Sci. Technol. 32, 99–105.

Jiang, R., Da, Y., Han, X., Chen, Y., Deng, Y., Hu, W., 2021. Ultrafast synthesis for functional nanomaterials. Cell Rep. Phys. Sci. 2, 100302.

Jinsoo, K., Jae Won, L., Tai Gyu, L., Suk Woo, N., Jonghee, H., 2005. Nanostructured titania membranes with improved thermal stability. J. Mater. Sci. 40 (7), 1797–1799.

Joerger, R., Klaus, T., Granqvist, C.G., 2000. Biologically produced silver–carbon composite materials for optically functional thin-film coatings. Adv. Mater. 12, 407–409.

Jones, A.C., Chalker, P.R., 2003. Some recent developments in the chemical vapour deposition of electroceramic oxides. J. Phys. D Appl. Phys. 36, 80.

Kalhapure, R.S., Sonawane, S.J., Sikwal, D.R., et al., 2015. Solid lipid nanoparticles of clotrimazole silver complex: an efficient nano antibacterial against *Staphylococcus aureus* and MRSA. Colloids Surf. B Biointerfaces 136, 651–658.

Khaleel, A., Richards, R.M., 2001. Ceramics. In: Klabunde, K.J. (Ed.), Nanoscale Materials in Chemistry. WileyInterscience, New York, pp. 85–120.

Khanna, P., Kaur, A., Goyal, D., 2019. Algae-based metallic nanoparticles: synthesis, characterization and applications. J. Microbiol. Methods 163, 105656.

Khodakarami, M., Bagheri, M., 2021. Recent advances in synthesis and application of polymer nanocomposites for water and wastewater treatment. J. Clean. Prod. 296, 126404.

Kim, C.S., Okuyama, K., Nakaso, K., Shimada, M., 2004. Direct measurement of nucleation and growth modes in titania nanoparticles generation by a CVD method. J. Chem. Eng. Jpn. 37 (11), 1379.

Kim, C., Nakaso, K., Xia, B., Okuyama, K., Shimada, M., 2005. A new observation on the phase transformation of TiO_2 nanoparticles produced by a CVD method. Aerosol. Sci. Technol. 39 (2), 104–112.

Kolen'ko, Y.V., Kovnir, K.A., Gavrilov, A.I., Garshev, A.V., Meskin, P.E., Churagulov, B.R., Bouchard, M., Colbeau Justin, C., Lebedev, O.I., Van Tendeloo, G., Yoshimura, M., 2005. Structural, textural, and electronic properties of a nanosized mesoporous $Zn_xTi_{1-x}O_{2-x}$ solid solution prepared by a supercritical drying route. J. Phys. Chem. B 109 (43), 20303–20309.

Kumari, P., Weqar, M., Siddiqi, A., 2019. Usage of nanoparticles as adsorbents for waste water treatment: an emerging trend. Sustain. Mater. Technol. 22, e00128.

Lee, E.J., Huh, B.K., Kim, S.N., Lee, J.Y., Park, C.G., Mikos, A.G., Choy, Y.B., 2017. Application of materials as medical devices with localized drug delivery capabilities for enhanced wound repair. Prog. Mater. Sci. 89, 392–410.

Leonidas, M., Qiuju, G., Stina, J., Ulrika, R., Paul, C., 2017. Green conversion of municipal solid wastes into fuels and chemicals. Electron. J. Biotechnol. 26, 69–83.

Li, X., Xu, H., Chen, Z.S., Chen, G., 2011. Biosynthesis of nanoparticles by microorganisms and their applications. J. Nanomater. 1–16.

Liang, H., Esmaeili, H., 2021. Application of nanomaterials for demulsification of oily wastewater: a review study. Environ. Technol. Innovat. 22, 101498.

Liang, C., Shimizu, Y., Sasaki, T., Koshizaki, N., 2004. Synthesis, characterization, and phase stability of ultrafine TiO_2 nanoparticles by pulsed laser ablation in liquid media. J. Mater. Res. 19 (5), 1551–1557.

Liu, F., Yang, J.H., Zuo, J., et al., 2014. Graphene-supported nanoscale zero-valent iron: removal of phosphorus from aqueous solution and mechanistic study. J. Environ. Sci. 26 (8), 1751–1762.

Luechinger, N.A., Grass, R.N., Athanassiou, E.K., Stark, W.J., 2009. Bottom-up fabrication of metal/metal nanocomposites from nanoparticles of immiscible metals. Chem. Mater. 22, 155–160.

Mathur, S., Veith, M., Sivakov, V., Shen, H., Huch, V., Hartmann, U., Gao, H.B., 2002. Phase-selective deposition and microstructure control in iron oxide films obtained by single-source CVD. Chem. Vap. Depos. 8, 277.

McNamee, K.P., Pitet, L.M., Knauss, D.M., 2013. Synthesis, assembly, and cross-linking of polymer amphiphiles in situ: polyurethane–polylactide core–shell particles. Polym. Chem. 4, 2546–2555.

Nagaveni, K., Hedge, M.S., Ravishankar, N., Subbanna, G.N., Madras, G., 2004a. Synthesis and structure of nanocrystalline TiO2 with lower band gap showing high photocatalytic activity. Langmuir 20, 2900–2907.

Nagaveni, K., Hegde, M.S., Madras, G., 2004b. Structure and photocatalytic activity of $Ti_{1-x}M_xO_2$; (M = W, V, Ce, Zr, Fe, and Cu) synthesized by solution combustion method. J. Phys. Chem. B 108 (52), 20204–20212.

Palencia, A., 2021. A critical analysis of environmental sustainability metrics applied to green synthesis of nanomaterials and the assessment of environmental risks associated with the nanotechnology. Sci. Total Environ. 793, 148524.

Parashar, V., Parashar, R., Sharma, B., Pandey, A.C., 2009. Parthenium leaf extract mediated synthesis of silver nanoparticles: a novel approach towards weed utilization. J. Nanomater. Biostruct. 4, 45–50.

Park, N.K., Han, G.B., Yoon, S.H., Ryu, S.O., Lee, T., 2010. Preparation and absorption properties of ZnO nanostructures for cleanup of H_2S contained gas. J. Int. J. Precis. Eng. Manuf. 11, 321.

Parveen, S., Sharma, G., Sharma, S.B., 2021. A review on synthesis and biological evaluation of plants based metallic nanoparticles. Global J. Res. Rev. 8 (2), 60.

Peng, F., Cai, L., Huang, L., Yu, H., Wang, H., 2008. Preparation of nitrogen-doped titanium dioxide with visible-light photocatalytic activity using a facile hydrothermal method. J. Phys. Chem. Solid. 69 (7), 1657–1664. CODEN: JPCSAW; ISSN:0022-3697.

Pierre, A.C., 1998. Introduction to Sol-Gel Processing. Kluwer Academic Publishers, Boston, p. 394.

Rao, G.P., Lu, C., Su, F., 2007. Sorption of divalent metal ions from aqueous solution by carbon nanotubes: a review. Separ. Purif. Technol. 58, 224–231.

Saleh, T.A., 2021a. Protocols for synthesis of nanomaterials, polymers, and green materials as adsorbents for water treatment technologies. Environ. Technol. Innovat. 24, 101821.

Saleh, T.A., 2021b. Nanomaterials: classification, properties, and environmental toxicities. Environ. Technol. Innovat. 20, 101067.

Saleh, T.A., 2022a. Experimental and analytical methods for testing inhibitors and fluids in water-based drilling environments. Trac. Trends Anal. Chem. 116543. https://doi.org/10.1016/j.trac.2022.116543.

Saleh, T.A., 2022b. Advanced trends of shale inhibitors for enhanced properties of water-based drilling fluid. Upstream Oil Gas Technol. 100069. https://doi.org/10.1016/j.upstre.2022.100069.

Saravanan, R., Sacari, E., Gracia, F., Khan, M.M., Mosquera, E., Gupta, V.K., 2016. Conducting PANI stimulated ZnO system for visible light photocatalytic degradation of coloured dyes. J. Mol. Liq. 221, 1029–1033.

Seifipour, R., Nozari, M., Pishkar, L., 2020. Green synthesis of silver nanoparticles using *Tragopogon collinus* leaf extract and study of their antibacterial effects. J. Inorg. Organomet. Polym. Mater. 30, 2926–2936.

Sudarshan, T.S., 2003. In coated powders—new horizons and applications, advances in surface treatment: research & applications (ASTRA). In: Proceedings of the International Conference, Hyderabad, India, pp. 412–422.

Tsyurupa, M.P.,, Davankov, V.A., 2002. Hypercrosslinked polymers: basic principle of preparing the new class of polymeric materials. React. Funct. Polym. 53 (2–3), 193–203.

Tsyurupa, M.P., Davankov, V.A., 2006. Porous structure of hypercrosslinked polystyrene: state-of-the-art mini-review. React. Funct. Polym. 66 (7), 768–779.

Van de Krol, R., Goossens, A., Schoonman, J., 1997. In situ X-ray diffraction of lithium intercalation in nanostructured and thin film anatase TiO_2. J. Electrochem. Soc. 144.

Xin, X., Wei, Q., Yang, J., Yan, L., Feng, R., Chen, G., et al., 2012. Highly efficient removal of heavy metal ions by aminefunctionalized mesoporous Fe_3O_4 nanoparticles. Chem. Eng. J. 184, 132–140.

Yan, J., Han, L., Gao, W., Xue, S., Chen, M., 2015. Biochar supported nanoscale zerovalent iron composite used as persulfate activator for removing trichloroethylene. Bioresour. Technol. 175, 269–274.

Yang, J., Mei, S., Ferreira, J.M.F., 2001. The effect of F^--doping and temperature on the structural and textural evolution of mesoporous TiO_2 powders. Mater. Sci. Eng. B C15.

Zeng, X., Huang, J., 2020. Anisole-modified hyper-cross-linked resins for efficient adsorption of aniline from aqueous solution. J. Colloid Interface Sci. 569, 177–183.

Chapter 6

Large-scale production of nanomaterials and adsorbents

1. Introduction

Nanomaterials and nanocomposites are extremely promising for a variety of technologies, including adsorption and separation, due to their enhanced surface properties. They promise to make a positive difference in our lives and the environment in a variety of ways. Nanomaterials' fundamental properties and practical applications are the subjects of extensive research worldwide. To bring such a beneficial breakthrough into the everyday lives of ordinary people, large-scale manufacturing of high-quality nanomaterials is required. While numerous production methods are available on a laboratory scale, only a few are being used on a large scale, and each approach can only produce a finite number of economically viable nanomaterials.

Nanoparticles of superior quality are increasingly being used in the industry, establishing them as the next generation of appealing resources with numerous potential applications. Without a doubt, the existing divide between fundamental research on nanomaterials and their application in real life will be bridged over the next decade.

This chapter discusses current trends in the production of nanomaterials and adsorbents on a large scale. Additionally, some fundamental aspects of product development are described to assist academic researchers and scientists who wish to commercialize their research findings.

2. Prerequisites of introducing nanomaterials

The selection of large-scale nanoparticle production methods for industrial use depends on three factors:

- ➢ Type of nanoparticles
- ➢ Type of approach used (bottom-up vs. top-down)
- ➢ Regulatory requirements

Surface Science of Adsorbents and Nanoadsorbents. https://doi.org/10.1016/B978-0-12-849876-7.00007-5

Other determining parameters such as the nature of precursors, temperature stability, and use of solvent may vary depending upon the above three factors.

Nanomaterials have been used in a variety of technological applications, including adsorption, separation, and environmental remediation, as a result of the development of new synthesis methods and the manipulation of the size, shape, and nanostructure of materials. Scale-up, in the context of nanomaterials manufacturing, refers to the transition from microscopic (molecular) laboratory production to industrial large-scale production on a macroscopic (bulk) scale.

There are some prerequisites for the introduction of any nanomaterials into use or the market. The prerequisites include the following:

- A certified large-scale manufacturing or production process must be available.
- The manufacturing process requires to be qualified, validated, and accepted by the regulatory authorities.
- The production process should result in a high-quality product with zero to acceptable contamination levels as defined by regulatory authorities or criteria, such as elemental contaminants from manufacturing equipment or residual solvent.
- Acceptable cost: Manufacturers seek cost-effective, easy, and consistent operation and production setup.

3. Manufacturing, industry, and academia

Nanomanufacturing is a term that refers to the manufacturing of nanomaterials. Nanomanufacturing is the process of scaling up, dependable, and cost-effectively producing nanoscale materials, structures, devices, and systems. Additionally, it encompasses research, development, and integration of top-down and more complex bottom-up or self-assembly processes. Simply put, nanotechnology enables the development of novel materials and products.

Nanotechnology is a multidisciplinary field that encompasses physics, chemistry, biology, materials science, and engineering. Collaboration between scientists from diverse fields will almost certainly result in the development of novel materials with tailored properties (Fig. 6.1). The success of nanomanufacturing is contingent on strong collaboration between academia and industry in order to stay informed about current and future demands and to design products that can be directly transferred to the industrial sector.

4. Terminologies used in the scale-up process

The scale-up process entails research and development, as well as laboratory work to optimize the parameters and conditions of chemical reactions required for the production of nanomaterials, as illustrated in Fig. 6.2. This is in

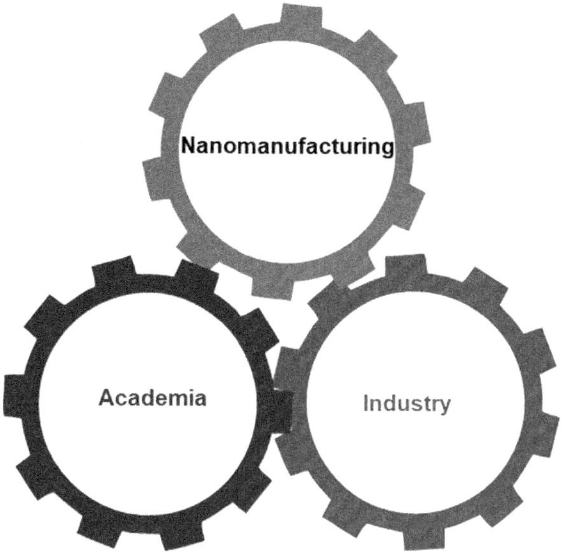
Nanomanufacturing
Academia
Industry

FIGURE 6.1 Nanomanufacturing dependent on academia and industry.

addition to the process design. If the results of the laboratory and bench scale synthesis of nanomaterials are promising, a pilot plant can be investigated. On the basis of the results and observations, a decision can be made regarding the feasibility of expanding nanomaterial production on a large scale. Fig. 6.3 illustrates several critical terms in this context.

A pilot plant can be used for the following:

- ➢ Evaluating the laboratory experimental results and making process and product modifications and enhancements
- ➢ Producing small quantities of materials for chemical and microbiological evaluations, restricted market testing, or providing samples to possible consumers and shelf-life and storage stability studies
- ➢ Providing data that can be utilized to decide whether or not to move forward with a full-scale manufacturing process; in case a favorable choice is made, planning and building a full-scale plant or altering an existing facility

5. Steps in scale-up production of a material

There are several steps involved in scaling up a new material's large production, as illustrated in Fig. 6.4. These activities include defining the new nanomaterial product, conducting additional laboratory work, and developing a scale-up strategy. The following steps involve defining the critical rate steps

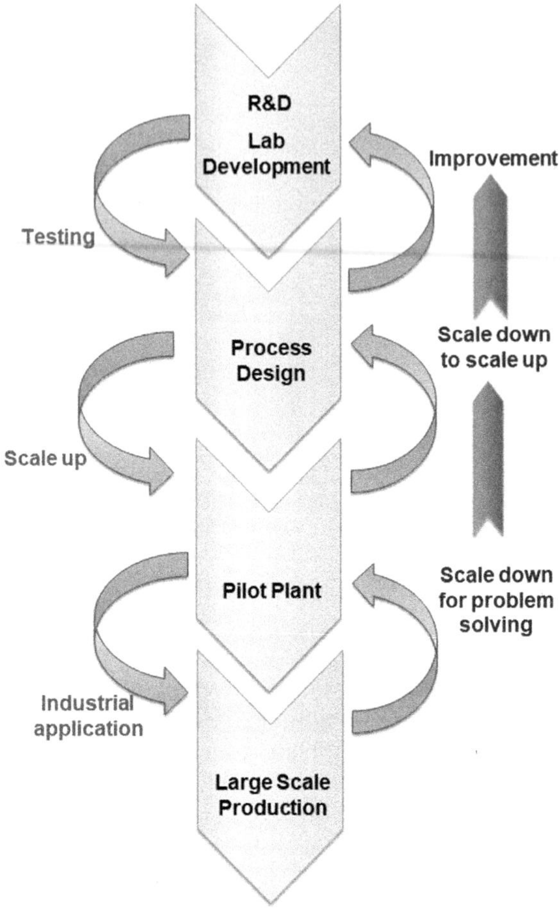
R&D
Lab
Development
Improvement
Testing
Process
Design
Scale down
to scale up
Scale up
Pilot Plant
Scale down
for problem
solving
Industrial
application
Large Scale
Production

FIGURE 6.2 Process of scale-up and main phases sequence for the large-scale production of nanomaterial.

and ramping up production. The following step is to develop a pilot plan, followed by an evaluation of the results, as illustrated in Fig. 6.5. As illustrated in Fig. 6.6, these are the steps to transfer the production of nanomaterials from laboratory to industrial scale.

5.1 Lab-scale, bench scale, pilot scale, and scale-up

The process of nanomaterial development tends to proceed in small steps when scaled up. It begins at the laboratory scale and progresses to the bench scale, then to the pilot scale, and finally to the production scale (Fig. 6.7). This will mitigate the risk associated with the larger investment in the next step.

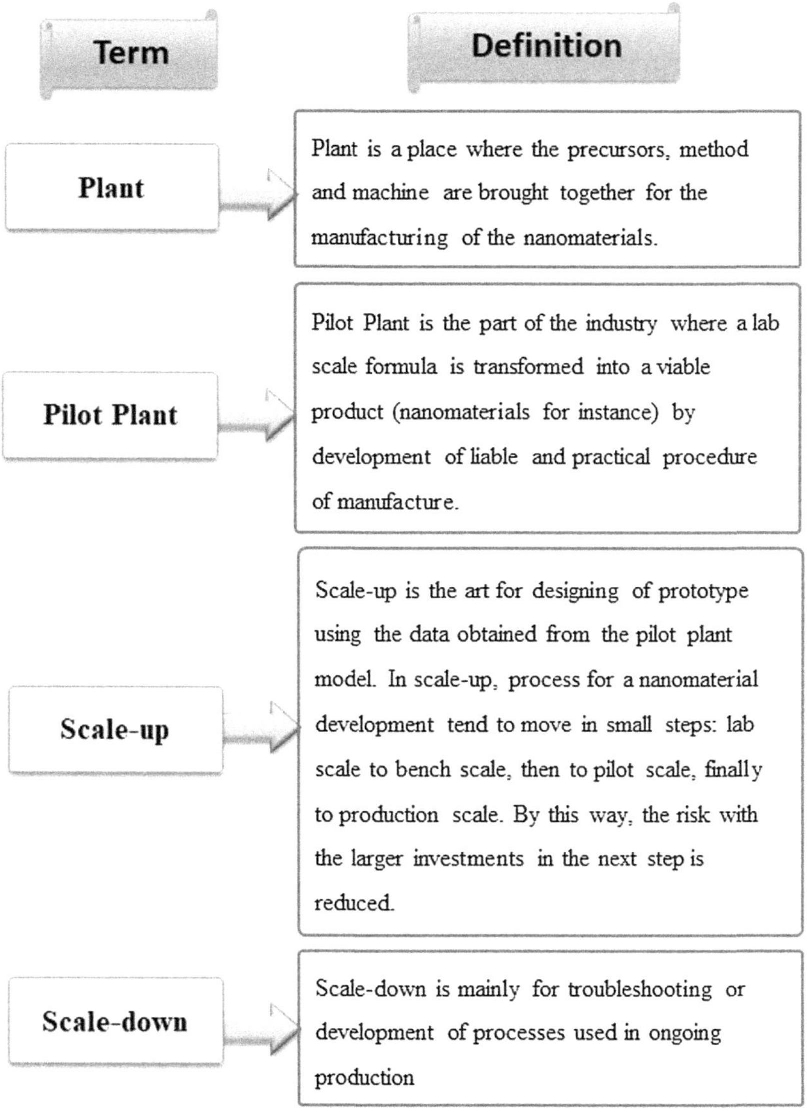
Term
Definition
Plant
Plant is a place where the precursors, method and machine are brought together for the manufacturing of the nanomaterials.
Pilot Plant
Pilot Plant is the part of the industry where a lab scale formula is transformed into a viable product (nanomaterials for instance) by development of liable and practical procedure of manufacture.
Scale-up
Scale-up is the art for designing of prototype using the data obtained from the pilot plant model. In scale-up, process for a nanomaterial development tend to move in small steps: lab scale to bench scale, then to pilot scale, finally to production scale. By this way, the risk with the larger investments in the next step is reduced.
Scale-down
Scale-down is mainly for troubleshooting or development of processes used in ongoing production

FIGURE 6.3 Some of the main terms commonly used in scale-up process.

5.2 Lab-scale

Lab-scale systems, alternatively referred to as laboratory-scale units, are the next step up from the glass beaker used to determine the potential of a particular chemistry. The laboratory-scale is used to screen an idea and optimize the parameters and process for the synthesis of a material or nanomaterial.

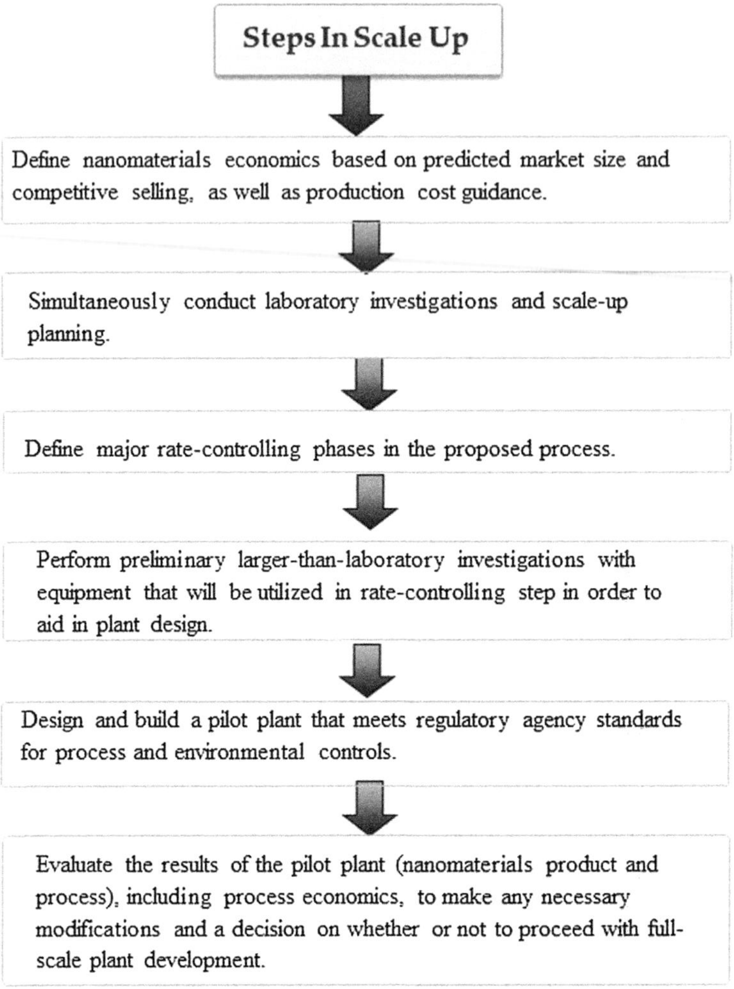
Steps In Scale Up
Define nanomaterials economics based on predicted market size and competitive selling, as well as production cost guidance.
Simultaneously conduct laboratory investigations and scale-up planning.
Define major rate-controlling phases in the proposed process.
Perform preliminary larger-than-laboratory investigations with equipment that will be utilized in rate-controlling step in order to aid in plant design.
Design and build a pilot plant that meets regulatory agency standards for process and environmental controls.
Evaluate the results of the pilot plant (nanomaterials product and process), including process economics, to make any necessary modifications and a decision on whether or not to proceed with full-scale plant development.

FIGURE 6.4 Steps to scale up nanomaterials production.

5.3 Bench scale

Bench scale synthesis is typically performed in a laboratory and employs systems incapable of handling small quantities of reagents and precursors used in the synthesis. It is a procedure that entails setting up a small reaction setup on a laboratory bench and carrying out the synthesis under optimal conditions. Additionally, it entails verifying the yield and purity of products such as nanomaterials.

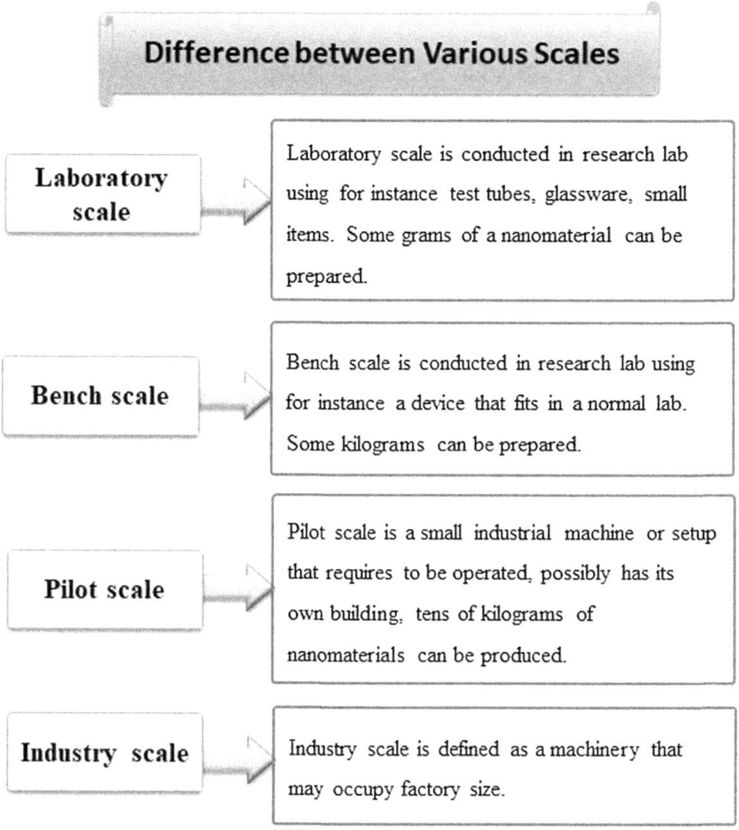
Difference between Various Scales
Laboratory scale
Laboratory scale is conducted in research lab using for instance test tubes, glassware, small items. Some grams of a nanomaterial can be prepared.
Bench scale
Bench scale is conducted in research lab using for instance a device that fits in a normal lab. Some kilograms can be prepared.
Pilot scale
Pilot scale is a small industrial machine or setup that requires to be operated, possibly has its own building, tens of kilograms of nanomaterials can be produced.
Industry scale
Industry scale is defined as a machinery that may occupy factory size.

FIGURE 6.5 Difference between various scales of materials production.

5.4 Pilot plant studies

A pilot scale is used to validate proof of concept before scaling up to produce larger quantities of a product or nanomaterial. Typically, these take place on-site at the location where the full-scale facility will be built and operated. The pilot scale manufactures a material in a manner that is fully representative of and comparable to the production scale process of increasing batch size/procedure for applying the same process to varying output volumes. Pilot plant studies are conducted to evaluate the production and processing of nanomaterials on a smaller scale before investing significant funds in full-scale production. The consequences of a multiplication of scale are frequently impossible to predict. It is impossible to build a large-scale processing facility successfully using only laboratory data.

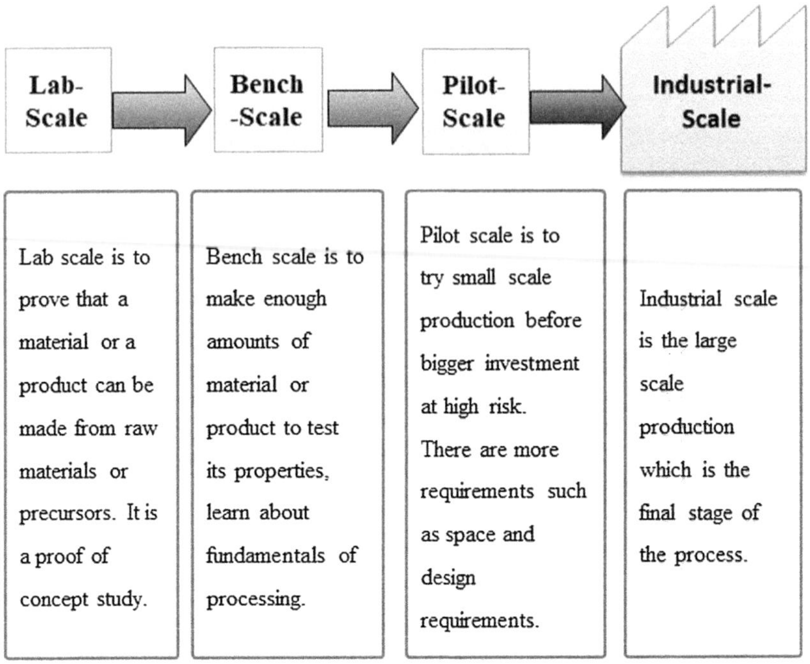
Lab-Scale
Bench-Scale
Pilot-Scale
Industrial-Scale
Lab scale is to prove that a material or a product can be made from raw materials or precursors. It is a proof of concept study.
Bench scale is to make enough amounts of material or product to test its properties, learn about fundamentals of processing.
Pilot scale is to try small scale production before bigger investment at high risk. There are more requirements such as space and design requirements.
Industrial scale is the large scale production which is the final stage of the process.

FIGURE 6.6 Steps from the lab to the industrial production of nanomaterials.

A pilot plant can be used to

- ➢ evaluate the findings of laboratory studies as well as make product and process improvements.
- ➢ produce small quantities of nanomaterials for chemical and microbiological evaluations, restricted market testing, or provision of samples to potential consumers, and shelf-life and storage stability studies.
- ➢ provide data that can be utilized to decide whether or not to move forward with a full-scale manufacturing process; and, if a favorable choice is made, planning and building a full-scale plant or altering an existing facility.

5.5 Industrial scale

The term "full scale" refers to the final design, construction, and operation of a system. It is the result of incorporating bench and pilot scale results in order to optimize the final design.

Full-scale manufacturing decisions are required for design, including the following:

What is the current state of market demand?

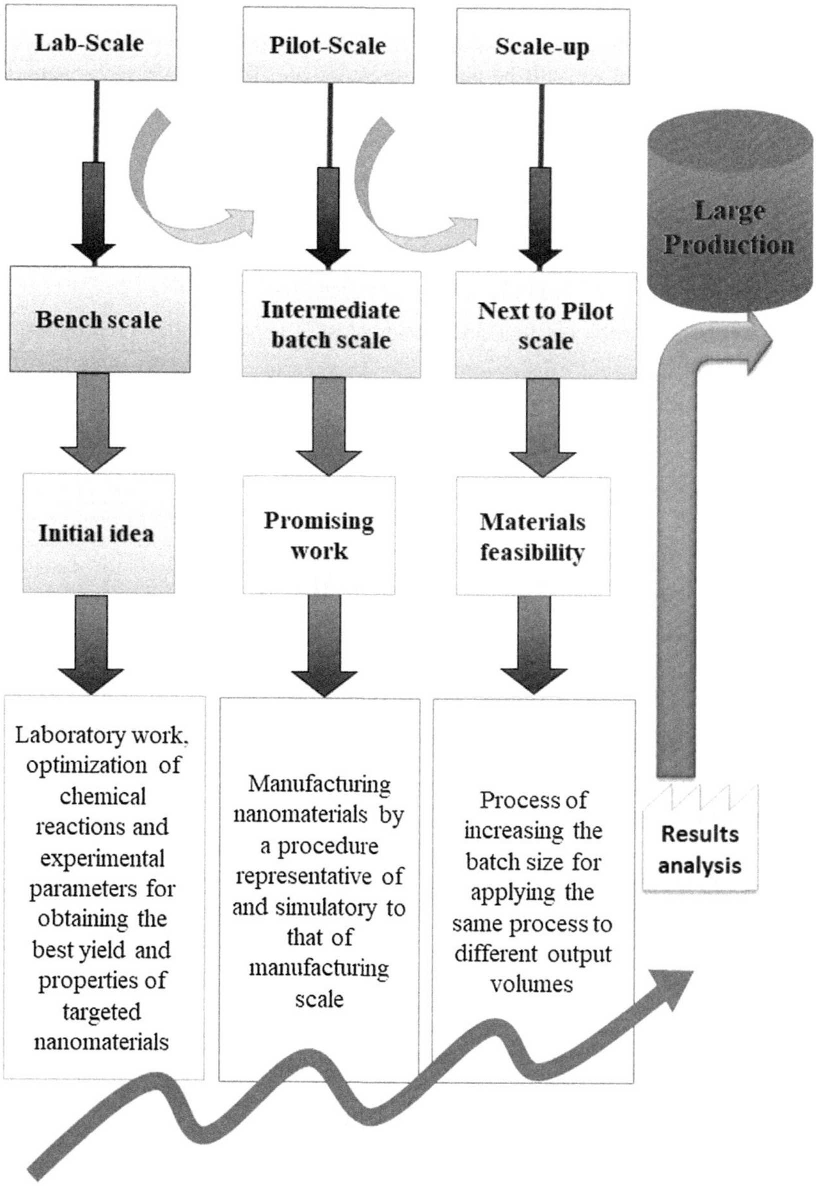
Lab-Scale
Pilot-Scale
Scale-up
Large Production
Bench scale
Intermediate batch scale
Next to Pilot scale
Initial idea
Promising work
Materials feasibility
Laboratory work, optimization of chemical reactions and experimental parameters for obtaining the best yield and properties of targeted nanomaterials
Manufacturing nanomaterials by a procedure representative of and simulatory to that of manufacturing scale
Process of increasing the batch size for applying the same process to different output volumes
Results analysis

FIGURE 6.7 Nanomaterial development phases from lab, pilot-scale, and scale-up to the industrial large scale production.

What is the forecast for it?
How many units should be manufactured?
Demand is determined by price, and production costs are determined by the quantity produced.

What are the available raw materials?
What are the estimates of cost, supply, and pricing, among other things?
Do you believe we will be able to increase our manufacturing capacity?
Which consequences of such an increase should be considered during the initial design phase?

5.6 Down-scaling

Scale-down is used mainly for cases such as troubleshooting and improvement of processes used in ongoing productions. It is used also to:

- ➢ solve some problems that are causing losses (poor quality or production interruption).
- ➢ advance productivity or product quality.
- ➢ evaluate lower cost raw materials, additives, consumables, and unit operations.
- ➢ assessment of new raw materials, test new mixer, heater, sensor, or control system.

6. Gas-, liquid-, and solid-based methods

The most promising methods for large-scale production can be classified into vapor, liquid, and solid methods.

6.1 Vapor-phase synthesis

This section discusses methods for directly converting gas to particles. Three fundamental steps are required to form nanoparticles using the vapor-phase synthesis method. It all began with the nucleation of clusters from supersaturated vapors, either homogeneous or heterogeneous. Homogeneous nucleation occurs when vapor molecules condense and form nuclei without the presence of foreign ions or particles. Heterogeneous nucleation, on the other hand, occurs on foreign nuclei, ions, or surfaces.

The nanocrystal nuclei condense from the constituent vapor and grow as a result of collision and coalescence. The primary particles coagulate and form aggregates as the temperature drops.

Numerous alternative vapor-phase synthesis approaches have been investigated in order to increase nanoparticle output. Vapor synthesis can be accomplished in a variety of ways, including flame synthesis, laser ablation, chemical vapor condensation, and arc discharge (Swihart, 2003; Yang et al., 2010; Khang and Lee, 2010; Ashkarran et al., 2009).

6.2 Liquid-phase synthesis

Another widely used method for producing nanoparticles is liquid-phase synthesis, also known as the wet chemical process. A typical liquid-phase synthesis procedure involves combining well-defined concentrations of various ions in a solution and subjecting it to controlled heat, pressure, and temperature to induce the formation of insoluble nanoparticles that precipitate out of the solution. In comparison to vapor-phase synthesis, liquid-phase synthesis is more energy efficient and is frequently performed at low temperatures and pressures below atmospheric. It does not require a great deal of heat, energy, or pressure. Additionally, liquid-phase synthesis enables more precise control over the stoichiometry composition of the end product, and nanoparticles can be conducted or functionalized throughout the synthesis method.

While liquid-phase synthesis has a number of advantages, it is still limited in terms of industrial applicability and scalability. Chemicals used in liquid-phase synthesis are frequently toxic, explosive, and corrosive, posing a significant risk to worker safety and the environment. Additionally, nanomaterials must be processed and purified following their creation. The purity of nanoparticles generated by liquid-phase synthesis is often lower than that of those synthesized by vapor-phase synthesis.

Chemical Precipitation (Cushing et al., 2004), Hydrothermal Method (Hayashi and Hakuta, 2010), and Sol–Gel Method are just a few examples (Tavakoli et al., 2007).

6.2.1 Solid-phase synthesis

In comparison to vapor-phase and liquid-phase synthesis, solid-phase synthesis is considered the first technology for the fabrication of nanomaterials. For solid-phase synthesis, physical crushing of coarse granules into finer powders is required. The majority of solid-phase synthesis is performed using planetary or rotating ball mills. These procedures have the advantage of being straightforward and requiring little expense in terms of equipment. However, contaminations, large particle size, and broad size distribution can make this sort of synthesis difficult. An example includes mechanical milling (Wang et al., 2002).

6.3 Comparison

Vapor-phase synthesis is a relatively clean method for synthesizing high-purity nanomaterials. Nonetheless, the nanomaterials generated in this manner are predominantly polycrystalline in nature and prone to agglomeration. Wet chemical synthesis, also referred to as liquid-phase synthesis, enables precise control of stoichiometry composition and is more easily scaled up than dry chemical synthesis because it does not require high temperatures or pressures. However, because stabilizers or capping agents are required to prevent agglomeration, the quality of the generated nanoparticles remains a critical issue. Contamination is a problem in solid-phase synthesis, as it is in liquid-phase synthesis. Nonetheless, it is a relatively efficient and straightforward method for increasing production capacity and can be used to produce alloy compounds at low temperatures (Su and Chang, 2018).

Examples

Production of carbon nanomaterials by catalytic chemical vapor deposition (Qi et al., 2010)
Green synthesis of nanomaterials (Huston et al., 2021)
Production of graphene by wet chemical methods (Tene et al., 2020)
Production of graphene by electrochemical methods (Parvez et al., 2019)
Production of graphene by hydrothermal methods (Vacacela Gomez et al., 2019)

7. Advantages of liquid-based wet chemical methods

Liquid-based wet chemical approaches have some advantages.

- ➢ They allow for the proper selection of reaction parameters for the desired product (e.g., kinds of solvents and/or surfactants, temperature, and duration).
- ➢ By adjusting the reaction's thermodynamic and kinetic parameters, the nucleation and development of NMs may be easily regulated.
- ➢ They allow for low temperature and relatively simple processing conditions.
- ➢ The cost is low because expensive devices are not needed. The cost can be decreased even more by carefully planning reactions and selecting ingredients.
- ➢ The product has good homogeneity and phase purity due to the liquid-phase conditions.
- ➢ By simply changing the reaction conditions, such as salt precursors, temperature, pH, solvent, hydrolyzing agents, and capping agent, different morphologies can be prepared. Examples of nanostructures include nanorods, nanosheets, nanoneedles, nanoflowers, nanoplates, and hierarchical (Gao et al., 2013).

8. Large-scale production of nanomaterials

Nanostructures are critical for technological advancements in a variety of fields, including engineering, water, energy, and electronics. While there are numerous applications for these nanoparticles in the literature, there is currently a small market for nanomaterial-based products. This disparity is exacerbated by a lack of effective and low-cost industrial processes, as well as businesses' unwillingness to adopt and invest in new infrastructure. Large-scale nanomaterial production should pass the same tests that industries use to determine the appropriateness of a technique. Nanostructures must be manufactured to exacting industrial specifications in order to achieve the following: (i) desired properties; (ii) sufficient quality and purity for intended applications; (iii) cost-effective and environmentally friendly methods; (iv) controllable and stable process; and (v) scalable procedures. Although nanostructures have entered the value chain, they remain rare in industrial goods. Carbon nanostructures, metal oxide nanoparticles, and others are mass produced in large quantities throughout the world (Charitidis et al., 2014; Moon et al., 2016). Numerous businesses have begun integrating nanomaterials into their manufacturing processes, but in small quantities due to their high cost. Around the world, inorganic nanomaterials are primarily produced on a large scale via liquid-phase and vapor-phase processes (Kumar and Sinha Ray, 2018) (Table 6.1).

8.1 Classification of methods

The two primary techniques for nanomanufacturing are depicted in Fig. 6.8: top-down and bottom-up. Top-down fabrication begins with a block of starting material and gradually removes it until the desired nanoscale output is obtained. Bottom-up design begins with atomic and molecular-scale components, providing engineers with additional building options.

Both techniques are applicable in the gas, liquid, supercritical fluid, and solid states, as well as in a vacuum. However, the fundamental attributes to manage remain the same: particle size, shape, size distribution, composition, and degree of particle agglomeration, among others (Table 6.2). The critical parameters to consider when selecting a technology on a large scale are depicted in Fig. 6.9.

8.2 The top-down approach

The top-down approach begins with bigger (macroscopic) beginning structures that can be manipulated externally during processing and transformed into nanostructures via severe plastic deformation induced by mechanical, chemical, or other types of energy. Two examples are wet media milling or bead milling (shear forces) and high-pressure homogenization (HPH) (cavitation force).

8.3 The bottom-up approach

The bottom-up technique entails synthesis or downsizing of materials components (down to the atomic level), followed by a self-assembly process to create nanostructures. Sol–gel processing, precipitation, supercritical, aerosol-based chemical vapor deposition (CVD), plasma or flame spraying synthesis, laser pyrolysis, and atomic or molecular condensation are all examples of this type of synthesis.

TABLE 6.1 Various synthesis processes are compared based on various parameters.

Method	Yield	Controllability	Efficacy	Cost	Comments
Wet chemical routes	Medium	Relatively high	Relatively high	Relatively low	High controllability, continuous monitoring, several stages, and a wide range of nanostructures
Ball milling	Relatively high	No	Relatively low	Relatively low	Impurities are easily introduced, particles clump together, and there is no morphology
Chemical vapor deposition (CVD) and physical vapor deposition (PVD), atomic layer deposition (ALD)	Relatively low	Relatively low	Relatively low	Relatively high	It involves the use of toxic gases and precursors, thin films and superlattices, high homogeneity
Deformation	Relatively high	No	Average	Average	Mechanical properties of bulk materials should be good
Etching	Relatively low	Average	Relatively low	Relatively high	Mainly nanoarrays
Molecular beam epitaxy (MBE) fast-quenching	Relatively high	No	Relatively low	Average	Ultrarapid cooling is required to avoid phase transition
Sputtering	Relatively low	Medium	Relatively low	Relatively high	Consider uniformly thick films, high temperatures, and vacuums, all of which necessitate the use of competent operators
Electrospinning	Acceptable	Relatively low	Average	Relatively high	Nanoribbons >10 nm (aligned or random) were mostly required, as well as a polymeric solution

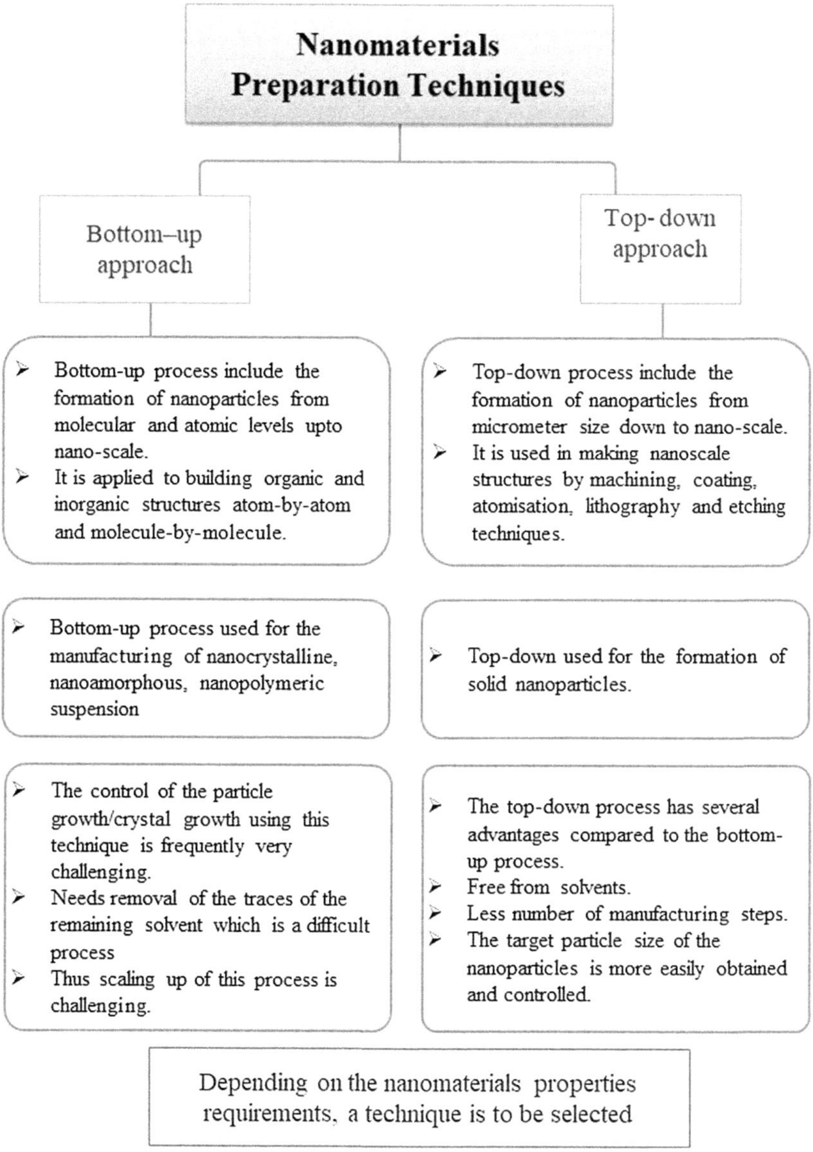
Nanomaterials Preparation Techniques
Bottom–up approach
Top-down approach
➢ Bottom-up process include the formation of nanoparticles from molecular and atomic levels upto nano-scale.
➢ It is applied to building organic and inorganic structures atom-by-atom and molecule-by-molecule.
➢ Top-down process include the formation of nanoparticles from micrometer size down to nano-scale.
➢ It is used in making nanoscale structures by machining, coating, atomisation, lithography and etching techniques.
➢ Bottom-up process used for the manufacturing of nanocrystalline, nanoamorphous, nanopolymeric suspension
➢ Top-down used for the formation of solid nanoparticles.
➢ The control of the particle growth/crystal growth using this technique is frequently very challenging.
➢ Needs removal of the traces of the remaining solvent which is a difficult process
➢ Thus scaling up of this process is challenging.
➢ The top-down process has several advantages compared to the bottom-up process.
➢ Free from solvents.
➢ Less number of manufacturing steps.
➢ The target particle size of the nanoparticles is more easily obtained and controlled.
Depending on the nanomaterials properties requirements, a technique is to be selected

FIGURE 6.8 Promising nanomaterial preparation techniques for large-scale production.

Bottom-up nanomanufacturing is made possible by advanced methods:

➢ Chemical vapor deposition: In this method, chemicals are mixed together to create ultrapure, high-performance films.
➢ Molecular beam epitaxy: This method is based on the deposition of single crystals and is particularly valuable in the semiconductor industry.

TABLE 6.2 A comparison analysis between top-down and bottom-up approaches.

Term	Top-down methods	Bottom-up methods
Solvents	There are no solvents used in the procedure	Solvents are used
Precursors	Poorly soluble precursors can be processed using the top-down process	Only precursors that have high solubility can be used
Preparation steps	Preparation steps are easy to adapt and reproduce	Preparation steps require precautions as any error during preparation lead to a change in the properties of the obtained material
Risk	The possibility of establishing an amorphous state or a polymorphic transition is minimal	The transition between amorphous and polymeric states is a prevalent problem
Size	Can be controlled	Easy to control
Scaling up	More feasibly	Requires more optimization
Modification of nanomaterials surface	Not possible during the synthesis. But it can be done in a separation additional chemical functionalization process	Can be performed by the chemical process

- Atomic layer epitaxy: It is a method of forming one-atom-thick layers on a surface.
- Dip pen lithography: It is a technique in which a chemical fluid is dipped into the tip of an atomic force microscope, which subsequently "writes" on a substrate surface.
- Nanoimprint lithography: It is a technique for imprinting or printing microscopic features onto a surface.
- Roll-to-roll processing: It is a technique for mass-producing nanoscale gadgets on a roll of ultrathin plastic or metal.
- Self-assembly: It is a process by which individual chemical or biological molecular structures naturally link together to form an organized structure without the need for outside instruction.

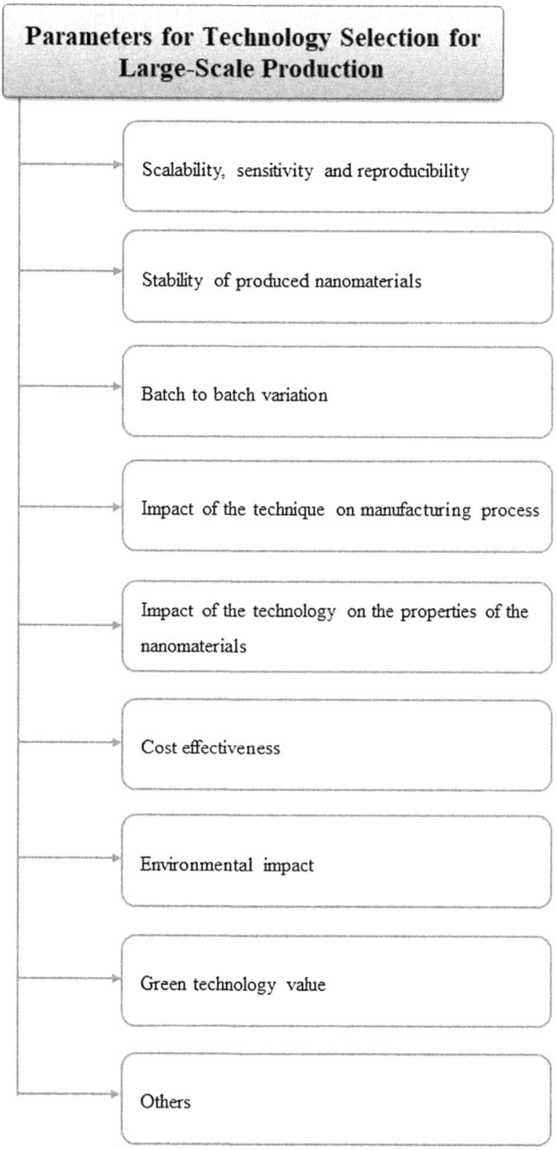
Parameters for Technology Selection for Large-Scale Production
Scalability, sensitivity and reproducibility
Stability of produced nanomaterials
Batch to batch variation
Impact of the technique on manufacturing process
Impact of the technology on the properties of the nanomaterials
Cost effectiveness
Environmental impact
Green technology value
Others

FIGURE 6.9 The most important factors to consider while choosing a technology on a wide scale.

9. Requirements for scale-up

There are some general requirements for scale-up nanomaterial production. These include the following:

➢ Reporting plan and personal responsibility

Scientists with prior experience in both pilot plant operations and actual production are preferred. They must comprehend the formulator's goal as well as the manufacturing staff's viewpoint. Engineering knowledge should be present in the group, and scaling-up also requires engineering principles.

- ➢ Space requirements

  This includes (i) administration and information processing, (ii) physical testing area, (iii) standard pilot plant equipment floor space, and (iv) storage area.
- ➢ Raw materials

  Raw materials used in the small-scale production cannot necessarily be representative of large-scale production. So, it is necessary to determine the required raw precursors for the production of nanomaterials.
- ➢ Equipment

  The simplest, most cost-effective, and most efficient equipment capable of generating nanomaterials within the required requirements is employed. The equipment should be large enough that the experimental trials done are comparable to production batches.
- ➢ Production rates

  While determining production rates, both current and future market trends/requirements are taken into account.
- ➢ Process evaluation

  It is necessary to perform the reactions and order of mixing of components and note the mixing speed, mixing time, and rate of addition of granulating agents, solvents, and solutions.
- ➢ Manufacturing procedures

  The important aspects include weight sheet, processing directions, and manufacturing procedure.
- ➢ Nanomaterial stability and uniformity

  The main goal of the pilot plant is to ensure that the nanomaterials are both physically and chemically stable. As a result, the stability of each pilot batch reflecting the nanomaterials formulation and production technique should be investigated. Stability tests should also be performed on completed packages.

10. Challenges to scale up nanomaterial production

There are still several challenges and difficulties in manufacturing scale-up and large-scale production of nanoparticles (Kaur et al., 2014; Ragelle et al., 2017) (Fig. 6.10). Some of the challenges are the following:

- ➢ Reproducibility: Changes in operating conditions can have a significant impact on the properties of the nanoparticles produced. Scale changes necessitate adjustments to process variables such as reaction time, agitation rate, mixing process, and equipment used. Environmental factors such as temperature, pH, and pressure can also affect reproducibility. However,

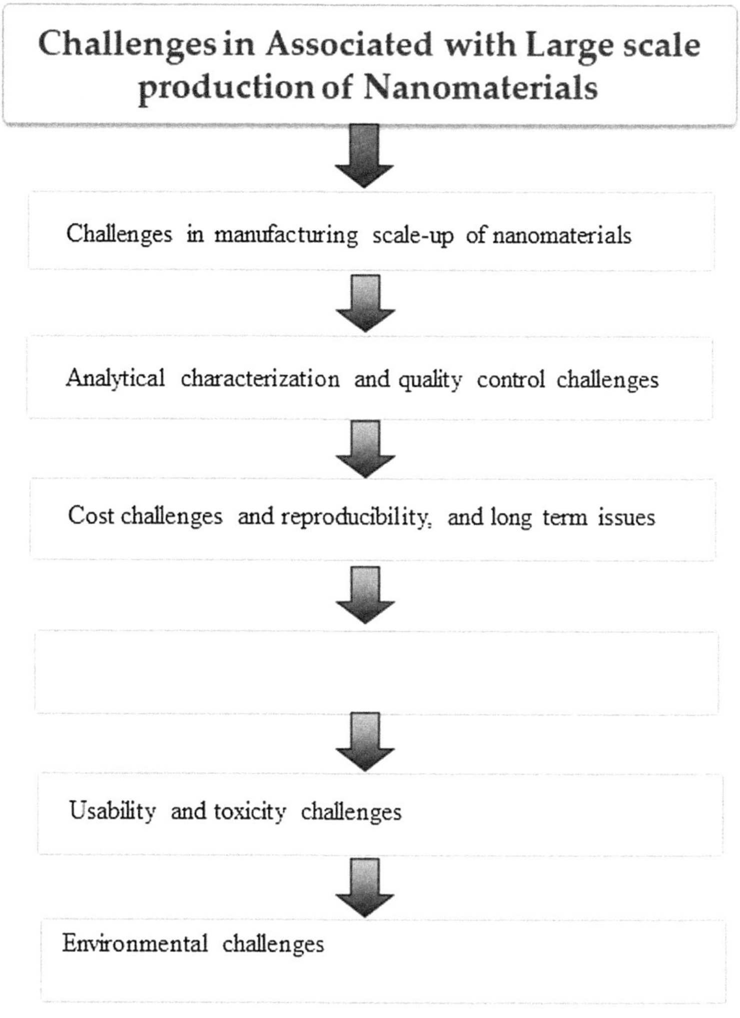
Challenges in Associated with Large scale production of Nanomaterials
Challenges in manufacturing scale-up of nanomaterials
Analytical characterization and quality control challenges
Cost challenges and reproducibility, and long term issues
Usability and toxicity challenges
Environmental challenges

FIGURE 6.10 Challenges associated with the large-scale production of nanomaterials.

reproducibility can be improved by implementing a robust quality control (QC) system and conducting in-process testing. Utilizing a quality-by-design approach, we identify critical quality properties of nanoparticles. To ensure batch-to-batch reproducibility, identify potentially harmful process parameters (Saleh, 2021).

➢ The physical stability of nanomaterials may be compromised during and after production. It is necessary to characterize processes and nanomaterials. To minimize physical changes, critical manufacturing parameters should be identified and analyzed during the development stage.

- ➢ Chemical transformations of the nanomaterials are expected during and after production. Thus, the addition of stabilizers or coatings may help to minimize chemical changes.
- ➢ Environmental protection: Nanomaterials produced on a large scale may have an effect on the environment. This necessitates toxicity testing and evaluation. By and large, nanoparticles formed exclusively in a liquid environment may have a significantly lower environmental impact.

11. Converting waste materials into adsorbents

There are several types of waste materials that can be converted into viable materials and adsorbents. Examples are:

- ➢ Household waste like carpeting composites, scrap tires, fruit waste, and coconut shells.
- ➢ Organic and inorganic waste materials like peat moss, algae, chitosan, and seafood processing waste (seaweed, etc.).
- ➢ Biomass waste includes agricultural products like bagasse pith, sawdust, cob, maize and rice husk, coconut shells, bark, chitosan, cyclodextrin, starch, peat, and cotton.
- ➢ Industrial waste like plastics, fibers, red mud, fly ash, blast-furnace slag and petroleum, fertilizer waste, sugar industry waste, and biosorbents (Gupta, 2009; Gupta and Saleh, 2013; Saleh, 2022).

11.1 Methods of conversion wastes to value-added products

The vast array of products of materials and adsorbents that can be generated from wastes and biomass using different types of conversion technologies, including pyrolysis, gasification, torrefaction, anaerobic digestion, and hydrothermal processing. To transform waste and biomass into value-added products, various thermochemical and biological methods have been used. Pyrolysis is the most convenient of these processes since it has various advantages in terms of storage, transportation, and flexibility in terms of solicitation, such as turbines, combustion appliances, boilers, engines, and so on.

11.2 Procedures of conversion

Due to the huge amount of trash and the necessity for cost-effective sorbents, the conversion of waste biomaterials to sorbents could be a dual solution for waste management and cost-effective products formation technologies. Through physical and chemical treatments, they can be transformed into adsorbents for the manufacture of value-added goods. The character of the derived materials is influenced by converting waste to viable products. The guidelines of the experimental process are presented in Fig. 6.11. The general

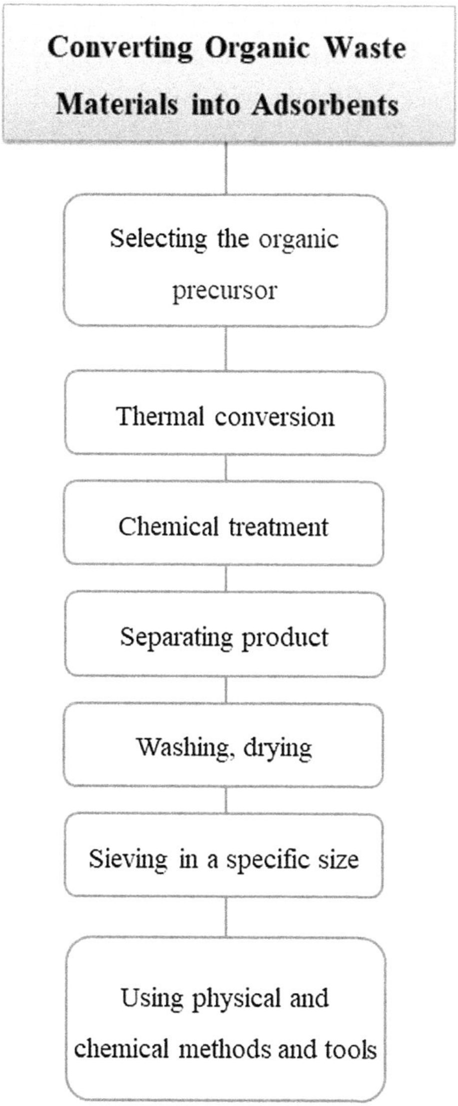
Converting Organic Waste Materials into Adsorbents
Selecting the organic precursor
Thermal conversion
Chemical treatment
Separating product
Washing, drying
Sieving in a specific size
Using physical and chemical methods and tools

FIGURE 6.11 Guidelines for converting organic waste materials into adsorbents.

protocol for converting waste materials into adsorbents is proposed in Fig. 6.12. It starts with the selection of the precursor or the type of waste material. After sample collection, to remove dirt and undesirable components, it is washed with water. After being washed and dried, the samples are pulverized or broken into little bits (Fig. 6.13). In the case of organic wastes, the thermal conversion process for solid organic materials or carbon-based materials can be carried out utilizing the following methods:

(i) Drying, which entails the evaporation of water.

FIGURE 6.12 Procedure for converting inorganic waste materials into adsorbents.

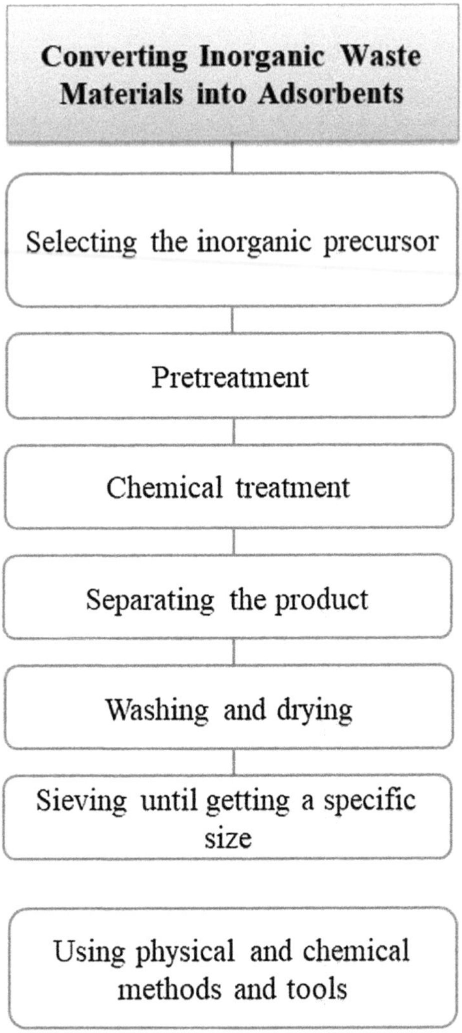
Converting Inorganic Waste Materials into Adsorbents
Selecting the inorganic precursor
Pretreatment
Chemical treatment
Separating the product
Washing and drying
Sieving until getting a specific size
Using physical and chemical methods and tools

(ii) Pyrolysis without oxygen utilizing thermal heat. Note that carbonization refers to the end product being largely carbon and generated after heating without air at temperatures exceeding 500°C, whereas pyrolysis refers to intense carbonization that occurs at a quick rate and takes much less time. Both operations are carried out in an inert atmosphere and are subjected to heat deterioration. Pyrolysis, on the other hand, can take many distinct forms (Akhil et al., 2021; Danmaliki and Saleh, 2016; Saleh and Danmaliki, 2016).

(iii) Gasification, which is the partial conversion of pyrolysis-dried products to fuel gases.

(iv) Heat is released by combustion or oxidation.

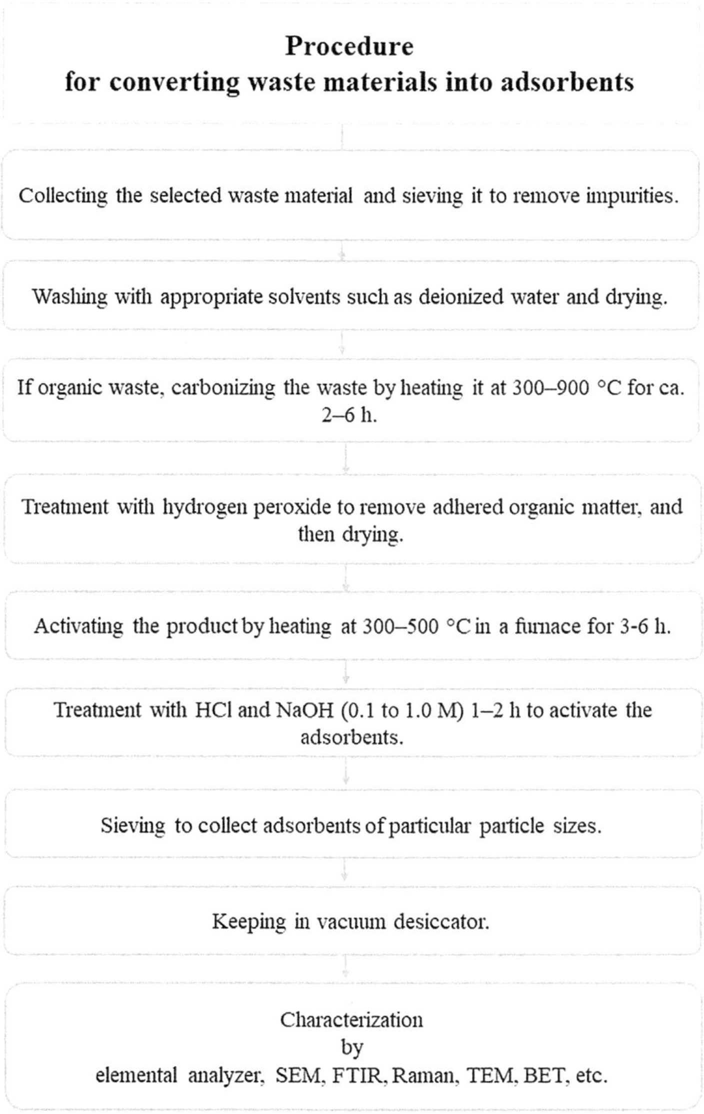
Procedure
for converting waste materials into adsorbents
Collecting the selected waste material and sieving it to remove impurities.
Washing with appropriate solvents such as deionized water and drying.
If organic waste, carbonizing the waste by heating it at 300–900 °C for ca. 2–6 h.
Treatment with hydrogen peroxide to remove adhered organic matter, and then drying.
Activating the product by heating at 300–500 °C in a furnace for 3-6 h.
Treatment with HCl and NaOH (0.1 to 1.0 M) 1–2 h to activate the adsorbents.
Sieving to collect adsorbents of particular particle sizes.
Keeping in vacuum desiccator.
Characterization
by
elemental analyzer, SEM, FTIR, Raman, TEM, BET, etc.

FIGURE 6.13 Procedures for turning waste materials into adsorbents.

(v) Microwave method: microwaves can help a moisture sample heat very quickly (Yang et al., 2010). Moisture is preferable for microwave absorbents to improve the early heating of biomasses by microwave radiation; therefore no predrying is required. Microwave heating also prevents undesirable reactions, resulting in a high-quality product (Ren et al., 2012). To address the limits of traditional furnace pyrolysis, microwave-assisted pyrolysis is utilized.

Gasification, pyrolysis, and torrefaction take place in the absence of oxygen or with less oxygen than is required for combustion (Basu, 2013). The operating conditions (temperature, heating rate, and oxygen source), as well as the yield of products (gas, oil/condensables, and char), differ between these processes. Pyrolysis favors the creation of liquid due to its rapid heating rate and mild temperature. Torrefaction produces chars due to its low temperature and lengthy residence time, while gasification produces chars mostly from noncondensable and condensable gases because of its high temperature and heating rate (Matsakas et al., 2017). By-products (bio-oil and biogas) can be gathered during these operations.

11.3 Pyrolysis

11.3.1 The pyrolysis' basic principles

The vast array of products that can be generated from garbage, plastic, or biomass has prompted extensive research into pyrolysis, gasification, torrefaction, anaerobic digestion, and hydrothermal processing, among other thermal conversion processes. Pyrolysis is more convenient since it offers numerous storage, transportation, and solicitation advantages, such as turbines, combustion appliances, boilers, engines, and so on. Pyrolysis is also a sustainable way to generate energy from waste, in addition to its advantage of converting waste into valuable materials such as adsorbents.

The pyrolysis' basic principles are that the thermal decomposition of lignocellulosic biomass or industrial wastes takes place in the absence of oxygen in an inert atmosphere. Argon or nitrogen gas flow is frequently required to create an inert atmosphere. The fundamental chemical process is multistep and extremely complex. Char, oil, and gases are the end products of biomass pyrolysis (Demiral and Sensoz, 2008). The principal by-products of the pyrolysis process are methane, hydrogen, carbon monoxide, and carbon dioxide. Without the presence of air/oxygen, the organic components contained in the bulk substrate begin to breakdown about 350–550°C and can continue until 700–800°C (Fig. 6.14).

11.3.2 Types and classifications of the pyrolysis process

Depending on the heating rate, the pyrolysis process can be characterized as slow or fast. The time it takes to heat the biomass substrate to pyrolysis temperature is longer than the time it takes to keep the substrate at the characteristic pyrolysis reaction temperature in a slow pyrolysis process. In fast pyrolysis, however, the precursors' initial heating time is shorter than the end retention period at the pyrolysis peak temperature. Pyrolysis can be classified into two categories based on the medium used: hydrous pyrolysis and hydropyrolysis. Slow pyrolysis and fast pyrolysis are normally carried out in an inert environment, whereas hydrous pyrolysis and hydropyrolysis are

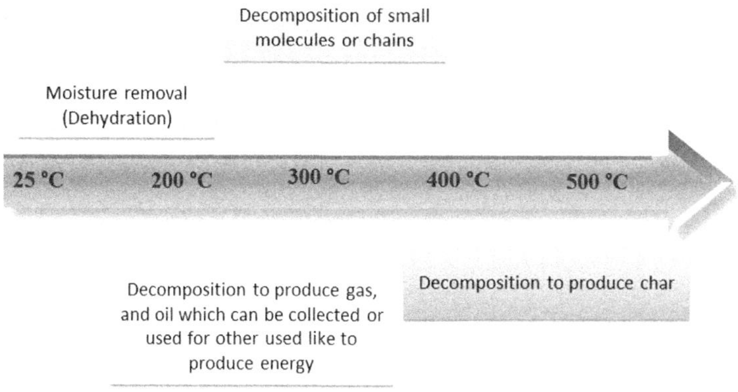
Decomposition of small molecules or chains
Moisture removal (Dehydration)
25 °C
200 °C
300 °C
400 °C
500 °C
Decomposition to produce gas, and oil which can be collected or used for other used like to produce energy
Decomposition to produce char

FIGURE 6.14 Decomposition behavior of the constituents of the waste at different temperature.

carried out in the presence of water and hydrogen, respectively. Fig. 6.15 lists different types of pyrolysis processes.

For slow pyrolysis, the vapor's residence duration in the pyrolysis medium is longer. This method is mostly employed in the manufacturing of char. Carbonization and conventional are two types of carbonization. The vapor residence time, on the other hand, is measured in seconds or milliseconds. This type of pyrolysis, which is generally used to produce bio-oil and gas, is divided into two categories: flash and ultrarapid.

The pyrolysis process is influenced by

- Treated material composition—each of the key parts of biomass and trash has a different thermal decomposition temperature, implying that they contribute to the process results in different ways. Because of the wide range of material compositions, pilot studies are always recommended to accurately predict the pyrolysis process performance.
- The temperature of the process has a significant impact on the treatment outcomes. Pyrolysis at higher temperatures produces more noncondensable gases (syngas, synthetic gas), but at lower temperatures, high-quality solid products are produced.
- The residence time of material in the pyrolysis chamber—has an impact on the degree of thermal conversion of the obtained solid product as well as the vapor residence time, which has an impact on the vapor composition (condensable/noncondensable phase).
- Particle size and physical structure—has an effect on the rate at which material is pyrolyzed. In general, lesser particle size materials are more quickly impacted by thermal decomposition, resulting in larger amounts of pyrolysis oil than bigger particle size materials.

FIGURE 6.15 Waste materials can be pyrolyzed in a variety of methods to produce adsorbents.

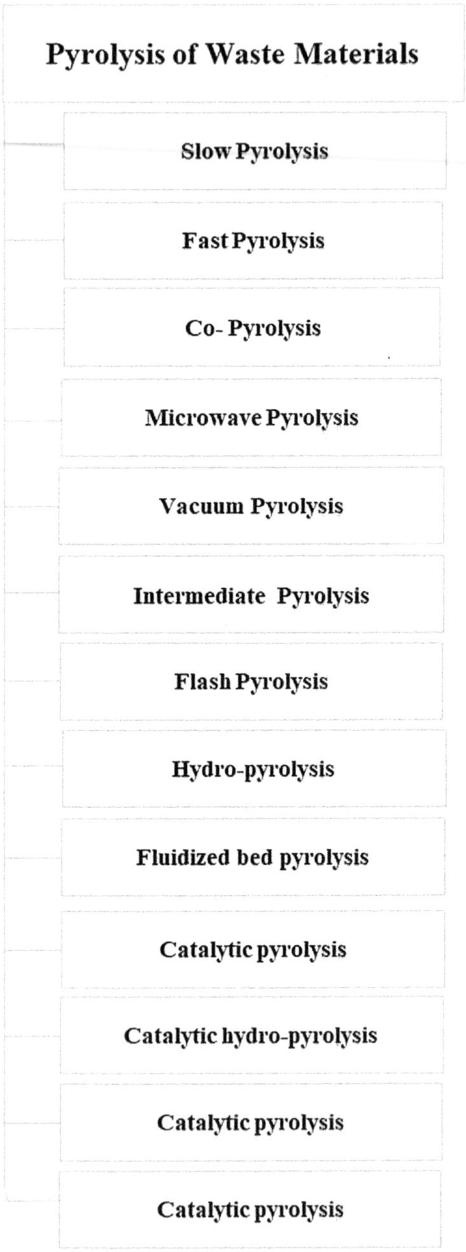
Pyrolysis of Waste Materials
Slow Pyrolysis
Fast Pyrolysis
Co- Pyrolysis
Microwave Pyrolysis
Vacuum Pyrolysis
Intermediate Pyrolysis
Flash Pyrolysis
Hydro-pyrolysis
Fluidized bed pyrolysis
Catalytic pyrolysis
Catalytic hydro-pyrolysis
Catalytic pyrolysis
Catalytic pyrolysis

11.3.3 Types of reactor

The significance of a suitable reactor in any pyrolysis process cannot be overstated. For a high bio-oil yield, reactors have been constructed to meet certain conditions, taking into account characteristics such as heating temperature, vapor product residence time, and needed pressure. In light of the foregoing, researchers have created a variety of reactors for specific applications.

There are several types of pyrolysis reactors (Zaman et al., 2017) including:

- ➢ Fixed bed reactor
- ➢ Fluidized bed reactor
- ➢ Bubbling fluidized bed reactor
- ➢ Circulating fluidized bed reactors
- ➢ Ablative reactor
- ➢ Vacuum pyrolysis reactor
- ➢ Rotating cone reactor
- ➢ PyRos reactor
- ➢ Auger reactor
- ➢ Plasma reactor
- ➢ Microwave reactor
- ➢ Solar reactor

11.4 Physical or chemical treatment

There are two methods for further treatment and modification of the resulting biochar: chemical and physical methods, as shown in Fig. 6.16.

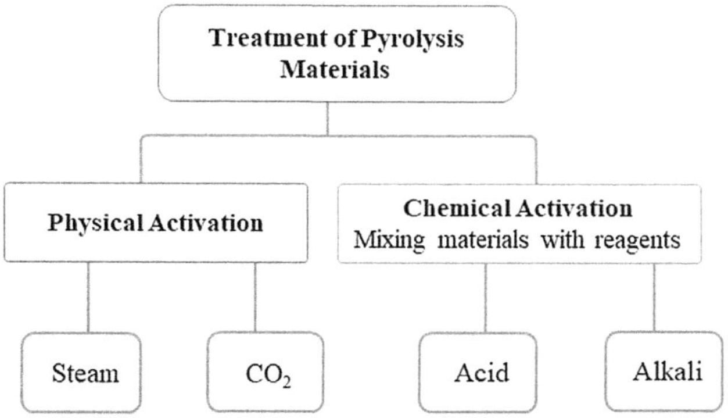
Treatment of Pyrolysis Materials
Physical Activation
Chemical Activation
Mixing materials with reagents
Steam
CO2
Acid
Alkali

FIGURE 6.16 Physical and chemical treatment of pyrolysis materials.

Following the thermal treatment, the produced materials are subjected to further chemical treatment, which includes acid or basic treatment. After that, the product is separated, rinsed, and dried. The resultant materials are sieved to a particular size to achieve a homogeneous size. To get a good knowledge of the materials' properties, characterization can be done using a variety of instruments including FTIR, Raman, BET, SEM, TEM, XRD, XRF, and XPS. Microporous materials (pore diameters D 2 nm) and mesoporous materials (pore diameters $D > 2$ nm) are the two types of porous materials (pore diameters 2 50 nm). Metal-organic frameworks and zeolites are examples of microporous structures. The kind of procedure and waste source determine the adsorption characteristics and pore-size distribution of the produced materials.

11.5 Some important considerations

The following should be taken into account throughout the conversion process:

(i) The mixing of various types of wastes might result in adsorbents with diverse properties; consequently, pollutant sorption capacity varies.
(ii) Different regions have different wastes. As a result, integration must be adaptable to varied conditions.
(iii) Contaminants in some waste materials should be eliminated before conversion.
(iv) Some of the materials obtained can be employed in a variety of applications, including cementitious applications, construction, and road construction (Nazia et al., 2020; Kasidit et al., 2021).
(v) The adsorptive capabilities of adsorbents are related to their molecular structures or surface active sites. As a result, different adsorbents have varying adsorption capabilities. Furthermore, surface area/porosity may alter adsorption capacity, necessitating more in-depth research. More thorough research is needed to determine which components in the precursor are primarily responsible for surface area, microporosity, and mesoporosity. The pore-size distribution of the materials, as well as the size of dye molecules and functional groups present on the adsorbent's surface, should be assessed.

12. Conclusion

The synthesis and processing of nanomaterials can be challenging at times due to the inherent difficulties in ensuring that nanoparticles possess the desired properties. The properties of nanomaterials vary according to the method used. Even if the particle size and crystal phase of those nanomaterials products appear to be very similar on a specification sheet, some critical properties for specific applications may vary significantly depending on the manufacturing processes used. While conducting research on nanomaterials applications, it is critical to choose synthesis techniques based on the properties required for specific applications as well as the technique's scalability.

The method for producing large-scale nanomaterials is determined by several factors, including the type of material, the synthesis approach, and regulatory requirements. Certain methods for synthesizing nanomaterials, as well as those for adsorbents derived from wastes, are promising for large-scale production. Certain existing synthesis technologies can be used or modified to meet quality, target material specifications, desired physicochemical properties, and other requirements.

Although the use of nanoparticles in large products has grown tremendously, the development of many groundbreaking materials remains on the laboratory bench. One of the most significant obstacles to commercializing the use of these nanomaterials is the lack of an adequate large-scale nanoparticle manufacturing process. It is critical to choose an acceptable method that combines the synthesis of nanomaterials with the desired properties and minimal contaminants with the technique's scalability.

Due to the absence of a comprehensive regulatory framework and guidance on safety criteria, their industrial application faces numerous obstacles; specific provisions have not yet been developed. Additionally, restrictions on intellectual property rights, as well as a lack of an appropriate framework for patent registration, contribute to product commercialization delays.

References

Akhil, D., Lakshmi, D., Kartik, A., et al., 2021. Production, characterization, activation and environmental applications of engineered biochar: a review. Environ. Chem. Lett. 19, 2261–2297.

Ashkarran, A.A., Iraji Zad, A., Mahdavi, S.M., et al., 2009. ZnO nanoparticles prepared by electrical arc discharge method in water. J. Mater. Chem. Phys. 118 (1), 6–8.

Basu, P., 2013. Biomass Gasification, Pyrolysis and Torrefaction—Practical Design and Theory, second ed. Elsevier Inc., CA, USA.

Charitidis, C.A., et al., 2014. Manufacturing nanomaterials: from research to industry. Manuf. Rev. 1, 11.

Cushing, B.L., Kolesnichenko, V.L., Charles, J., O'Connor, C.J., 2004. Recent advances in the liquid-phase syntheses of inorganic nanoparticles. Chem. Rev. 104 (9), 3893–3946.

Danmaliki, G.I., Saleh, T.A., 2016. Influence of conversion parameters of waste tires to activated carbon on adsorption of dibenzothiophene from model fuels. J. Clean. Prod. 117, 50–55.

Demiral, I., Sensoz, S., 2008. The effects of different catalysts on the pyrolysis of industrial wastes (olive and hazelnut bagasse). Bioresour. Technol. 99, 8002–8007.

Gao, M.-R., et al., 2013. Nanostructured metal chalcogenides: synthesis, modification, and applications in energy conversion and storage devices. Chem. Soc. Rev. 42 (7), 2986–3017.

Gupta, V.K.,S., 2009. Application of low-cost adsorbents for dye removal—a review. J. Environ. Manag. 90 (8), 2313–2342.

Gupta, V.K., Saleh, T.A., 2013. Sorption of pollutants by porous carbon, carbon nanotubes and fullerene—an overview. Environ. Sci. Pollut. Res. 20, 2828–2843.

Hayashi, H., Hakuta, Y., 2010. Hydrothermal synthesis of metal oxide nanoparticles in supercritical water. Materials 3 (7), 3794.

Huston, M., DeBella, M., DiBella, M., Gupta, A., 2021. Green synthesis of nanomaterials. Nanomaterials 11 (8), 2130. https://doi.org/10.3390/nano11082130.

Kasidit, J., Yimponpipatpol, A., Ngamthanacom, N., Panomsuwan, G., 2021. Conversion of industrial carpet waste into adsorbent materials for organic dye removal from water. Clean. Eng. Technol. 100150.

Kaur, I.P., et al., 2014. Issues and concerns in nanotech product development and its commercialization. J. Contr. Release 193, 51–62.

Khang, Y., Lee, J., 2010. Synthesis of Si nanoparticles with narrow size distribution by pulsed laser ablation. J. Nanoparticle Res. 12 (4), 1349–1354.

Kumar, N., Sinha Ray, S., 2018. Synthesis and functionalization of nanomaterials. In: Sinha Ray, S. (Ed.), Processing of Polymer-Based Nanocomposites, Springer Series in Materials Science, vol 277.

Matsakas, L., Qiuju, G., Stina, J., Ulrika, R., Paul, C., 2017. Green conversion of municipal solid wastes into fuels and chemicals. Electron. J. Biotechnol. 26, 69–83.

Moon, J.-W., et al., 2016. Manufacturing demonstration of microbially mediated zinc sulfide nanoparticles in pilot-plant scale reactors. Appl. Microbiol. Biotechnol. 100 (18), 7921–7931.

Nazia, H., Bhuiyan, M.A., Kumar Pramanik, B., Nizamuddin, S., Griffin, G., 2020. Waste materials for wastewater treatment and waste adsorbents for biofuel and cement supplement applications: a critical review. J. Clean. Prod. 255, 120261.

Parvez, K., Worsley, R., Alieva, A., Felten, A., Casiraghi, C., 2019. Water-based and inkjet printable inks made by electrochemically exfoliated graphene. Carbon 149, 213–221.

Qi, X., Qin, C., Zhong, W., Au, C., Ye, X., Du, Y., 2010. Large-scale synthesis of carbon nanomaterials by catalytic chemical vapor deposition: a review of the effects of synthesis parameters and magnetic properties. Materials 3, 4142–4174.

Ragelle, H., et al., 2017. Nanoparticle-based drug delivery systems: a commercial and regulatory outlook as the field matures. Expet Opin. Drug Deliv. 14 (7), 851–864.

Ren, S., Lei, H., Wang, L., Bu, Q., Wei, Y., Liang, J., Liu, Y., Julson, J., Chen, S., Wu, J., Ruan, R., 2012. Microwave torrefaction of Douglas fir sawdust pellets. Energy Fuel 26, 5936–5943.

Saleh, T.A., 2021. Protocols for synthesis of nanomaterials, polymers, and green materials as adsorbents for water treatment technologies. Environ. Technol. Innovat. 24, 101821.

Saleh, T.A., 2022. Experimental and Analytical methods for testing inhibitors and fluids in water-based drilling environments. Trac. Trends Anal. Chem. 116543. https://doi.org/10.1016/j.trac.2022.116543.

Saleh, T.A., Danmaliki, G.I., 2016. Influence of acidic and basic treatments of activated carbon derived from waste rubber tires on adsorptive desulfurization of thiophenes. J. Taiwan Inst. Chem. Eng. 60, 460–468.

Su, S.S., Chang, I., 2018. Review of production routes of nanomaterials. In: Brabazon, D., et al. (Eds.), Commercialization of Nanotechnologies—A Case Study Approach. Springer, Cham. https://doi.org/10.1007/978-3-319-56979-6_2.

Swihart, M.T., 2003. Vapor-phase synthesis of nanoparticles. Curr. Opin. Colloid Interface Sci. 8 (1), 127–133.

Tavakoli, A., Sohrabi, M., Kargari, A., 2007. A review of methods for synthesis of nanostructured metals with emphasis on iron compounds. Chem. Pap. 61 (3), 151–170.

Tene, T., Tubon Usca, G., Guevara, M., Molina, R., Veltri, F., Arias, M., Caputi, L.S., Vacacela Gomez, C., 2020. Toward large-scale production of oxidized graphene. Nanomaterials 10, 279.

Vacacela Gomez, C., Tene, T., Guevara, M., Tubon Usca, G., Colcha, D., Brito, H., Molina, R., Bellucci, S., Tavolaro, A., 2019. Preparation of few-layer graphene dispersions from hydrothermally expanded graphite. Appl. Sci. 9, 2539.

Wang, Z., Liu, Y., Zhang, Z., 2002. Handbook of nanophase and nanostructured materials-synthesis. In: Wang, Z., Liu, Y., Zhang, Z. (Eds.), Handbook of Nanophase and Nanostructured Materials-Synthesis. Kluwer Academic Plenum Publishers, USA.

Yang, S., Jang, Y.H., Kim, C.H., et al., 2010. A flame metal combustion method for production of nanoparticles. Powder Technol. 197 (3), 170–176.

Zaman, C., Pal, K., Yehye, W.A., Sagadevan, S., Tawab Shah, S., Abimbola Adebisi, G., Marliana, E., Faijur Rafique, R., Bin Johan, R., 2017. Pyrolysis: A Sustainable Way to Generate Energy from Waste. https://doi.org/10.5772/intechopen.69036.

Chapter 7

Characterization and description of adsorbents and nanomaterials

1. Introduction

Nanotechnologies are the design, characterization, production, and application of structures, devices, and systems by controlling shape and size at the nanometer scale. Nanoscience can be defined as the study of phenomena and material manipulation at the atomic, molecular, and macromolecular scales, where properties vary dramatically from those at larger scales. Scanning microscopes can determine the positions of substrate and adsorbate atoms on surfaces, control the initial quantum states of molecular beams impacting a surface, and evaluate desorbing reaction products state-by-state. This creates an ideal environment for developing a microscopic theoretical description that can either explain experimental results or, in the case of theoretical predictions, be tested experimentally. Indeed, the microscopic theoretical treatment of surfaces and processes on surfaces has progressed remarkably in recent years. While phenomenological thermodynamic approaches were popular decades ago, microscopic concepts currently predominate in the study of surface processes (Saleh, 2018, 2020a,b, 2021). Without using any empirical parameters, a variety of surface qualities can be explained from basic principles (Table 7.1).

This chapter focuses on the discussion of the characterization of materials and nanomaterials as adsorbents and nanoadsorbents. It discusses the structural, surface characterization, elemental analysis, crystallinity characterization, morphology, pore structure characterization, surface charge, biological evaluation, and genera highlights on the tools used to evaluate their performance.

Surface Science of Adsorbents and Nanoadsorbents. https://doi.org/10.1016/B978-0-12-849876-7.00008-7

TABLE 7.1 Definition of some important terms.

Term	Definition
Material characterization	Material characterization is the process of defining the properties of one or more components of a substance or material. It is the process of measuring and determining the physical, chemical, mechanical, and microstructural properties of materials. It allows the higher level of understanding required to resolve significant issues, including developing new materials, solving failure and process-related problems. Characterization allows the manufacturer to make critical products and decisions.
Topography	Topography: The surface futures of an object of how it looks, its texture, a direct relation between these features and materials properties.
Morphology	Morphology: The shape and size of the particles making up the object, a direct relation between these structures and materials properties.
Composition	Composition: The elements and compounds that the object is composed of and the relative amounts of each, direct relationship between compositions and materials properties.
Crystallographic	Crystallographic information: How the atoms are arranged in the object, direct relation between these arrangements and materials properties.
Thermogravimetric analysis (TGAs)	Thermogravimetric analysis (TGAs) technique is a technique used to measure the weight change that occurs as a sample is heated at a consistent rate to assess a material's thermal stability and fraction of volatile components.

2. Properties to be determined

Several properties of any new nanomaterials can be determined. These can be classified into chemical, physical, mechanical, and biological.

➢ Chemical properties:
- Molecular structure
- Composition, including purity, and known impurities or additives
- Surface chemistry
- Functional groups
- Active sites
- Attraction to media like water molecules or oils and fats

➢ Physical properties:
- Size, shape, specific surface area, and the ratio of width and height
- Whether it is held in a solid, liquid, or gas

- Whether they stick together
- Number size distribution
- How smooth or bumpy their surface is
- Structure, including crystal structure and any crystal defects
- How well they dissolve

➢ Mechanical properties:

- Hardness and elastic modulus
- Stiffness and strength
- Interfacial adhesion and friction
- Movement law
- Thermal stability
- Other size-dependent characteristics
- The mechanical properties of materials usually increase with decreasing size (not always true)
- Mechanical strength begins at the micron meter scale, which is markedly different from the size dependency of other properties.

3. Characterization of nanomaterials as adsorbents

The process of studying the morphological and functional qualities of a material is known as material characterization. It is a fundamental tool to ensure the highest quality in the synthesis, design, and manufacturing processes of new materials. The objective is to know important information about the material, such as its structure, chemical, physical and mechanical properties, and degree of resistance and reliability or their possible applications (Saleh, 2022). Knowing the material's properties is essential to predict its performance and estimate its useful function and stability based on the expected environmental exposure conditions.

The adsorbent material is characterized to clarify details about the bulk crystal structure, molecular and electronic structure, surface properties, and pore structure. Fig. 7.1 depicts some of the most common characterization approaches. In the following section, several instruments and techniques used for characterization are discussed including:

➢ Structural Characterization
➢ Surface Characterization
➢ Elemental Analysis
➢ Crystallinity nature
➢ Morphology of the materials
➢ Pore Structure Characterization
➢ Surface Charge
➢ Thermal stability
➢ Biological Evaluation
➢ Performance

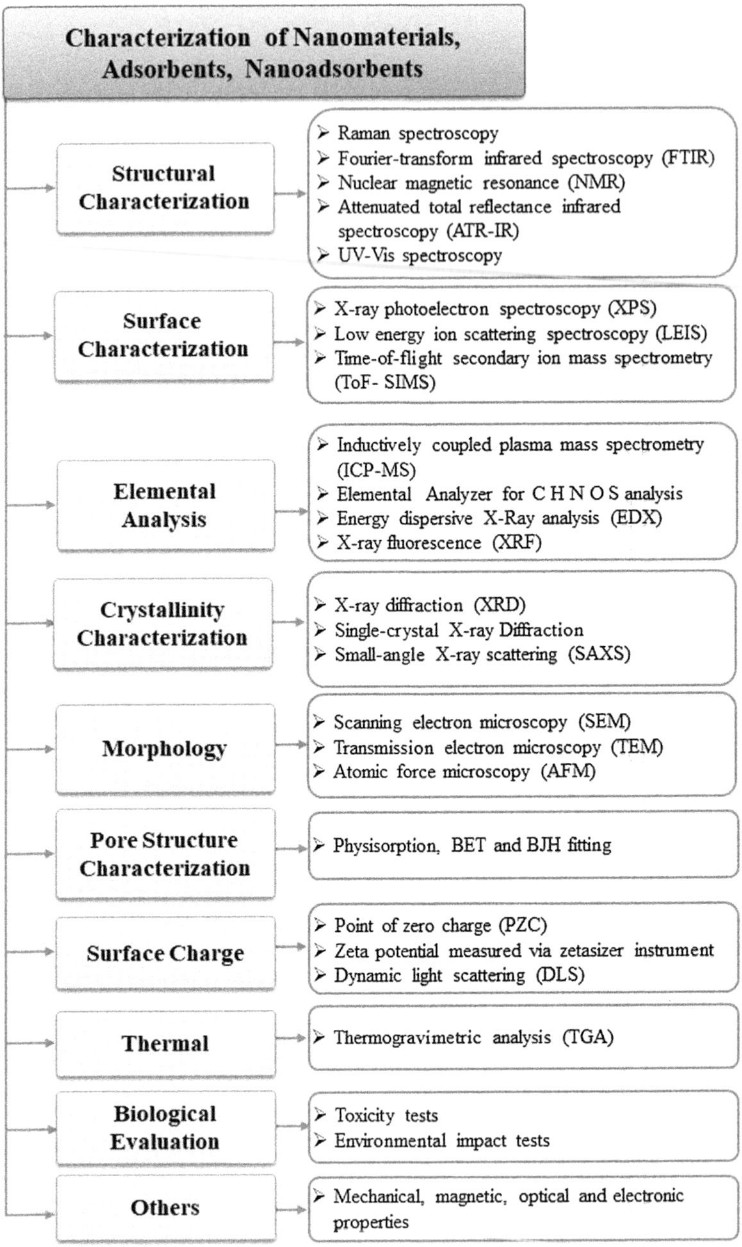
Characterization of Nanomaterials, Adsorbents, Nanoadsorbents
Structural Characterization
➢ Raman spectroscopy
➢ Fourier-transform infrared spectroscopy (FTIR)
➢ Nuclear magnetic resonance (NMR)
➢ Attenuated total reflectance infrared spectroscopy (ATR-IR)
➢ UV-Vis spectroscopy
Surface Characterization
➢ X-ray photoelectron spectroscopy (XPS)
➢ Low energy ion scattering spectroscopy (LEIS)
➢ Time-of-flight secondary ion mass spectrometry (ToF- SIMS)
Elemental Analysis
➢ Inductively coupled plasma mass spectrometry (ICP-MS)
➢ Elemental Analyzer for C H N O S analysis
➢ Energy dispersive X-Ray analysis (EDX)
➢ X-ray fluorescence (XRF)
Crystallinity Characterization
➢ X-ray diffraction (XRD)
➢ Single-crystal X-ray Diffraction
➢ Small-angle X-ray scattering (SAXS)
Morphology
➢ Scanning electron microscopy (SEM)
➢ Transmission electron microscopy (TEM)
➢ Atomic force microscopy (AFM)
Pore Structure Characterization
➢ Physisorption, BET and BJH fitting
Surface Charge
➢ Point of zero charge (PZC)
➢ Zeta potential measured via zetasizer instrument
➢ Dynamic light scattering (DLS)
Thermal
➢ Thermogravimetric analysis (TGA)
Biological Evaluation
➢ Toxicity tests
➢ Environmental impact tests
Others
➢ Mechanical, magnetic, optical and electronic properties

FIGURE 7.1 Types of characterization techniques of nanomaterials and adsorbents.

4. Structural characterization

The composition and nature of bonding materials are primarily studied using structural characteristics. Structural characterization provides important information regarding nanomaterial qualities.

4.1 Raman spectroscopy

Raman spectroscopy is a nondestructive chemical examination technique that can provide details about chemical structure, phase and polymorphy, crystallinity, and molecular interactions. Raman scattering is a light-scattering technique in which a molecule scatters incident light from a high-intensity laser. Fourier-transform infrared spectroscopy (FTIR) and Raman spectroscopies are commonly used to investigate the vibrational characteristics of nanomaterials. Compared to other elemental analytical procedures, these are the most established and viable techniques. The fingerprint region, which provides the signature information about the material, is the most important range for nanomaterials (NMs).

The functionalization of nanomaterials and their interaction with other substrates can be studied via the FTIR and X-ray photoelectron spectroscopy (XPS) technique. FTIR is used to confirm the functionalization as it showed characteristic peaks, such as the signature vibrational bands of carboxylated C–O 2050 cm^{-1} and the O–H band at 3280 cm^{-1}. The degree of functionalization can be revealed from the red or blue shift values of the FTIR bands (Dablemont et al., 2008).

Due to its signal improved capacity via the surface plasmon resonance (SPR) phenomenon, surface-enhanced Raman spectroscopy (SERS) is increasingly being used as a vibrational conformational tool. One study used the SERS technology to investigate the vibrational characteristics of nanostructured metals such as silver and gold. It concluded that plasmonic resonances in nanomaterials are responsible for the increased spectra (Saleh, 2018).

4.2 Fourier-transform infrared spectroscopy

FTIR is a technique employed to obtain an infrared spectrum of absorption, emission, and photoconductivity of solid, liquid, and gas. It can be used to detect various functional groups in nanomaterials. FTIR spectra can be recorded between 4000 and 400 cm^{-1}. This technique utilizes the mathematical process (Fourier transform) to translate the raw data (interferogram) into the actual spectrum.

Other spectroscopies used to report surface information include diffuse reflectance infrared Fourier-transform spectroscopy (DRIFTS) and Raman spectroscopy (Tran et al., 2017). However, considering the laser penetration depths utilized in both, it is crucial to highlight that unless the overlayer is probed carefully, both DRIFTS and Raman spectroscopy offer bulk and surface information (Bañares and Wachs, 2010). Given the ability to use liquid-phase samples and the lack of interference by water as seen in infrared spectroscopy, Raman spectroscopy is excellent for studying adsorption processes in situ (Kumar et al., 2019).

4.3 Attenuated total reflectance infrared spectroscopy

Attenuated total reflectance infrared spectroscopy (ATR-IR) can study the interface of aqueous adsorption reactions using flow cells (Young and McQuillan, 2009; Elzinga and Sparks, 2007). The ATR-IR sensitivity toward the protonation of the functional groups on nanoadsorbents enables the discrimination of surface species using the nondegenerate symmetric stretching ν_1 and the triply degenerate symmetric stretching ν_3.

4.4 Nuclear magnetic resonance

Nuclear magnetic resonance (NMR) is a vital technique that uses a magnetic field and radiofrequency pulses to quantify the resonant frequency of an atomic nucleus in relation to its chemical or environmental surroundings (e.g., the most common stable isotopes 1H, 13C, and 15N). Nuclei in a strong constant magnetic field are agitated by a weak oscillating magnetic field (in the near field) and respond by emitting an electromagnetic signal with a frequency characteristic of the magnetic field at the nucleus. NMR is a highly comprehensive chemical investigation approach for nanomaterials. It can indicate how many hydrogen atoms are in a molecule and where they are in the carbon chain. Each hydrogen atom's nucleus acts like a tiny magnet, aligning with an applied magnetic field most of the time, i.e., if we add energy, the tiny magnet can turn around and align itself with the magnetic field. When the external energy is eliminated, the magnetic nucleus must realign with the magnetic field and release the additional energy it has stored, which is detected and used to collect information about the excited hydrogen.

With atomic-level resolution, NMR can irreversibly evaluate a wide variety of structural and chemical features of nanomaterials in both solid and liquid states. Hence, the motivation for NMR spectroscopy as a unique and primary method for nanomaterial characterization is justified.

4.5 UV–Vis spectroscopy and photoluminescence

Optical characteristics are used in photocatalytic studies to figure out how the photochemical process works. The Beer–Lambert rule, as well as basic lighting principles (Swinehart, 1962), are used in these descriptions. These approaches can determine the absorbance, reflectance, luminescence, and phosphorescence properties of NMs. It is common knowledge that NMs, especially metallic and semiconductor NMs, have different colors. Thus, they are best suited for photo-related applications. Thus, knowing the values of absorption and reflectance of nanomaterials is important to understanding the underlying photodegradation mechanisms.

UV−Vis spectroscopy (or spectrophotometry) is a quantitative technique for determining the amount of light absorbed by a chemical material. This is accomplished by comparing the intensity of light passing through a sample to that passing through a reference sample or blank. The null ellipsometer, UV−Vis, and photoluminescence (PL) are optical devices that can be employed to explore the nanomaterial's optical properties.

5. Surface characterization

The surface of the adsorbent is characterized cautiously, as the surface sensitivity of various methods varies greatly. High sensitivity low-energy ion-scattering (HS-LEIS) probes the first few monolayers of a nanomaterial and assesses elemental compositions while also allowing for static depth profiling (Brongersma et al., 2007; Druce et al., 2014). The surface of nanoadsorbents utilized in the aqueous media, on the other hand, undergoes dissolution and restructuring, necessitating a comprehension of the surface region's alterations. X-ray photoelectron spectroscopy (XPS) as well as time-of-flight secondary ion mass spectroscopy (ToF-SIMS) are used to analyze this surface region of complicated reconstructed surfaces. The XPS method is frequently used to determine the chemical state of elements at the surface of nanoadsorbents or materials. Because these are vacuum systems for ex situ research, the impact of hydration on structured surfaces can be widely studied.

5.1 X-ray photoelectron spectroscopy

XPS is the most sensitive technique for determining the exact elemental ratio and bonding behavior of the elements in NMs, and it is frequently utilized. It is a surface-sensitive approach that can be utilized in depth-profiling research to figure out the general composition and its changes with depth. The number of electrons on the Y-axis plotted against the binding energy (eV) of the electrons on the X-axis is a typical XPS spectrum based on basic spectroscopic principles. Each element has its own binding energy fingerprint, resulting in a unique set of XPS peaks. Electronic configurations such as 1s, 2s, 2p, and 3s correspond to the peaks (Oprea et al., 2015; Wang et al., 2016)

5.2 Low-energy ion-scattering spectroscopy

It is a surface-sensitive analytical technique for determining a material's chemical and structural makeup. Directing a stream of charged particles known as ions toward a surface and observing the locations, velocities, and energy of the ions interacting with the surface is what constitutes LEIS. The information gathered in this way can be used to deduce information about the material. This includes the relative positions of atoms in a surface lattice and their elemental identities.

5.3 Time-of-flight secondary ion mass spectrometry

Time-of-flight secondary ion mass spectrometry (ToF-SIMS) is a surface-sensitive analytical approach that removes molecules from the sample's very outermost surface using a pulsed ion beam (Cs or microfocused Ga). The particles are eliminated from the surface's atomic monolayers (secondary ions). The mass of these particles is measured by monitoring the exact moment at which they reach the detector after accelerating through a "flight tube" (i.e., time-of-flight). Surface spectroscopy, surface imaging, and depth profiling are the three operational modes possible with ToF-SIMS.

This technique extracts mass spectral data and image data in the XY dimension across a sample and depth profile data in the Z dimension from nanomaterials. This ability to distinguish absorbed coatings or trace materials on the surface frequently aids in the identification of a foreign particle.

6. Elemental analysis

Elemental analysis of materials can be done by the following techniques:

6.1 Inductively coupled plasma mass spectrometry

Mass spectrometry (MS) techniques are useful to analyze the elemental composition of nanoparticles. This technique identifies atoms or molecules in a sample by measuring the mass-to-charge ratio of ions in the interfacial layer. MS methods ionize the sample using a variety of procedures, causing the molecules to charge and fragment. These molecules are subsequently separated into groups based on their mass-to-charge ratio, using one of many mass selection procedures. The elemental makeup of the specimen may now be determined by analyzing the newly separated groupings of molecules.

ICP spectroscopy (inductively coupled plasma) spectroscopy is an analytical method for detecting and measuring elements in chemical samples. It involves ionizing a sample with a very hot plasma, which commonly comprises argon gas. ICP refers to the process of producing electrically neutral plasma from argon gas to interact with the material. Single-particle inductively coupled plasma mass spectrometry (spICP-MS) resembles ICP-MS; however, testing can be performed by time-of-flight (TOF) and quadrupole (Q) analyzers. TOF analyzers determine the time taken by nanoparticles to reach into the detector after being accelerated by an electric field. A quadrupole analyzer creates a radiofrequency quadrupole (one sequence of electrical current) field between four rods set parallel to each other, which efficiently filters different elements through the instrument (Bustos and Winchester, 2016; Naasz et al., 2018).

6.2 Elemental analyzer for C H N O S analysis

The amount of carbon (C), hydrogen (H), nitrogen (N), sulfur (S), and oxygen (O) present in a sample is determined by elemental analysis of CHNSO, also known as organic elemental analysis or elemental microanalysis. It is a dependable and cost-effective method for determining the chemical composition and purity of materials and may be used for a variety of sample types, including viscous substances: solid, liquid, and volatile. Knowing the organic element composition also aids analysts in determining the structure of the sample substance. Organic compound chemical characterization is utilized in both research and quality control (QC).

6.3 Energy-dispersive X-ray analysis

EDX, EDS, and EDXS refer to the same technique: energy-dispersive X-ray spectroscopy. It is utilized to figure out what a material's elemental make-up is. Materials and product research, troubleshooting, deformulation, and other applications are all possible.

The imaging capability of the microscope identifies the specimen of interest in EDX systems, which are attachments to the electron microscopy equipment (scanning electron microscopy [SEM] or transmission electron microscopy [TEM]). EDX analysis produces spectra with peaks corresponding to the elements that really compose the sample under investigation. Image analysis and element mapping of a sample are also options.

6.4 X-ray fluorescence

The elemental composition of materials is determined using XRF (X-ray fluorescence). By detecting the fluorescence (or secondary) X-ray released by a sample when it is excited by the main X-ray source, XRF analyzers may determine the chemistry of a sample. XRF spectroscopy is perfect equipment for the quantitative and qualitative study of the composition of material because each element in the material creates a set of characteristic (a fingerprint) fluorescent X-rays which are unique to the element.

The advantage of XRF is that it is a nondestructive method for determining elemental compositions from small (tens of milligrams) to large (several grams) sample amounts. It may be used to produce representative data from practically any type of material, including nanopowders, liquid nanoparticle dispersions, and solid nanocomposites.

6.4.1 The X-ray fluorescence process

1. High-energy X-rays from a controlled X-ray tube are used to irradiate a solid or liquid sample.

2. An electron from one of the atom's inner orbital shells is expelled when an atom in the sample is targeted by an X-ray of sufficient energy (higher than the atom's K or L shell binding energy).
3. The atom regains stability, with an electron from one of the atom's higher energy orbital shells filling the vacancy in the inner orbital shell.
4. The electron emits a fluorescent X-ray when it falls to a lower energy state. The particular difference in energy between two quantum states of the electron is equal to the energy of this X-ray. XRF analysis is based on the measurement of this energy.

Most atoms have many-electron orbitals (K shell, L shell, M shell, for example). XRF peaks with variable intensities are formed when X-ray energy forces electrons to transfer in and out of these shell levels, and they appear in the spectrum, which graphically represents X-ray intensity peaks as a function of energy peaks. The element's peak energy is used to identify it, and the peak height/intensity is used to estimate its concentration.

7. Crystallinity characterization

XRD, energy-dispersive X-ray (EDX), and small-angle X-ray scattering are the common techniques used to study the crystalline properties of nanomaterials.

7.1 X-ray diffraction

Bulk characterization procedures are critical for identifying the adsorbent's material species and bulk crystalline or amorphous phase. Powder X-ray diffraction is one of the most used techniques for determining crystalline phases in a solid powder material (XRD). XRD has been used to analyze the bulk structure of adsorbents (Kiani et al., 2018; Cullity and Stock, 2001). XRD is used to reveal the structural properties of nanomaterials. It provides information on a nanomaterial's crystallinity and phase. It also employs the Debye–Scherrer formula to roughly estimate particle size.

XRD can also assess crystallite size by using the Debye–Scherrer equation on the peak widening. XRD has the advantage of analyzing the average of all grains present in a powder sample, whereas microscopy provides only local information.

XRD, on the other hand, does not provide adequate surface-specific data. It may not give information about a surface's adsorption properties. Electron microscopy techniques including scanning electron microscopy (SEM) as well as transmission electron microscopy (TEM) can be used to investigate additional bulk information including morphology and particle size. EDS is a type of energy-dispersive X-ray spectroscopy that provides elemental analysis spectra in conjunction with electron microscope examination. EDS can detect bulk components with concentrations of more than 0.1 wt% and other important components; however, it is unsuited for detecting trace elements (under 0.1 wt%) (Goldstein et al., 2003; Nasrazadani and Hassani, 2016).

X-ray absorption spectroscopy (XAS) as well as 31P nuclear magnetic resonance spectroscopy (NMR) are two techniques that can be used to analyze bulk materials and offer essential coordination information for phosphorous in the adsorbent (Feng et al., 2013; Kim et al., 2011).

7.2 Single-crystal X-ray diffraction

Single-crystal X-ray diffraction is considered a nondestructive analytical technique that offers good information on the internal lattice of crystalline substances, such as unit cell dimensions, bond lengths, bond angles, and site-ordering characteristics. Single-crystal refinement is closely linked, in which the data provided by X-ray analysis are interpreted and refined to obtain the crystal structure.

Strengths

- No separate standards required
- Nondestructive
- Detailed crystal structure, include unit cell dimensions, bond lengths, bond angles, and site-ordering information
- Determining of crystal-chemical controls on mineral chemistry
- With specialized chambers, structures of high pressure and/or temperature phases can be determined
- Powder patterns can also be derived from single-crystals by the use of specialized cameras

Limitations

- Must have a single, robust (stable) sample, generally between 50 and 250 microns in size
- Optically clear sample
- Twinned samples can be handled with difficulty
- Data collection generally requires between 24 and 72 h

7.3 Small-angle X-ray scattering

Small-angle X-ray scattering (SAXS) is an analytical method for determining the intensity of X-rays scattered by a sample as a function of the scattering angle. When a material possesses structural characteristics on the nanometer scale, typically in the range of 1–100 nm, an SAXS signal is observed.

8. Morphology

Because morphology controls most of the properties of materials, morphological features always attract a lot of attention. Various techniques for morphological characterization exist, although microscopic techniques such as scanning electron microscopy (SEM), transmission electron microscopy (TEM), atomic force microscopy (AFM), and polarized optical microscopy (POM) are the most common.

Electron microscopes are scientific tools that study objects on a very tiny scale using a beam of highly powerful electrons. This investigation may reveal the following details:

- Topography: the surface futures of an object, its looks, texture, and the direct relation between these features and the material's properties.
- Morphology: the shape and size of the particles making up the object, and the direct relationship between these structures and the material's properties.
- Composition: the elements and compounds that the object is composed of, the relative amounts of each, and the direct relationship between the composition's and material's properties.
- Crystallographic information: the arrangement of atoms in the object and the direct relationship between these arrangements and the material's properties.

Some of the several types of techniques are in Fig. 7.2.

8.1 Scanning electron microscopy

The scanning electron microscope (SEM) uses a diversity of signals at the surface of the solid material using a concentrated beam of high-energy electrons (Fig. 7.3). The signals created by electron-sample interactions offer information on the sample's exterior morphology (texture), chemical composition, crystalline structure, and orientation of the materials that make up the sample, among other things. In general, data are collected across a specific area of the sample's surface, and a two-dimensional picture is created to depict the spatial variations in these qualities.

Essential components commonly of all SEMs include the following:

- Electron Source ("Gun")
- Electron Lenses
- Sample Stage
- Detectors for all signals of interest
- Display/Data output devices
- Infrastructure Requirements:
  - Power Supply
  - Vacuum System
  - Cooling system
  - Vibration-free floor
  - Room free of ambient magnetic and electric fields

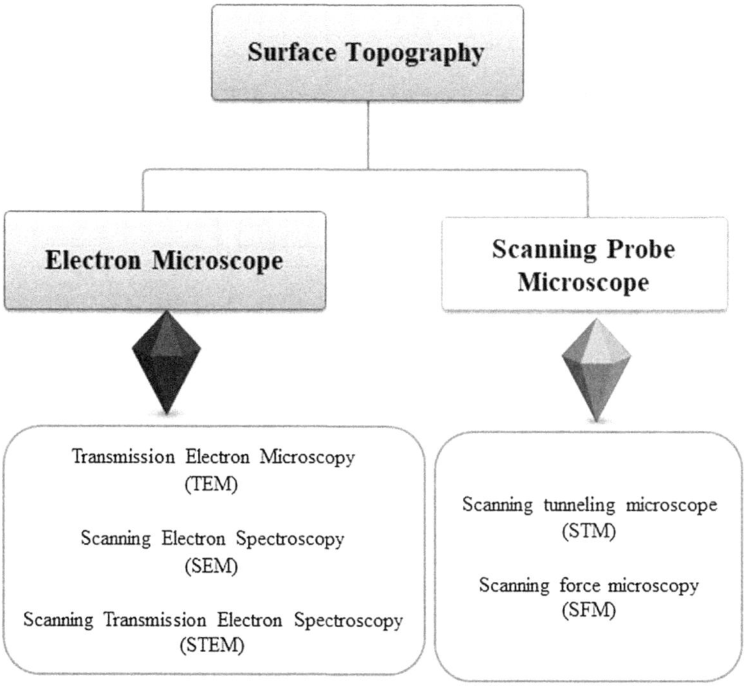
Surface Topography
Electron Microscope
Scanning Probe Microscope
Transmission Electron Microscopy (TEM)
Scanning Electron Spectroscopy (SEM)
Scanning Transmission Electron Spectroscopy (STEM)
Scanning tunneling microscope (STM)
Scanning force microscopy (SFM)

FIGURE 7.2 Types of morphology techniques used for materials characterization.

8.1.1 Guideline of sample (specimen) preparation for SEM measurements

8.1.1.1 Cleaning

Cleaning the surface of the specimen: to remove impurities (dust, soil, mud, etc.).

8.1.1.2 Stabilizing

Stabilizing the specimen: some samples require chemical fixation to preserve and stabilize their structure. Hard, dry materials including shells, and wood, can be inspected with little additional treatment. Fixatives are commonly used for stabilization. Fixation is accomplished by immersing the sample in a buffered chemical fixative solution, such as glutaraldehyde, which is sometimes used in conjunction with formaldehyde and other fixatives. Aldehydes, osmium tetroxide, tannic acid, and thiocarbohydrazides are some of the fixatives that can be utilized.

8.1.1.3 Rinsing

Rinsing the specimen: The sample is to be rinsed to remove excessive fixatives.

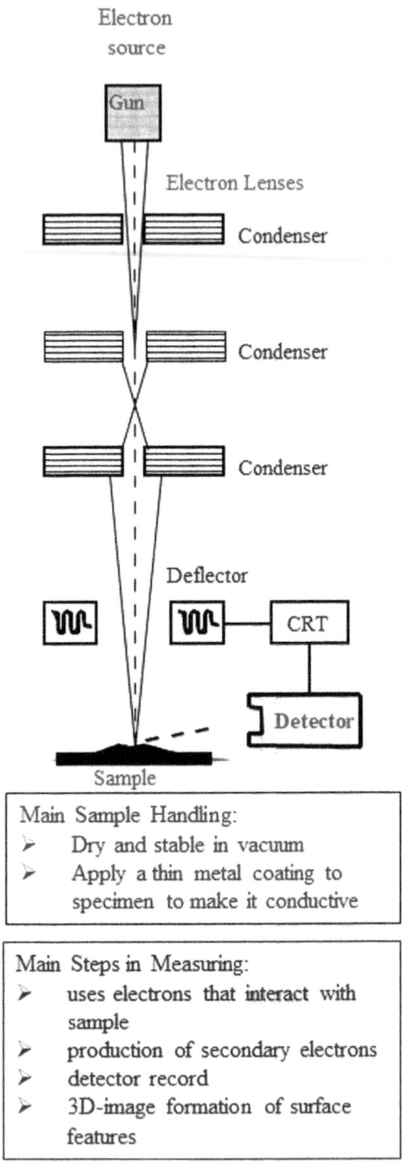
Electron source
Gun
Electron Lenses
Condenser
Condenser
Condenser
Deflector
CRT
Detector
Sample
Main Sample Handling:
Dry and stable in vacuum
Apply a thin metal coating to specimen to make it conductive
Main Steps in Measuring:
uses electrons that interact with sample
production of secondary electrons
detector record
3D-image formation of surface features

FIGURE 7.3 Components of scanning electron microscopy (SEM) and the main function steps.

8.1.1.4 Dehydrating

Dehydrating the specimen: Remove water. Air-drying causes collapse and shrinkage. This is usually precipitated by water replacement in cells with organic solvents such as acetone or ethanol. The dehydration is conducted several times with acetone or ethanol.

8.1.1.5 Drying

Drying the specimen: The specimen shall be thoroughly dry before use.

8.1.1.6 Mounting

Mounting the specimen: The specimen must be placed in the holder. A sample stub is a material holder rigidly attached to the specimen. Using an adhesive such as epoxy glue or electrically conductive double-sided sticky tape, mount a dry specimen on a specimen stub. The phenomenon of charge-up can be avoided by coating the nonconductor sample with metal (conductor).

8.1.1.7 Coating

The purpose of coating the specimen is to avoid build-up by letting the charge on the specimen surface to pass through the coated conductive film to the ground to increase the conductivity of the specimen while also preventing a high voltage charge on it. The conductive metal coating is thin, i.e., 20–30 nm. All metals are conductive and do not need to be prepared before usage. A device known as a "sputter coater" makes nonmetals conductive. Gold, platinum, gold–palladium alloys, osmium, tungsten, iridium, chromium, and graphite are all conductive materials (Zarraoa et al., 2019; Braet et al., 1997; Tsotsas and Mujumdar, 2011).

8.1.1.8 Electron beam-sample interactions

The incoming electron beam scatters both elastically and inelastically in the material. Hence, we can detect a variety of signals. The volume of the interaction grows as the acceleration voltage increases and reduces as the atomic number increases. Consequently, a beam scanning across the sample surface is synchronized with a cathode-ray-tube (CRT) beam. A secondary electron detector does attract the scattered electrons and registers different degrees of brightness on a monitor based on the number of electrons that reach the detector.

A material with a low atomic weight does not emit as many backscattered electrons as one with a high atomic weight. In actuality, the image represents the material surface's density. As images can only be processed in grayscale, SEM pictures are always black and white. These photos can be colored with a feature-detection software or by manually editing with a hand graphic editor, frequently for aesthetic reasons, to clarify the structure, or to provide the sample a more realistic appearance (Hawes, 2012; Stadtländer, 2005).

The electron scanning principle underpins the SEM technique, which delivers all available information about NMs at the nanoscale level. This technique researches not only the morphology of their nanomaterials but also the dispersion of NMs in the bulk or matrix. Fig. 7.4 displays an SEM image of carbon nanotubes with a uniform formation structure while Fig. 7.5 presents the SEM image of the graphene structure. The dispersion of nanoparticles in

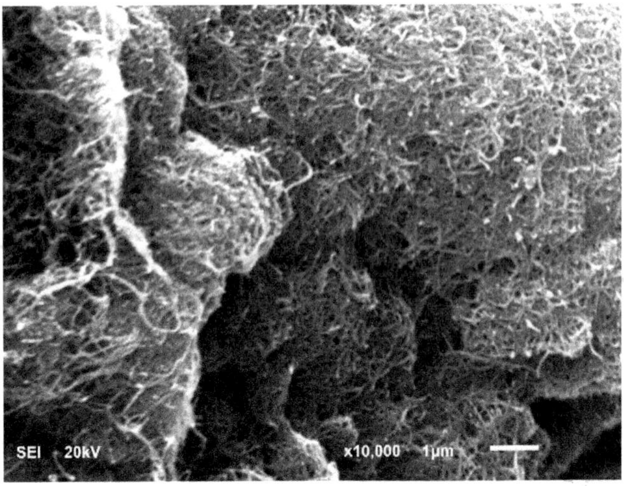
SEI 20kV
x10,000 1μm

FIGURE 7.4 SEM image of carbon nanotubes with a uniform formation structure as an example.

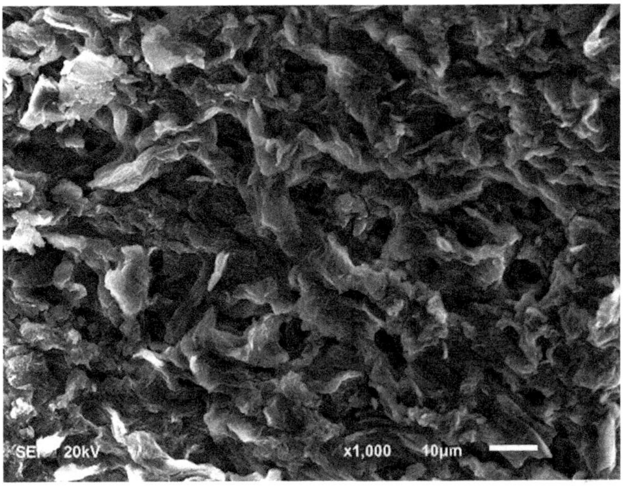
20kV
x1,000 10μm

FIGURE 7.5 SEM image of graphene with a uniform formation structure as an example.

the polymeric composite is shown in Fig. 7.6 as an example. The image indicates the homogenous dispersion of nanoparticles in the composite.

8.1.1.9 Merits

No other instrument likely compares to the SEM in terms of its scope of applications in the study of solid materials. SEM is essential in any field that

SEI 20kV

FIGURE 7.6 SEM image of carbon nanofiber modified with metal oxide nanoparticles as an example.

requires solid material characterization. While the focus of this contribution is on geological applications, it is crucial to emphasize that these are only a small portion of the scientific and industrial uses for this apparatus. Most SEMs are rather simple to use, having user-friendly "intuitive" interfaces. Many applications have just only a modest amount of sample preparation. Data capture is quick for many applications (less than 5 min per picture for SEI, BSE, and spot EDS studies). Modern SEMs produce data in easily transferable digital formats.

In summary, SEM advantages include:

- ➢ SEM provides detailed 3D and topographical images with versatile information garnered from different detectors.
- ➢ SEM works relatively fast.
- ➢ Modern SEMs provide a chance to obtain data in digital form.
- ➢ Most SEM samples require minimal preparation processes.

Limitations

Some of the SEM limitations are:

- SEMs are expensive and large. It is limited to solid samples.
- Special training is required to operate an SEM.
- SEM may carry a small risk of radiation exposure related with the electrons that scatter from beneath the sample surface.
- Solid samples must fit into the microscope chamber. The maximum size in horizontal dimensions is usually around 10 cm, while vertical dimensions are usually much less, seldom exceeding 40 mm. Samples must be stable in a vacuum of about 10^{-5} to 10^{-6} torr for most equipment. Those that are

liable to outgas at low pressures (rocks saturated with hydrocarbons, "wet" samples such as coal, organic compounds, or swelling clays, and samples that degrade at low pressure) are unsuitable for inspection in traditional SEMs.

- There are, however, "low vacuum" and "environmental" SEMs, and several of these types of samples can be investigated effectively in these specialist devices. Various equipment cannot detect elements with atomic numbers less than 11, and EDS detectors on SEMs cannot identify very light elements (H, He, and Li) (Na) (Goldstein et al., 2003; Clarke, 2002; Reimer, 1998).
- Unlike most electron probe microanalyzers, which use wavelength dispersive X-ray detectors (WDS), most SEMs use a solid-state X-ray detector (EDS). While these detectors are quick and simple to use, their energy resolution and sensitivity to low-abundance elements are limited. Unless the instrument can operate in a low vacuum mode, an electrically conductive coating is added to the electrically insulating specimen for investigation in typical SEMs.

8.2 Transmission electron microscopy

In electron microscopy, transmission electron microscopy (TEM) is the most widely used technique for characterizing nanomaterials. It provides information on the bulk material from low to high magnification. TEM offers crucial information about materials with two or more layers. It provides chemical information and images of nanomaterials with a spatial resolution equal to that of atomic dimensions. When an electron beam interacts with a thin foil material transmitting incident light, the beam is changed into elastically or inelastically dispersed electrons as in Fig. 7.7.

TEM can provide images with a resolution of several angstroms (0.019 nm). It uses the interaction of electrons passing through a material to scan thin (100 s/nm^{-1}) samples. The study of nanoscale morphological and chemical properties of materials down to near-atomic levels is possible because of the detection of a variety of secondary signals.

8.2.1 Sample preparation for TEM measuring

Processing, embedding, and polymerization are the three main phases in sample preparation. Fixation, rinsing, postfixation, dehydration, and infiltration are all steps in the processing process.

(i) Fixation preserves the sample and prevents additional deterioration.
(ii) Rinsing: The sample is washed with a buffer to maintain the pH and prevent extra acidity.
(iii) Postfixation: A secondary fixation with osmium tetroxide (OsO_4) increases the stability and contrast of fine structure.

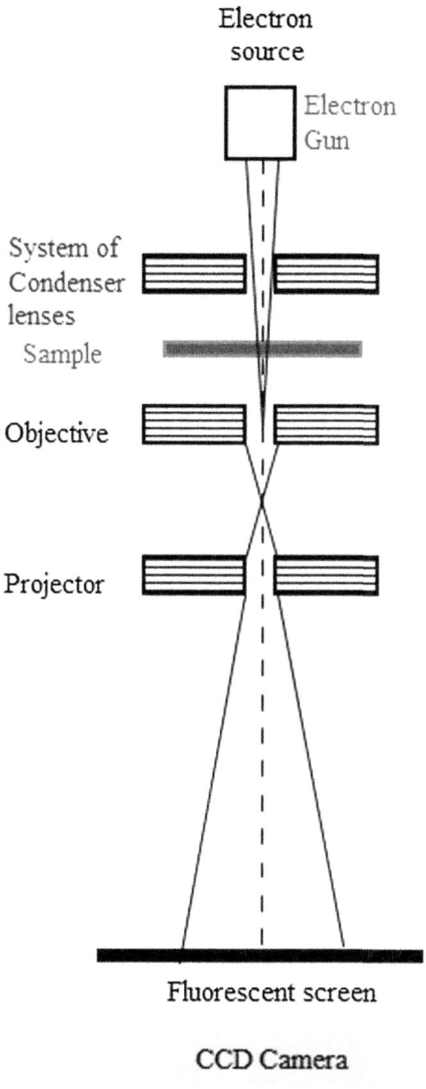
Electron
source
Electron
Gun
System of
Condenser
lenses
Sample
Objective
Projector
Fluorescent screen
CCD Camera

FIGURE 7.7 The main components of TEM.

(iv) Dehydration: The water content in the tissue sample should be replaced with an organic solvent, as the epoxy resin used in the infiltration and embedding step is not miscible with water.

(v) Infiltration: To enter the cells, epoxy resin is employed. It penetrates the cells and fills the spaces to create a strong plastic material that can withstand cutting pressure. The next phase after processing is embedding. Flat molds are used for this.

Polymerization: The resin is then allowed to cure overnight in an oven at a temperature of 60°C for polymerization.
Sectioning: For electron microscopy, the specimen must be sliced into very thin sections so that the electrons are semitransparent to them.

Fig. 7.8 shows the TEM image of graphene prepared from graphite by the Hummers' method. Fig. 7.9 shows the TEM image of carbon nanotubes (CNTs) prepared by the chemical vapor deposition (CVD) method as examples.

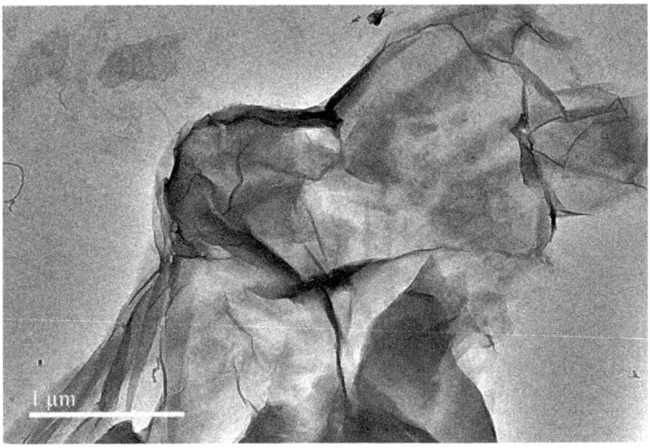
1 μm

FIGURE 7.8 TEM image of graphene nanosheets as an example.

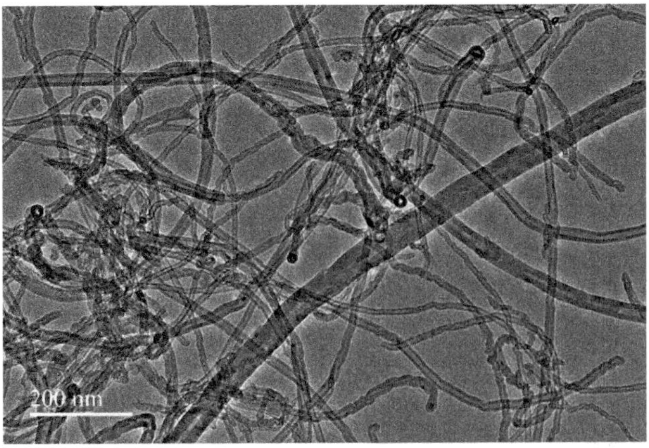
200 nm

FIGURE 7.9 TEM image of carbon nanotubes.

SEM and TEM have the following similarities:

- Working principle: electrons are used to acquire images of samples.
- Their main components, housed inside a chamber that is under high vacuum, are as follows:
  - An electron source
  - A series of electromagnetic and electrostatic lenses to control the shape and trajectory of the electron beam
  - Electron apertures

Differences between SEM and TEM:

- SEM makes images by detecting reflected or knocked-off electrons, whereas TEM creates images by using transmitted electrons (electrons traveling through the sample).
- SEM provides information on the sample's surface and composition, while TEM provides information on the sample's interior structure, such as crystal structure, shape, and stress state.

8.3 Scanning tunneling microscopy

With the introduction of the scanning probe microscope (STM), the family of scanning probe microscopes was born. While working at IBM Zurich Research Laboratories in Switzerland, Gerd Binnig and Heinrich Rohrer created the first operational STM (Fig. 7.10). Binnig and Rohrer were awarded the Nobel Prize in Physics in 1986 for their apparatus. Only conducting or semiconducting surfaces can be imaged by STM (Saleh and Gupta, 2016).

8.4 Atomic force microscopy

AFM is a sort of scanning probe microscopy (SPM) that has demonstrated resolution on the scale of fractions of a nanometer, which is more than 1000 times greater than the optical diffraction limit (Fig. 7.11). The AFM can image nearly any form of surface, including polymers, ceramics, composites, glass, and biological samples. A mechanical probe is used to touch the surface and collect data.

Concept of how AFM works:

- The physical parameter probed is a force resulting from various interactions.
- Therefore, the AFM image is formed when the force changes as the probe (or sample) is scanned in the x and y directions.
- The specimen is mounted on a piezoelectric scanner, ensuring three-dimensional positioning with high resolution.
- The force is measured by attaching the probe to a pliable cantilever, acting as a spring, and measuring the bending or deflection of the cantilever.

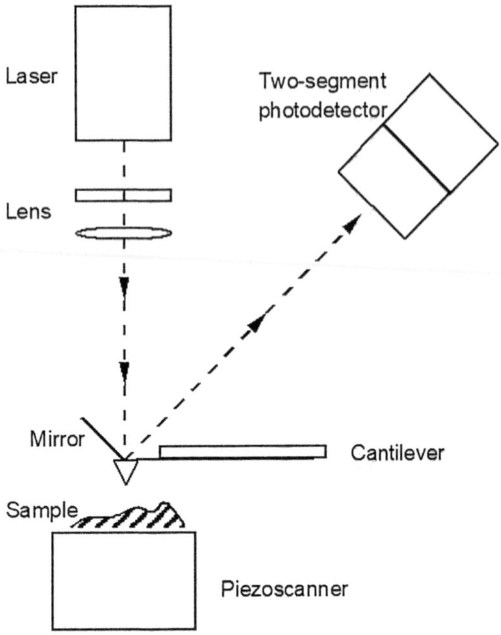
Laser
Two-segment photodetector
Lens
Mirror
Cantilever
Sample
Piezoscanner

FIGURE 7.10 The main components of STM

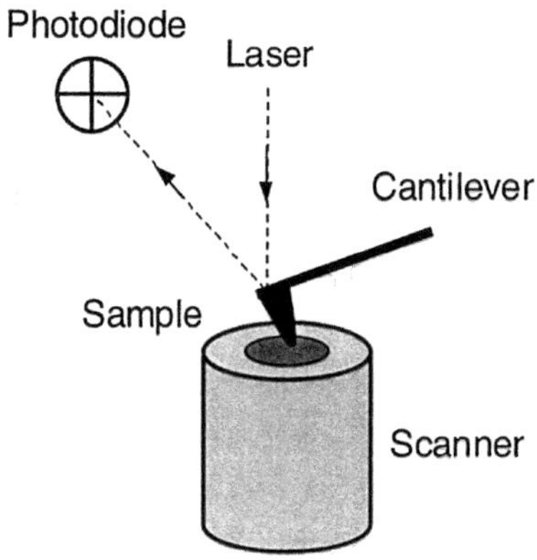
Photodiode
Laser
Cantilever
Sample
Scanner

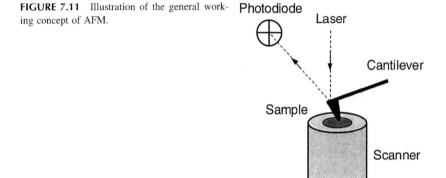

FIGURE 7.11 Illustration of the general working concept of AFM.

AFM is made up of a cantilever with a sharp tip (probe) that is used to scan the surface of the material. Silicon or silicon nitride cantilevers with a tip radius of curvature in nanometers are commonly used. Hooke's law states that when the tip contacts a sample surface, forces between the tip and the sample deflect the cantilever. Mechanical contact force, van der Waals forces, capillary forces, chemical bonding, and electrostatic forces are among the forces detected in AFM, depending on the scenario (Table 7.2).

TABLE 7.2 List of some commonly used other imaging techniques.

Technique	Description
Phase imaging	In dynamic mode, recording the phase difference between the driving signal and the AFM cantilever oscillation yields additional material properties such as elasticity, adhesion, and so on.
AFM in liquids	Both contact and tapping modes can be used to measure biological and other materials in aqueous solutions. Silicon nitride probes are extensively utilized in liquid applications. When measuring in liquids with soft AFM cantilevers, the AFM tip does not damage the sample surface.
High-speed scanning (HSS)	Dynamic processes in bioresearch can be observed by raising the scanning speed and feedback speed up to video rates using AFM cantilevers with megahertz resonance frequencies and high-speed electronics.
Ultrahigh vacuum (UHV) AFM	By insulating the sample surface from water film formation and contamination, AFM measurements under UHV enable atomic resolution imaging.
Force modulation (FM) microscopy	A periodic signal mechanically pushes the AFM cantilever in the vertical direction while scanning in contact mode (Maivald et al., 1991). The amplitude of AFM cantilever modulation varies depending on the elastic characteristics of the sample, allowing the elastic response of the sample to be mapped.
Lateral force microscopy (LFM)	Using a four-segment photodetector and scanning in contact mode, the microscope can detect the vertical deflection of the AFM cantilever, and its torsional twisting owing to lateral forces (typically friction) acting on the AFM probe tip.
Electrostatic force microscopy (EFM)	With applying a voltage bias on intermediately stiff AFM probes with an electrically conductive coating, the variations of the electric field gradient across the material surface are evaluated in a noncontact mode.

The ability of an AFM to achieve near-atomic-scale resolution depends on the three essential components:

- a cantilever with a sharp tip,
- a scanner that controls the x-y-z position, and
- the feedback control and loop.

AFM imaging can be performed in various modes.

8.4.1 Contact mode

The tip is dragged across the sample's surface, and the surface contours are measured either directly using the cantilever's deflection or, more often, using the feedback signal required to hold the cantilever in place. Fig. 7.12 represents AFM image of polyamide membrane used for water treatment.

Low-stiffness cantilevers are utilized to achieve a great enough deflection signal while keeping the contact force low because static signal measurement is prone to noise and drift.

Therefore, contact mode is used. AFM is almost always performed at a depth where the overall force is repulsive, i.e., when the solid surface is in firm contact. This mode provides

- High resolution
- Damage to sample
- Measurement of frictional forces

8.4.2 Noncontact mode

➢ The tip of the cantilever does not touch the sample surface in noncontact mode; instead, the cantilever is oscillated at its resonant frequency or just above it (amplitude modulation), with the amplitude typically a few nanometers (10 nm) down to a few picometers.

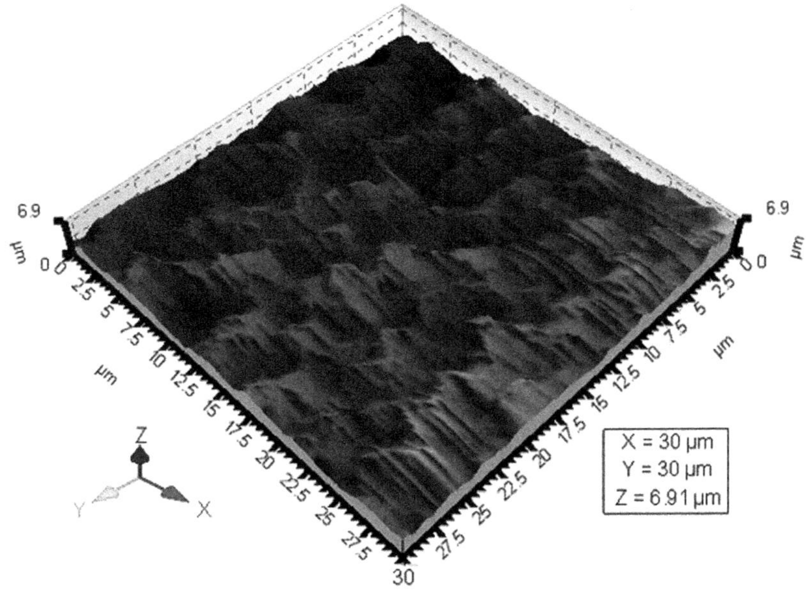
6.9
μm
0
6.9
μm
0
X = 30 μm
Y = 30 μm
Z = 6.91 μm
30
Z
Y
X

FIGURE 7.12 Example of an AFM image of polyamide membrane used for water treatment.

- The van der Waals forces, which are highest between 1 and 10 nm above the surface, act to reduce the cantilever's resonance frequency.
- Measuring the tip-to-sample distance at each (x,y) data point makes it possible for the scanning software to construct a topographic image of the material surface.
- The mode of noncontact AFM does not suffer from the tip or sample degradation effects that can occur with contact AFM after several scans. For soft samples, including biological samples and organic thin films, noncontact AFM is superior to contact AFM. This mode provides
  - Lower resolution
  - No damage to the sample

8.4.3 *Tapping mode*

- The cantilever is driven to oscillate up and down at or near its resonance frequency.
- The amplitude of this oscillation usually varies from several nm to 200 nm.
- The interaction of forces acts on the cantilever when the tip comes close to the surface, electrostatic forces, dipole−dipole interactions, Van der Waals forces, etc.
- The amplitude of the cantilever's oscillation changes (usually decreases) as the tip approaches the sample.
- This amplitude is fed into the electronic servo that controls the cantilever's height above the sample.
- As a result, a tapping AFM picture is created by imaging the force of the tip's intermittent contacts with the sample surface. This mode provides
  - Better resolution
  - Minimal damage to the sample

Advantages of AFM

- Compared to the scanning electron microscope, the AFM has various advantages (SEM).
- The AFM gives a real three-dimensional surface profile, unlike the electron microscope, which provides a two-dimensional projection or image of a sample.
- Furthermore, AFM-viewed materials do not require any specific treatments (such as metal/carbon coatings) that modify or damage the material irreparably.
- While an electron microscope requires a costly vacuum condition to operate well, most AFM modes may operate in ambient air or even a liquid environment, allowing researchers to investigate biological macromolecules and even living creatures.
- AFM, in theory, can provide better resolution than SEM.

Disadvantages

- ➢ The picture size of an AFM is smaller than that of an SEM.
- ➢ The SEM has a depth of field of millimeters and can image a region on the order of millimeters by millimeters.
- ➢ The AFM can only picture objects with a maximum height of a few micrometers and a scanning area of roughly 150 by 150 μm.
- ➢ Another drawback is that the quality of a picture at high resolution is restricted by the radius of curvature of the probe tip, and choosing the wrong tip for the needed resolution can result in image artifacts.

9. Pore structure, size, and surface area

9.1 Physisorption, BET, and BJH fitting

The identification of pore texture information, which has ramifications for the adsorbent's diffusion properties, necessitates pore structure characterization. The Barrett–Joyner–Halenda theory and the Brunauer–Emmett–Teller theory may quantify surface area and pore structure information, respectively, from nitrogen physisorption data (Barrett et al., 1951; Brunauer et al., 1938). The optimum method for determining the surface area of NMs is by using BET. The adsorption and desorption concepts, as well as the BET theorem, are used in this technique. BET forms several types of isotherms including Type-I, Type-II, Type-III, and Type-IV (Fagerlund, 1973).

MgO has been observed to have a poor surface area and porosity when used without support, such as charcoal (Zhu et al., 2020) or diatomite (Xia et al., 2016) to attain high dispersion of MgO porosity and sites. High reaction rates can be achieved. Furthermore, in the instance of MgO, it has been demonstrated that increasing the surface area and pore volume by constructing mesoporous MgO structures is conceivable (Zhou et al., 2011). The transport of phosphate through the material is dictated by the adsorbent's pore properties, which impact the rate of adsorption. Improving the surface area and pore volume with the use of support materials and mesoporous particles manufacturing, as stated above, is of interest to improve the rate of adsorption. However, while BJH and BET data are obtained on dry powder, the sorbent might react inversely in the aqueous media owing to dissolving and possible pore collapse. Characterization before and after the use in water can reveal whether phosphate has caused significant area loss and pore-clogging.

9.2 Dynamic light scattering

The dynamic light scattering (DLS) technique is used to characterize particle sizes in suspensions and emulsions. It is based on the Brownian motion of particles—this states that small particles move fast, while large particles move more slowly in a liquid. The light scattered by particles comprises information on the diffusion speed and therefore on the size distribution. DLS enables the analysis of particles in a size. It provides information on particle size distribution.

9.3 Other techniques

The magnitude of the NMs can be estimated using a variety of methods. SEM, TEM, XRD, AFM, and DLS are a few examples. SEM, TEM, XRD, and AFM can provide a more accurate picture of particle size (Kestens et al., 2016). The NMs size can be determined at a very low level using the zeta potential size analyzer/DLS. Sikora et al. employed DLS to explore the size variation of silica NMs in relation to serum protein absorption in one research. The size of the sample grew as the protein layer was acquired. However, in the event of agglomeration and hydrophilicity, DLS may not be capable of precise assessment. Thus we should rely on the high-resolution differential centrifugal sedimentation (DCS) technique instead (Sikora et al., 2016). Nanoparticle tracking analysis (NTA) is used in the case of biological systems such as proteins and DNA. NMs are visualized in liquid media that relates the Brownian motion rate to particle size. This method can be used to determine the size distribution profile of NMs in a liquid media with diameters ranging from 10 to 1000 nm. When compared to DLS, this technique produced good results and was found to be very precise for size monodisperse and polydisperse samples, with significantly greater peak resolution. Gross et al. determined the particle size and concentration of different sized NMs in polymer and protein suspensions, as well as provided an overview of the effect of experimental and data assessment parameters (Gross et al., 2016).

10. Surface charge

10.1 Point of zero charge

The strength of Coulombic interactions between the phosphate ions and the adsorbent is determined by surface charge characterization, while the surface composition and molecular structures are significant in influencing the adsorption mechanisms. Functional groups such as $-CO_3^{2-}$, $-OH$, and $-HCO_3^-$ groups on an adsorbent impact the surface charge by producing charge deficits. The zeta potential measures a particle's surface charge in an aqueous phase affected by the pH and ionic strength of the solution. The zeta potential of the particle surface as a function of pH was reported in a prior study on employing MgO for phenol ozonation (Wang et al., 2017). Wang et al. show that beyond pH = 9, MgO is positively charged, which is the region of concern for phosphate sorption and struvite formation. In the pH range of 2.0–10.0, MgO–diatomite is positively charged, with the isoelectric point at pH = 11. This demonstrates that phosphate sorption occurs at a variety of pH levels (Li et al., 2019).

10.2 Zeta potential measured via Zetasizer instrument

Colloid–electrolyte interactions are studied using the zeta potential. The basic premise is that particles with oppositely charged surfaces (counterions) attract one other, whereas similarly charged particles (co-ions) repel each other. The Zetasizer equipment is used to determine the zeta potential. The zeta potential must be determined in order to estimate flocculation. The velocity of flocculation is also affected by the surface charge of the droplets.

11. Thermal stability

Thermal stability determines the material's ability to retain its properties at required temperatures over extended service time. In addition to temperature and time, thermal stability is affected by load and environmental conditions. The physical properties of a substance are evaluated as a function of temperature. This may include some reaction products, and there is a need for the temperature to be controlled.

There are some methods to determine thermal stability:

- Thermogravimetry
- Differential thermal analysis
- Differential scanning calorimetry

Thermogravimetry is a thermal analytical procedure used widely in the investigation of polymeric systems. It involves monitoring the weight loss of the sample in a given atmosphere as a function of temperature. Gases used include N_2, O_2, air, and He to measure the change from ambient to 1000°C. The first derivative of the mass loss can be recorded.

The temperature difference between nanomaterials and reference material is measured as a function of temperature. Nanomaterials and reference material are subjected to the same controlled temperature. Differences in heat flow are measured as a function of temperature.

12. Biological evaluation

The nanomaterials are to be evaluated and studied for cell toxicity, immunotoxicity, and genotoxicity. Various methods are available for the toxicity assessment imposed by nanoparticles on the organisms. The methods for toxicity assessment can be categorized as in vitro and in vivo.

12.1 In vitro assessment methods

One of the most significant ways is to investigate nanoparticle toxicity in vitro. The benefits include lower costs, faster processing, and fewer ethical concerns. Proliferation tests, apoptosis assays, necrosis assays, oxidative stress assays, and DNA damage assays are all types of assessments.

12.2 In vivo toxicity assessment methods

Animal models such as mice and rats are commonly used to investigate in vivo toxicity. Biodistribution, clearance, hematology, chemistry, and histopathology are among the approaches used to measure in vivo toxicity. Biodistribution studies investigate how nanoparticles find their way into a tissue or organ. Radiolabels are used to detect nanoparticles in dead or live animals (Kim et al., 2001). The measurement of nanoparticle excretion and metabolism at various periods following exposure is used to determine their clearance. Examining changes in serum chemistry and cell type following nanoparticle exposure is another way of determining in vivo toxicity. The toxicity level generated by a nanoparticle is determined by the histopathology of the cell, tissue, or organ following exposure (Lei et al., 2008). Nanoparticles have been detected in exposed tissues such as the lung, eyes, brain, liver, kidneys, heart, and spleen (Baker et al., 2008). Microelectrochemistry and microfluidics are being used to advance toxicity evaluation (Ewing et al., 1983; Kumar et al., 2017).

13. Mechanical properties

Researchers can use NMs' unique mechanical properties to find new applications in a variety of domains, including tribology, surface engineering, nanofabrication, and nanomanufacturing. To determine the actual mechanical nature of NMs, many mechanical parameters such as hardness, stress, strain, elastic modulus, adhesion, and friction can be examined. Various tools can be used to evaluate the mechanical properties of nanomaterials (Wu et al., 2020).

14. Magnetic properties

The magnetic properties of NMs depend on the synthesis method and chemical structure. Investigators from a diverse range of disciplines are interested in magnetic NMs, including heterogeneous and homogeneous catalysis, biomedicine, magnetic fluids, data storage, magnetic resonance imaging (MRI), and environmental remediation such as water purification. The magnetic property of NMs can be measured using a vibrating-sample magnetometer, such as VSM, LakeShore 7307.

15. Benefits of nanomaterials characterization for industry

Nanomaterial characterization has significant benefits for the industry. In forensic engineering, several analysis techniques are utilized to collect the exact data on the structural features and behavior of all types of substances under diverse environmental circumstances and stimuli.

Material characterization techniques are utilized in this way to select the best nanomaterials, prevent defects in the structure, and improve the design and manufacturing efficiency, thus extending the useable life of items and maximizing the resources available in the firm. Table 7.3 lists some of the commonly used instruments and techniques used for the determination of the characteristics (properties) of materials and nanomaterials.

TABLE 7.3 Instruments and techniques used for the determination of the characteristics (properties) of materials and nanomaterials.

Characteristics (properties)	Instrument used
Chemical structure and chemical bonds	FTIR, NMR, UV, Raman, XPS
Elemental-chemical composition	ICP, SEM with EDX, XRD, XPS
Crystallinity and structures	XRD, HRTEM, electron diffraction
Shape	SEM, TEM, AFM, ferromagnetic resonance spectroscopy (FMR), elliptically polarized light scattering (EPLS), 3D-tomography
Single-particle properties	Sp-ICP-MS, HRTEM
3D visualization	3D-tomography, AFM, SEM
Surface area, specific surface area	BET
Surface charge	Zeta potential, electrophoretic mobility (EPM)
Structural defects	HRTEM, electron backscatter diffraction (EBSD)
Size	SEM, TEM, XRD, DLS, AFM
Size distribution	DCS, DLS
Detection of nanoparticles	TEM, SEM, STEM, EBSD, magnetic susceptibility
Agglomeration state	Zeta potential, DLS, DCS, UV–Vis, SEM, Cryo-TEM, TEM
Density	DCS, resonant mass measurement (RMM)
Chemical state and oxidation state	XPS, X-ray absorption spectroscopy (XAS), electron energy-loss spectroscopy (EELS), mössbauer spectroscopy, wet chemical analysis, indirect stoichiometry based calculations from electron probe analysis data, redox titration, thermogravimetric analysis

TABLE 7.3 Instruments and techniques used for the determination of the characteristics (properties) of materials and nanomaterials.—cont'd

Characteristics (properties)	Instrument used
Optical properties	UV−Vis-NIR, Photoluminescence (PL), EELS-STEM
Magnetic properties	Vibrating-sample magnetometers and superconductor quantum interference devices (SQUIDs), vibration sample magnetometry (VSM), mössbauer, MFM, FMR, X-ray magnetic circular dichroism (XMCD), magnetic susceptibility
Dispersion of nanoparticles in matrices or supports	TEM, SEM, AFM
Concentration of nanoparticles in a media	ICP-MS, UV−Vis, DCS, tunable resistive pulse sensing (TRPS), particle tracking analysis (PTAs)

16. Challenges with nanomaterials

Although NMs are gaining interest and being investigated by several researchers, there are some challenges that limit their possible applications. The challenges include:

- Difficult to detect without sophisticated equipment
- Difficult to predict how particles will behave in the environment (dispersed/clumped)
- Small size may result in particles passing into the body more easily (inhalation, ingestion, absorption)
- More reactive due to surface area to volume ratio
- Potential to adsorb toxic chemicals
- Persistence—The longevity of particles in the environment and body are unknown
- Adverse health effects in humans from deliberate or accidental exposure
- Adverse effects on the environment from deliberate or accidental exposure
- Potentially explosive properties of nanostructures
- Some structures are likely to have a unique toxicological profile
- Standardized terminology agreed recently
- Particle size could be less important than the surface characteristics of the material
- Standard dose−response tests may not be appropriate

17. Conclusions

In short, the characterization of nanomaterials is a critical research tool for improving the efficiency of manufacturing processes. Nanomaterials with sizes ranging from a few nanometers to 500 nm have been discovered using various characterization techniques such as SEM, TEM, and XRD. The morphology, on the other hand, can be controlled. Nanomaterials have a vast surface area due to their small size, making them a good contender for a variety of applications. Furthermore, at that size, the optical properties are prominent, thereby enhancing the usefulness of these materials in photocatalytic applications.

References

Baker, G.L., et al., 2008. Inhalation toxicity and lung toxicokinetics of C60 fullerene nanoparticles and microparticles. Toxicol. Sci. 101 (1), 122–131.

Bañares, M.A., Wachs, I.E., 2010. Encyclopedia of Analytical Chemistry. John Wiley & Sons, Ltd, Chichester, UK.

Barrett, E.P., Joyner, L.G., Halenda, P.P., 1951. The determination of pore volume and area distributions in porous substances. I. Computations from nitrogen isotherms. J. Am. Chem. Soc. 73, 373–380.

Braet, F., Zanger, R., Wisse, E., 1997. Drying cells for SEM, AFM and TEM by hexamethyldisilazane: a study on hepatic endothelial cells. J. Microsc. 186, 84–87.

Brongersma, H.H., Draxler, M., de Ridder, M., Bauer, P., 2007. Surface composition analysis by low-energy ion scattering. Surf. Sci. Rep. 62, 63–109.

Brunauer, S., Emmett, P.H., Teller, E., 1938. Adsorption of gases in multimolecular layers. J. Am. Chem. Soc. 60, 309–319.

Bustos, A.R.M., Winchester, M.R., 2016. Single-particle-ICP-MS advances. Anal. Bioanal. Chem. 408 (19), 5051–5052.

Clarke, A.R., 2002. Microscopy Techniques for Materials Science. CRC Press.

Cullity, B.D., Stock, S.R., 2001. Elements of X-Ray Diffraction. Prentice-Hall.

Dablemont, C., Lang, P., Mangeney, C., Piquemal, J.-Y., Petkov, V., Herbst, F., Viau, G., 2008. FTIR and XPS study of Pt nanoparticle functionalization and interaction with alumina. Langmuir 24, 5832–5841.

Druce, J., Téllez, H., Burriel, M., Sharp, M.D., Fawcett, L.J., Cook, S.N., McPhail, D.S., Ishihara, T., Brongersma, H.H., Kilner, J.A., 2014. Surface termination and subsurface restructuring of perovskite-based solid oxide electrode materials. Energy Environ. Sci. 7, 3593–3599.

Elzinga, E.J., Sparks, D.L., 2007. Phosphate adsorption onto hematite: an in situ ATR-FTIR investigation of the effects of pH and loading level on the mode of phosphate surface complexation. J. Colloid Interface Sci. 308, 53–70.

Ewing, A.G., Bigelow, J.C., Wightman, R.M., 1983. Direct in vivo monitoring of dopamine released from two striatal compartments in the rat. Science 221 (4606), 169–171.

Fagerlund, G., 1973. Determination of specific surface by the BET method. Mater. Constr. 6, 239–245.

Feng, X., Li, W., Yan, Y., Sparks, D.L., Phillips, B.L., 2013. Solid-state NMR spectroscopic study of phosphate sorption mechanisms on aluminum (hydr)oxides. Environ. Sci. Technol. 47, 8308–8315.

Goldstein, J., Newbury, D.E., Joy, D.C., Lyman, C.E., Echlin, P., Lifshin, E., Sawyer, L., Michael, J.R., 2003. Scanning Electron Microscopy and X-Ray Microanalysis. Springer US, 689 pp.

Gross, J., Sayle, S., Karow, A.R., Bakowsky, U., Garidel, P., 2016. Nanoparticle tracking analysis of particle size and concentration detection in suspensions of polymer and protein samples: influence of experimental and data evaluation parameters. Eur. J. Pharm. Biopharm. 104, 30−41.

Hawes, C., 2012. Electron Microscopy of Plant Cells. Academic Press.

Kestens, G.R., Herrmann, J., Jämting, Å., Coleman, V., Minelli, C., Clifford, C., De Temmerman, P.-J., Mast, J., Junjie, L., Babick, F., Cölfen, H., Emons, H., 2016. Challenges in the size analysis of a silica nanoparticle mixture as candidate certified reference material. J. Nano Res. 18, 171.

Kiani, D., Sheng, Y., Lu, B., Barauskas, D., Honer, K., Jiang, Z., Baltrusaitis, J., 2018. Transient struvite formation during stoichiometric (1:1) NH^{4+} and PO_4^{3-} adsorption/reaction on magnesium oxide (MgO) particles. ACS Sustain. Chem. Eng. 7, 1545−1556.

Kim, S.C., et al., 2001. In vivo evaluation of polymeric micellar paclitaxel formulation: toxicity and efficacy. J. Contr. Release 72 (1), 191−202.

Kim, J., Li, W., Philips, B.L., Grey, C.P., 2011. Phosphate adsorption on the iron oxyhydroxides goethite (α-FeOOH), akaganeite (β-FeOOH), and lepidocrocite (γ-FeOOH): a 31P NMR Study. Energy Environ. Sci. 4, 4298−4305.

Kumar, V., Sharma, N., Maitra, S.S., 2017. In vitro and in vivo toxicity assessment of nanoparticles. Int. Nano Lett. 7, 243−256.

Kumar, N., Wondergem, C.S., Wain, A.J., Weckhuysen, B.M., 2019. In situ nanoscale investigation of catalytic reactions in the liquid phase using Zirconia-protected tip-enhanced Raman spectroscopy probes. J. Phys. Chem. Lett. 10, 1669−1675.

Lei, R., et al., 2008. Integrated metabolomic analysis of the nano-sized copper particle-induced hepatotoxicity and nephrotoxicity in rats: a rapid invivo screening method for nanotoxicity. Toxicol. Appl. Pharmacol. 232 (2), 292−301.

Li, J., Wang, X., Wang, J., Li, Y., Xia, S., Zhao, J., 2019. Simultaneous recovery of microalgae, ammonium and phosphate from simulated wastewater by MgO modified diatomite. Chem. Eng. J. 362, 802−811.

Maivald, P., Butt, H.J., Gould, S.A.C., Prater, C.B., Drake, B., Gurley, J.A., Elings, V.B., Hansma, P.K., 1991. Using force modulation to image surface elasticities with the AFM. Nanotechnology 2, 103−106. https://doi.org/10.1088/0957-4484/2/2/004.

Naasz, S., Weigel, S., Borovinskaya, O., Serva, A., Cascio, C., Undas, A.K., Simeone, F.C., Marvin, H.J.P., Peters, R.J.B., 2018. Multi-element analysis of single nanoparticles by ICP-MS using quadrupole and time-of-flight technologies. J. Anal. Atomic Spectrom. 33 (5), 835−845.

Nasrazadani, S., Hassani, S., 2016. Handbook of Materials Failure Analysis with Case Studies from the Oil and Gas Industry. Elsevier Inc., pp. 39−54

Oprea, B., Martínez, L., Román, E., Vanea, E., Simon, S., Huttel, Y., 2015. Dispersion and functionalization of nanoparticles synthesized by gas aggregation source: opening new routes toward the fabrication of nanoparticles for biomedicine. Langmuir 31, 13813−13820.

Reimer, L., 1998. Scanning Electron Microscopy: Physics of Image Formation and Microanalysis. Springer, 527 pp.

Saleh, T.A., 2018. Simultaneous adsorptive desulfurization of diesel fuel over bimetallic nanoparticles loaded on activated carbon. J. Clean. Prod. 172, 2123−2132.

Saleh, T.A., 2020a. Nanomaterials: classification, properties, and environmental toxicities. Environ. Technol. Innovat. 20, 101067.

Saleh, T.A., 2020b. Characterization, determination and elimination technologies for sulfur from petroleum: toward cleaner fuel and a safe environment. Trends Environ. Analyt. Chem. 25, e00080.

Saleh, T.A., 2021. Protocols for synthesis of nanomaterials, polymers, and green materials as adsorbents for water treatment technologies. Environ. Technol. Innovat. 24, 101821.

Saleh, T.A., 2022. Experimental and Analytical methods for testing inhibitors and fluids in water-based drilling environments. Trac. Trends Anal. Chem. 116543. https://doi.org/10.1016/j.trac.2022.116543.

Saleh, T.A., Gupta, V., 2016. Nanomaterial and Polymer Membranes: Synthesis, Characterization, and Applications, ISBN 9780128014400.

Sikora, A., Shard, A.G., Minelli, C., 2016. Size and ζ-potential measurement of silica nanoparticles in serum using tunable resistive pulse sensing. Langmuir 32, 2216–2224.

Stadtländer, C.T.K.-H., 2005. Dehydration and rehydration issues in biological tissue processing for electron microscopy. Micros. Today 13, 32–35.

Swinehart, D.F., 1962. The Beer-Lambert law. J. Chem. Educ. 39, 333.

Tran, H.N., You, S.J., Hosseini-Bandegharaei, A., Chao, H.P., 2017. Mistakes and inconsistencies regarding adsorption of contaminants from aqueous solutions: a critical review. Water Res. 120, 88–116.

Tsotsas, E., Mujumdar, A.S., 2011. Modern Drying Technology. In: Product Quality and Formulation, vol. 3. John Wiley & Sons.

Wang, Y.-C., Engelhard, M.H., Baer, D.R., Castner, D.G., 2016. Quantifying the impact of nanoparticle coatings and nonuniformities on XPS analysis: gold/silver core-shell nanoparticles. Anal. Chem. 88, 3917–3925.

Wang, B., Xiong, X., Ren, H., Huang, Z., 2017. Preparation of MgO nanocrystals and catalytic mechanism on phenol ozonation. RSC Adv. 7, 43464–43473.

Wu, Q., Miao, W.-S., Zhang, Y.-D., Gao, H.-J., Hui, D., 2020. Mechanical properties of nanomaterials: a review. Nanotechnol. Rev. 9 (1), 259–273.

Xia, P., Wang, X., Wang, X., Song, J., Wang, H., Zhang, J., 2016. Struvite crystallization combined adsorption of phosphate and ammonium from aqueous solutions by mesoporous MgO–loaded diatomite. Colloids Surf., A 506, 220–227.

Young, A.G., McQuillan, A.J., 2009. Adsorption/desorption kinetics from ATR-IR spectroscopy. Aqueous oxalic acid on anatase TiO_2. Langmuir 25, 3538–3548.

Zarraoa, L., González, M.U., Paulo, Á.S., 2019. Imaging low-dimensional nanostructures by very low voltage scanning electron microscopy: ultra-shallow topography and depth-tunable material contrast. Sci. Rep. 9, 16263.

Zhou, J., Yang, S., Yu, J., 2011. Facile fabrication of mesoporous MgO microspheres and their enhanced adsorption performance for phosphate from aqueous solutions. Colloids Surf., A 379, 102–108.

Zhu, D., Chen, Y., Yang, H., Wang, S., Wang, X., Zhang, S., Chen, H., 2020. Synthesis and characterization of magnesium oxide nanoparticle-containing biochar composites for efficient phosphorus removal from aqueous solution. Chemosphere 247, 125847.

Chapter 8

Properties of nanoadsorbents and adsorption mechanisms

1. Introduction

Nanotechnology, a new field of study, has the potential to replace traditional micron technologies by allowing functional materials to have size-dependent features. The interest in nanoscience, also known as low-dimensional systems science, is a realization of Feynman's famous phrase, "There's Plenty of Room at the Bottom." Based on Feynman's idea (1959), Drexler proposed "molecular nanotechnology" in his book *Engines of Creation: The Coming Era of Nanotechnology* in 1986, where he proposed employing molecular structures to guide and activate the synthesis of bigger molecules in a machine-like manner (Drexler, 1986).

When a material's dimension is lowered from a high size, the qualities remain constant at first, then modest changes occur until the size drops below 100 nm; at this point, major changes in properties occur. A quantum well is a three-dimensional nanostructure with only one dimension of nanoscale; a quantum wire is a three-dimensional nanostructure with two dimensions of nanometer scale, while a quantum dot is a three-dimensional nanostructure with all three dimensions of nanometer scale. Henceforth, a quantum dot is the ultimate example of nanomaterials because it possesses all three dimensions in the nanoscale. Because quantum mechanics causes changes in attributes, the word quantum is related with these three types of nanostructures.

2. Comparison between properties of adsorbents and nanoadsorbents

There are several types of adsorbents that are either already in use or under testing and evaluation. Adsorbents can be classified generally into bulk adsorbents and nanoadsorbents (Fig. 8.1). Under each class, the adsorbents can

Surface Science of Adsorbents and Nanoadsorbents. https://doi.org/10.1016/B978-0-12-849876-7.00010-5

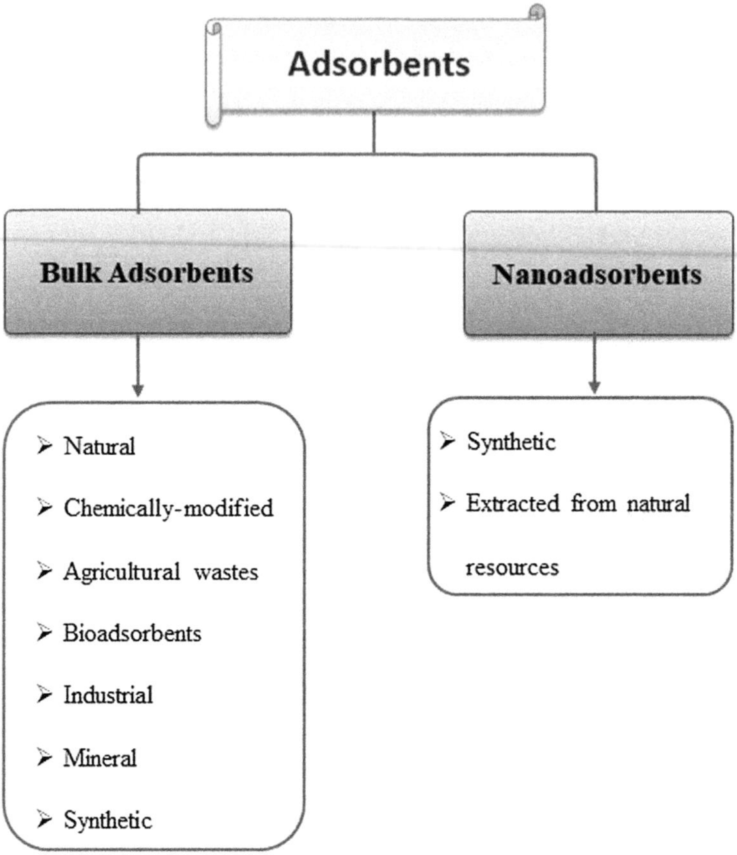
Adsorbents
Bulk Adsorbents
Nanoadsorbents
➢ Natural
➢ Chemically-modified
➢ Agricultural wastes
➢ Bioadsorbents
➢ Industrial
➢ Mineral
➢ Synthetic
➢ Synthetic
➢ Extracted from natural resources

FIGURE 8.1 General classification of adsorbents into bulk adsorbent and nano-adsorbents.

be categorized primarily by their involvement in adsorption applications, which is determined by their (i) inherent properties and (ii) subsequent external functionalization.

Bulk materials are particles that are larger than 100 nm in all dimensions (Fig. 8.2). Physical qualities are independent of size in bulk materials; however, distinct physical properties can be dependent on the size and form of NMs (Fig. 8.3). Some of the characteristics of bulk materials are:

- ➢ Bulk materials are particles that have their size above 100 nm in all dimensions.
- ➢ Can be seen by a simple microscope, or the naked eye.
- ➢ Low surface-to-volume ratio leads to better performance such as in catalysis, solar veils, and gas sensors.

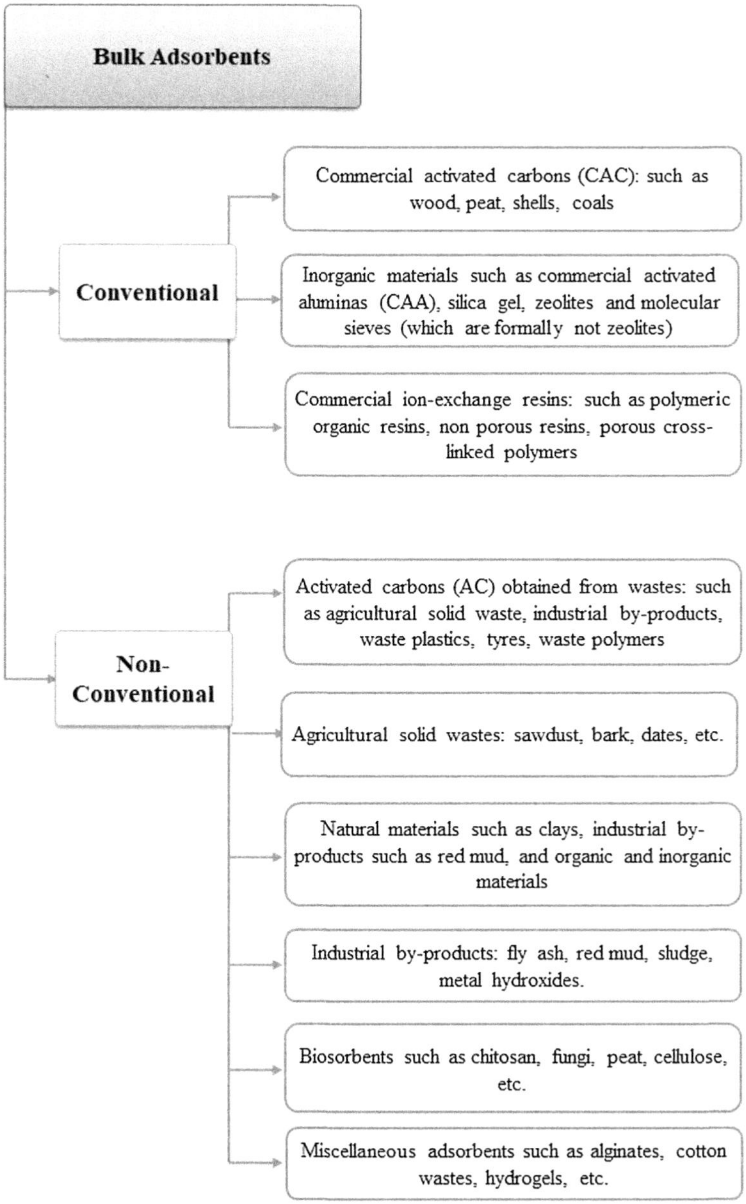
Bulk Adsorbents
Conventional
Commercial activated carbons (CAC): such as wood, peat, shells, coals
Inorganic materials such as commercial activated aluminas (CAA), silica gel, zeolites and molecular sieves (which are formally not zeolites)
Commercial ion-exchange resins: such as polymeric organic resins, non porous resins, porous cross-linked polymers
Non-Conventional
Activated carbons (AC) obtained from wastes: such as agricultural solid waste, industrial by-products, waste plastics, tyres, waste polymers
Agricultural solid wastes: sawdust, bark, dates, etc.
Natural materials such as clays, industrial by-products such as red mud, and organic and inorganic materials
Industrial by-products: fly ash, red mud, sludge, metal hydroxides.
Biosorbents such as chitosan, fungi, peat, cellulose, etc.
Miscellaneous adsorbents such as alginates, cotton wastes, hydrogels, etc.

FIGURE 8.2 Classification of bulk adsorbents into conventional and nonconventional.

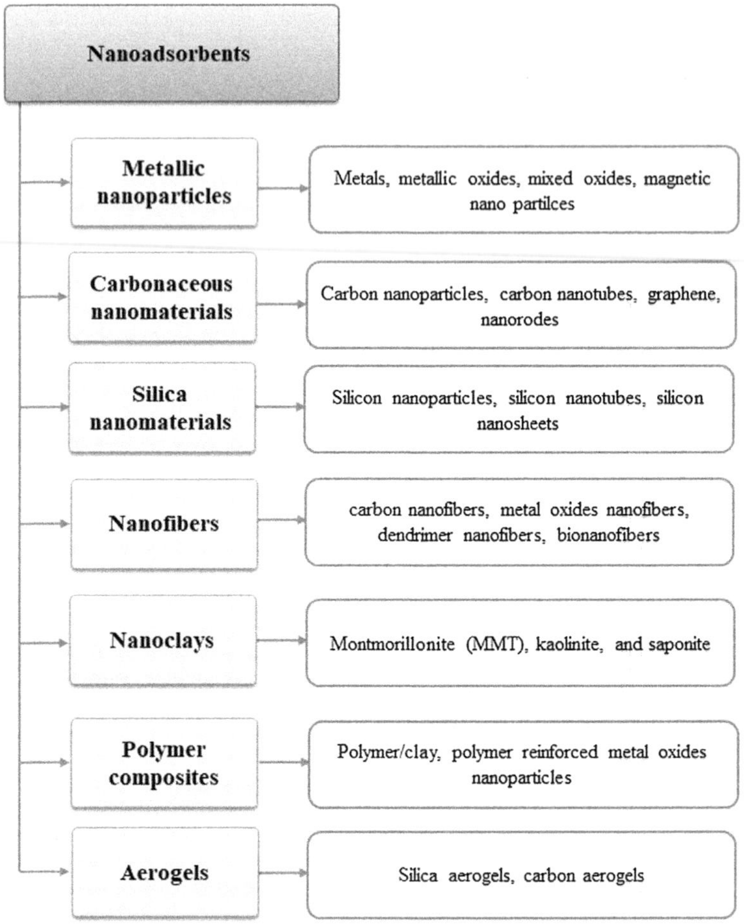
Nanoadsorbents
Metallic nanoparticles
Metals, metallic oxides, mixed oxides, magnetic nano partilces
Carbonaceous nanomaterials
Carbon nanoparticles, carbon nanotubes, graphene, nanorodes
Silica nanomaterials
Silicon nanoparticles, silicon nanotubes, silicon nanosheets
Nanofibers
carbon nanofibers, metal oxides nanofibers, dendrimer nanofibers, bionanofibers
Nanoclays
Montmorillonite (MMT), kaolinite, and saponite
Polymer composites
Polymer/clay, polymer reinforced metal oxides nanoparticles
Aerogels
Silica aerogels, carbon aerogels

FIGURE 8.3 Classification of nanoadsorbents inot metallic, carbonaceous, silica, nanofibers, clays, polymers and gels.

- ➢ The low percentage of atoms or molecules on the surface leads to their properties.
- ➢ Bulk forces are not as important as surface forces.
- ➢ Metal bulk have normal scattering properties.
- ➢ Semiconductor bulk may not exhibit confined energy states in the electronic bandstructure.
- ➢ Their chemical and physical properties cannot be tuned.
- ➢ Adsorption and absorption of molecules (gas or liquid phases) are low and slow.
- ➢ Examples include sand, cement, alumina, ore, salts, etc.

Material reduced to the nanoscale displays a range of enhanced properties, compared to what they display on the microscale. This can be explained due to several reasons and scenarios. Some of the characteristics of nanomaterials are:

- ➢ Nanomaterials (NMs) are chemical substances or materials that are of size, at least in one dimension, in nanoscale 1–100 nm.
- ➢ Cannot be seen by a simple microscope, or the naked eye. Advanced microscopic techniques are used.
- ➢ Large surface-to-volume ratio leads to better performance such as in catalysis, solar veils, and gas sensors.
- ➢ Particles with a very high aspect ratio, and hence larger surface area. A larger surface area enables better adhesion with the matrix/surface.
- ➢ Improvement in the mechanical performance of the parent material.
- ➢ The high percentage of atoms or molecules on the surface leads to unique properties.
- ➢ Surface forces are very important.
- ➢ Metal nanoparticles have unique scattering properties.
- ➢ Semiconductor nanoparticles may exhibit confined energy states in the electronic bandstructure.
- ➢ Their chemical and physical properties are unique and change by size and shape.
- ➢ NMs properties can be 'tuned' by varying the size of the particle (e.g., changing the fluorescence color so a particle can be identified).
- ➢ NMs complexity offers a variety of functions to products.
- ➢ Adsorption and absorption of molecules (gas or liquid phases) are high and fast.
- ➢ Examples are nanosilica, nanotitania, nanoalumina, etc.

Materials lowered to the nanoscale exhibit a variety of increased properties when compared to their microscale counterparts (Macwan et al., 2011). This is due to some factors as:

(i) The surface effects can be interpreted as follows:
- ➢ having more surface atoms than inner atoms,
- ➢ having more free energy surface available (as a result of the increased surface area and surface atoms, the surface energy associated with the particles increases), and
- ➢ the concept that increasing a substance's surface area enhances the rate of a chemical reaction, in general.

(ii) The volume effects that are due to:
- ➢ low wavelength and high frequency and energy,
- ➢ the blue shift of atoms for optical absorption spectra,
- ➢ superparamagnetism occurs when the nanoparticle is smaller than the magnetic domain in a material, and

- ➢ the fact that average energy spacing increases as the number of atoms decreases in a free-electron model, which improves nanoparticle catalytic characteristics (Wigginton et al., 2007).

(iii) The surface chemistry such as surface functionalization and charges.
(iv) The agglomeration state, shape, and fractal dimensions.
(v) The chemical composition and crystal structure, and the solubility.

3. Why nanomaterials?

Nanoadsorbents are capable of delivering ultrahigh adsorption capacities, rapid adsorption kinetics, high adsorption efficiency, and selectivity for a wide range of micropollutants (Valenzuela et al., 2020). Nanoadsorbents could be candidates for selective and reversible coronavirus adsorption from various media, including liquid and gas. They are even potential candidates for selective and reversible adsorption. These remarkable adsorption qualities and performances for micropollutants adsorption have been documented mostly under controlled experimental circumstances. Despite the limits of laboratory-scale adsorption studies, a substantial amount of scientific knowledge about the synthesis, characterization, and testing of nanoadsorbents for their potential to sequester micropollutants has been accumulated in the literature. As a result, the scientific, technical, and technological development communities have a lot of room to fine-tune their efforts and tap into the gold mine of nanoadsorbents for micropollutant adsorption in purification and separation facilities (Mudhoo and Sillanpää, 2021).

Some key characteristics are sought in an ideal nanoadsorbent (e.g., mesoporous nanoparticles, hydrogel, polymeric nanoparticles, aerogel, or carbon nanotube—type materials) for adsorbing various targets micropollutants from contaminated waters and wastewaters under a variety of chemical, physical, biological, and microbiological conditions. These include, for example (Perera and Ayres, 2020; Shen et al., 2019; Xiong et al., 2020):

- ➢ Adsorption and separation capabilities are both high
- ➢ Chemical and thermal stability, as well as sufficient selectivity
- ➢ High adsorbent recovery rate, regeneration, and recyclability
- ➢ Porosity tunability that is adequate
- ➢ The potential for specific types of functionalization to change the surface chemistry
- ➢ High mechanical strength, structural integrity, and possibility for shape recovery
- ➢ Self-healing and self-cleaning characteristics
- ➢ The capability of being mass-produced via green synthetic ways
- ➢ The ability to be integrated into large-scale water/wastewater treatment systems
- ➢ Bulk manufacturing and regeneration at a low cost

4. Properties of nanomaterials as adsorbents

The characteristics of matter at the nanoscale differ significantly from their bulk counterparts as size-dependent effects become more significant. Au solution, for example, appears yellow in bulk but purple or crimson at the nanoscale level. Nanomaterial characteristics can be tweaked by changing the size of the nanomaterial (Jose et al., 2019). When compared to bulk materials, electrical characteristics vary dramatically at the nanoscale. Boron, for example, is not considered a metal in its bulk form, but borophene, 2D network of boron, seems to be good 2D metals.

The mechanical characteristics of nanoparticles have significantly improved over their bulk counterparts due to gains in crystal perfection and reduction in crystallographic imperfections (Tomar et al., 2020a).

The following important features can be acquired by modifying the sizes and morphologies of nanomaterials, among a variety of other unique properties.

4.1 Innate (inherent) surface properties

Nanomaterials' intrinsic compositions, apparent sizes, and extrinsic surface structures are all linked to their physical, material, and chemical properties. It is well acknowledged that, as particle size drops to the nanoscale, physical and chemical properties differ from those associated with their bulk form. Recognition that a major portion of the atoms in nanoparticles is at or near the surface of the particles is equally essential and widely acknowledged; it, however, appears to be less understood (Baer et al., 2010).

The nature of active sites on nanostructured surfaces, as well as their distribution, is a tough problem to tackle. The basic intrinsic elements that govern nanoparticles' function as adsorbents in solution or substrate are listed below.

- ➢ Location of the most atoms on the surface.
- ➢ High surface area, and chemical activity.
- ➢ High adsorptive capacity, and surface binding energy.
- ➢ Lack of internal diffusional resistance.

Such properties increase the potential impact of surface accessibility and affinity, surface enrichment, number of active sorption sites, and, consequently, surface energy toward specific analytes as well by causing a significant fraction of atoms or molecules to be associated with surfaces and interfaces. The high surface area present in nanoparticles retains a solvent in unexpected circumstances due to capillary and sorption effects. It can be challenging to characterize the characteristics of actual nanoparticle surfaces even with surface tools as they have proven to be outstanding adsorbents in several applications including sorption, separation, and preconcentration (Valcarcel and Simonet, 2011; Saleh, 2022).

4.2 External functionalization

A variety of modifications in the surface characteristics of nanomaterials are observed when different functional groups are used. Excellent adsorption capabilities can be achieved by combining a wide range of nanomaterials with various external functionalization processes. The surface is further functionalized to prevent nanoparticles (NPs) from aggregating and to increase their selectivity. Coatings on nanoparticles may have a major impact on a range of properties. Adsorbents with functionalized groups have key properties including high absorption capacity (typically quantified as the breakthrough volume for a flowing system) and quick desorption (Jiménez-Soto et al., 2010; Saleh, 2021, 2018, 2015).

The search for functionalized groups is a critical component in improving analytical parameters, including selectivity, adsorption capacity, and affinity. It is accomplished by introducing numerous organic donor atoms into the surface of the nanomaterial, thereby increasing the interactions with the analytes of interest, like polarity or hydrophilicity. Electrostatic interactions have been found to allow amino and oxygen groups to align with transition metals. One such example is functionalizing graphene with oxygen-containing groups (Fig. 8.4). This enhances the graphene performance for sorption, for instance. Similarly, carbon nanotubes (CNTs) modified with amines show better sorption of mercury than CNTs without functionalization.

4.2.1 Oxidation of materials

When a material is oxidized, carbonyl, carboxyl, and hydroxyl groups are introduced at defect locations. Such groups preserve adsorbates when the pH is above the isoelectric point of the material. Methods of oxidation that are commonly used include reflux with oxidizing agents like nitric acid, hydrogen peroxide, or permanganate (Datsyuk et al., 2008; Peng and Liu, 2006; Zhang et al., 2002). By eliminating contaminants and generating surface flaws, such oxidation procedures can enhance surface area and allow the insertion of oxygen-containing functional groups. The particles' ends are more reactive than the surface, and, when oxidized, they tend to open up, thus resulting in defects such as pentagons and carboxyl as well as hydroxyl groups, which are responsible for the increased adsorption capabilities (Britto et al., 1999; Saleh, 2011, 2017, 2020a, b).

4.2.2 Selective functionalization

The selective functionalization of specific locations inside nanomaterials is of interest to control the surface properties. A good example of this is the

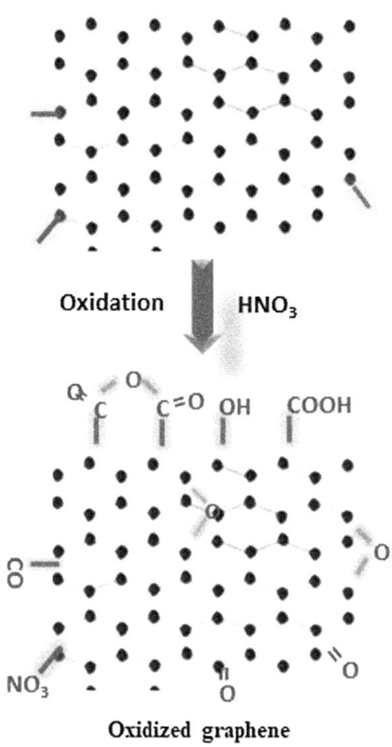
Oxidation
HNO$_3$
C=O
OH
COOH
O
CO
NO$_3$
Oxidized graphene

FIGURE 8.4 Functionalization of graphene with oxygen-containing groups.

functionalization of the inner surface in micropores and mesopores of silica. The surface could be functionalized with trimethylchlorosilane after partial cleavage of the surfactant, but the micropores remained inaccessible until a further heating treatment destroyed the residual surfactant. After that, the micropore surface is functionalized with trivinylchlorosilane and reacted with a palladium complex to produce a final product with accessible mesopores and nanoparticles within the micropores.

Another example is the formation of a passivation layer which can be done in neat hexamethyldisilazane (HMDS) followed by Soxhlet extraction with ethanol and heat treatment of about 250°C for 6 h. Increased passivation times from 8 to 24 h leads to materials displaying textural properties consistent with significant pore sialylation and use of alternate extraction conditions, including an acetone Soxhlet, resulting in materials with larger amounts of residual surfactants. The material is then treated with the required functionality in order to enhance its performance for the targeted application (Webb, 2015). Fig. 8.5 presents general steps of modification with pore protection passivation selective grafting protocol.

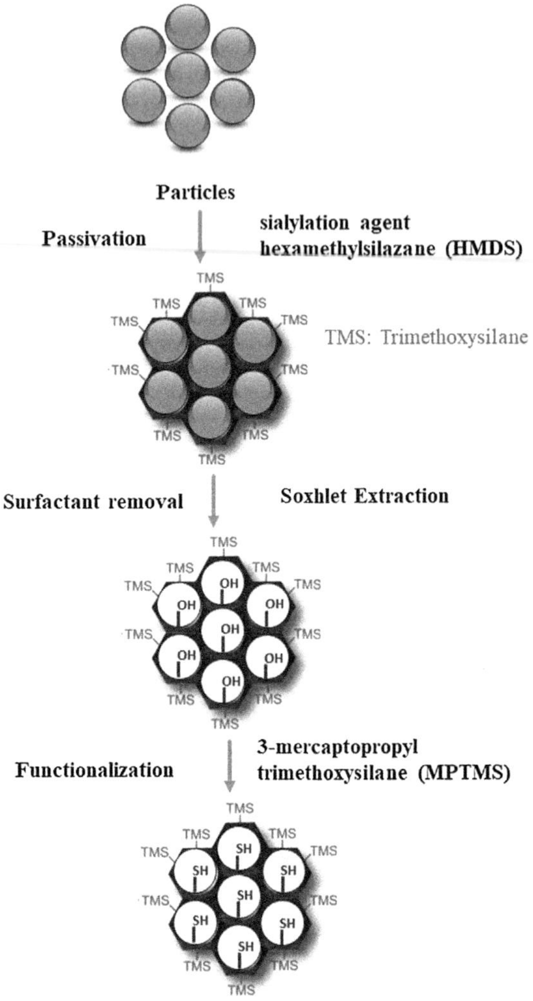
Particles
Passivation
sialylation agent
hexamethylsilazane (HMDS)
TMS
TMS: Trimethoxysilane
Surfactant removal
Soxhlet Extraction
OH
Functionalization
3-mercaptopropyl
trimethoxysilane (MPTMS)
SH

FIGURE 8.5 Modification with pore protection passivation selective grafting protocol.

4.2.3 Covalent conjugation

Covalent functionalization allows diverse chemical species to be attached to nanomaterials by forming covalent connections.

4.2.4 Noncovalent binding

Supramolecular or noncovalent interactions and functionalization require both attractive and repulsive forces between molecules. Chemisorption and/or physisorption can be used to adsorb guest molecules onto nanomaterials.

Many synthetic and natural systems used for recognition, interaction, and detection contain noncovalent interactions.

4.2.5 Intrinsic surface engineering

Intrinsic surface engineering refers to changes to the crystal structure of nanomaterials at the atomic level. Heteroatom inclusion or flaw engineering can be used to accomplish this. The electrical, catalytic, and structural properties of nanomaterials are altered by the incorporation or doping of electron-withdrawing and electron-donating elements (N, S, O, P, B, and other metals) (Wang, 2014).

4.2.6 Nanoparticle coating

Nanoparticles coating can be amorphous or crystalline coating. Amorphous, unlike crystalline, denotes that it has no consistent form. An amorphous solid, unlike a crystalline solid, does not have structural patterns and has a relatively unpredictable structural makeup. It can be performed by several methods like dip coating and electrochemical coating.

Groups of nanocomposite coating include

- ➢ Organic/inorganic nanocomposite coatings
- ➢ Organic/organic nanocomposite coatings
- ➢ Inorganic/organic nanocomposite coatings
- ➢ Inorganic/inorganic nanocomposite coatings

4.2.7 Other functionalization approaches

Various synthesis techniques for the functionalization of nanomaterials include:

- ➢ Physisorption (milling and/or mixing) and chemisorption (e.g., thiol groups linked to the surface of nanoparticles) are also used.
- ➢ Negatively charged nanoparticles attract positively charged small molecules and vice versa due to electrostatic interactions.
- ➢ Covalent interactions: utilizing conjugation chemistry to make use of functional groups on both nanoparticles and molecules.
- ➢ Supramolecular affinity is based on weak interactions between receptors and ligands.

4.3 Zeta potential

Zeta potential is typically established as a basic parameter to determine the surface charge of colloids and nanomaterials distributed in a liquid, which means electrophoretic mobility, velocity, when subjected to electric fields. The isoelectric point (IEP) is the point at which the zeta potential equals zero. IEP is used to quantify the surface charge of materials.

Functionalization of nanomaterials has a significant impact on adsorbent retention efficacy and causes nanoparticles' isoelectric point to fluctuate in relation to the pH of the solutions. For example, the negative charge on the surface provides electrostatic attractions that facilitate adsorbing cations easier when the pH of the solution is greater than the isoelectric point of nano-adsorbents. Whether the pH is above or below the isoelectric point, the regime can be altered. Surface charge, σ, computed as a function of pH, might be determined using potentiometric titrations of nanoparticles and bulk particles.

4.4 Point of zero charge

The adjustment of the surface charge is frequently preceded by changes in the media's pH and ionic strength. The point of zero charge (PZC) is defined as the point the pH at which positive and negative charges are balanced and no net charges are available on materials surfaces, allowing a colloid or nanoparticle to move in an electric field. In reality, the PZC dictates whether materials are hydrophilic in an aqueous suspension in addition to its mobility. Because surface charge influences nanomaterial stabilization and aggregation, nano-material dispersion near PZC nanomaterials is the least stable and has the largest inclination to aggregate owing to the lack of surface charge repulsion.

4.5 Surface area

Nanomaterials' surface area is often much larger than those of their bulk counterparts. Nanoparticles, too, display unique properties owing to their high surface area to volume ratio (Tomar et al., 2020b). A spherical particle has a diameter (D) of 100 nm. This provides an estimated surface area to volume ratio of $>107{:}1$, which is considerably larger than a macroparticle. The greater the surface area, the bigger its influence on reactions on adsorbent surfaces.

4.6 Mechanical properties

Mechanical attributes refer to a material's mechanical characteristics in various conditions and under diverse external loads. Mechanical qualities vary depending on the substance. Metals' mechanical properties, like those of classical materials, are made up of 10 components: plasticity, hardness, brittleness, strength, toughness, yield stress, rigidity, fatigue strength, elasticity, and ductility. The majority of nonmetallic materials are commonly brittle, lacking characteristics such as toughness, ductility, elasticity, and plasticity. Furthermore, some organic materials are flexible but lacking in characteristics such as brittleness and rigidity.

Because of their volume, surface, and quantum influence, nanomaterials offer exceptional mechanical properties superior to those of their macroscopic counterparts. When nanomaterials are introduced into a material, they refine the grain to some extent, generating an intragranular or intergranular structure, which improves the grain boundary and improves the mechanical characteristics of the material. Concrete's compressive strength, bending strength, and splitting tensile strength can all be improved by adding 3 wt/% nano-SiO_2 (Ghabban et al., 2018). The addition of 3% nano oil palm empty fruit string filler to kenaf epoxy composites improves tensile strength, elongation at break, and impact strength significantly (Saba et al., 2016).

4.7 High thermal stability

Thermal stability can be described as the ability of a material to resist breaking down under heat stress. The maximum usage temperature is the highest temperature to which a material may be heated without breaking down or degrading. When compared to bulk analogues, remarkable thermal and electrical stability are demonstrated at the nanosize due to the nature of the nanomaterial. Graphene, which is made from graphite, has very high thermal stability (Krishnan et al., 2019).

4.8 Support surface

The use of nanomaterials such as 2D sheets made of different components allows for effective dispersion of active nanoparticles, significantly improving its performance (Zhu et al., 2020).

4.9 Antimicrobial activity

Several nanomaterials contain antiviral, antibacterial, and antifungal characteristics, making them ideal for treating pathogen-related illnesses. They can also be used to adsorb some bacteria or work as antibacterial agents (Kotb et al., 2020).

5. Factors affecting properties and performance of nanomaterials

There are several factors affecting the properties and performance of nanomaterials. Examples are:

- ➢ Nanoparticle selection: This term refers to the improvement phase in the nanomaterials.
- ➢ Production process: The processing parameters primarily represent the effect of the manufacturing process on the mechanical characteristics of a nanomaterial. The mechanical characteristics of nanomaterials are

influenced, to some extent, by temperature, technique, treatment duration, nanoparticle dosage, and ratio.

- Grain size: Nanograins and grain boundaries make up the majority of nanomaterials. As a result, grain size is one of the most important elements influencing nanomaterial mechanical properties.
- Grain boundary structure: The grain boundaries of a nanomaterial have a higher volume fraction than microscale and standard materials due to the size effect.

6. Types of nanoadsorbents according to their properties

The qualities and behavior of many types of nanomaterials as sorbents are discussed here. A wide range of nanomaterials is being utilized for adsorptive, separation, and analytical processes. Adsorbents should be classified based on their inherent surface qualities and further functionalization.

6.1 Metal nanoparticles

Metal nanoparticles are nanomaterials that are made up of only one element. Individual atoms or groups of numerous atoms can exist. Au, Ag, Pt, Cu, Pd, Re, Zn, Ru, Co, Cd, Al, Ni, and Fe are some of the most commonly produced nanoparticles. Simple approaches, such as bio-assisted method, hydrothermal method, and microwave-assisted method, are used to make metal nanoparticles in the form of colloidal fluids or solid nanoparticles. They have remarkable properties including localized surface plasmon resonance (LSPR), high reactivity, and broad electromagnetic spectrum absorption. Metal nanoparticles are very fascinating materials for a variety of practical applications due to their improved optical, optoelectrical, catalytic, antimicrobial/cancer/viral characteristics (Chakraborty and Pradeep, 2017; Zaleska-Medynska et al., 2016).

6.2 Metal oxide

Metal oxides are thought to be one of the stable substances found in nature. They are created when electronegative oxygen reacts with an electropositive metal. Due to the presence of anionic oxygen, they have polar surfaces and are insoluble in most organic solvents owing to the strong interaction between the metal and oxygen. In the oxidative nature of the Earth, metal oxides are the lowest free energy states for metals. Because of their high natural abundance, chemical stability, tunable bandgap/band edge positions, and outstanding thermal/electrical conductivity, they are currently commonly used nanomaterials. Semiconductors, superconductors, and even insulators are among their applications. Various varieties of adaptable metal oxides, such as alumina, titania, iron oxides, silica, and zinc oxides, can be prepared in response to increased industrial demand for use in environmental remediation, water treatment, cosmetics, biomedical, and energy.

To fulfill the severe needs of good characteristics and efficiency, the metal oxides are altered by doping, which results in the formation of heterostructures. Metal hydroxides, also known as (layered) metal hydroxides, are a type of inorganic nanomaterial having a variety of properties that can be tailored by changing the composition and structure. These materials can be found in hydroxides that have neutral layers but no intercalated molecules, or in hydroxides that have a cationic layer but intercalated molecules.

6.3 Nanoparticles coatings

The creation of particles of a specific size and shape is an important aspect of their synthesis (scatter of sizes is small and can be controlled, at least). The ability to control the geometry of anisotropic nanostructures and synthesize them is particularly significant. Nanoparticles are frequently isolated from others using a covering around the particle to eliminate (or greatly reduce) interparticle interaction.

6.4 Metal chalcogenides

Metal chalcogenides (O_2, S_2, Se_2, and Te_2) are a large class of inorganic materials that comprise at least one metal and one chalcogen anion. Owing to the unusual behavior of oxygen (gaseous and outstanding nonmetallic characteristics) and polonium, they are mainly constrained to telluride, selenide, and sulfide nanostructures instead of oxide and polonium structures (strong metallic properties) (Xu et al., 2017; Butler et al., 2013; Lipatov et al., 2015).

6.5 Magnetic nanoparticles

Magnetic characteristics are one area where the difference between a huge (bulk) material and nanostructures is very noticeable. The type and chemical composition, as well as the degree of defectiveness of the crystal lattice, are important factors in determining the magnetic properties of nanomaterials. Other factors include particle size, shape, morphology, particle interactions with the surrounding matrix, and other particles. These properties can be controlled, to some extent, by modifying their structure, composition, size, and shape.

The coprecipitation method is a simple and effective way to make magnetic iron oxide nanoparticles. The iron salt, ferrous/ferric ratio, reaction pH, temperature, and ionic strength all impact the composition, size, and shape of nanoparticles.

6.6 Nanoporous materials

Porous materials are defined as any solid that has void space(s), that is, space that is not occupied by the primary framework of atoms that make up the

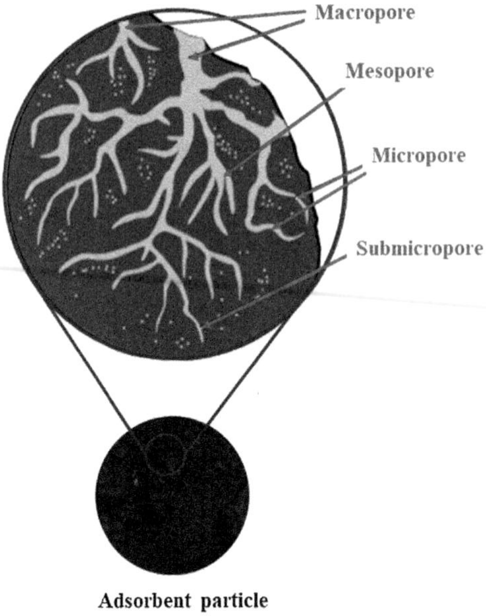
Macropore
Mesopore
Micropore
Submicropore
Adsorbent particle

FIGURE 8.6 Types of pores in porous materials.

solid's structure. They are a type of material that has a low density, a large specific surface area, and a variety of innovative physical, mechanical, and thermal properties, among others. Activated charcoal is an example of a porous material that has been used for many years. Porous solids' technical and scientific significance stems from their capacity to interact with molecules, atoms, and ions, at outer surfaces and also allow interior access (Mishra et al., 2018; Pal, 2020). When the pore diameter of porous materials is less than 100 nm, they are called nanoporous materials. Porous materials are commonly categorized into mesoporous, macroporous, and microporous (Fig. 8.6).

6.6.1 *Macroporous materials*

A porous material is defined as a macroporous material when it has a pore size >50 nm.

6.6.2 *Mesoporous materials*

A porous material is defined as a mesoporous material when it has pores in 2–50 nm size.

6.6.3 *Microporous materials*

Microporous material is described as a microporous material when it has pore sizes in the range of 2 nm or less. Examples are mesoporous carbon nanotube and silica nanoparticle. Mesoporous silica comprises honeycomb-like porous structures with pore size and outer particle diameter in the nanometer range. This type of material has hundreds of empty channels which are capable of encapsulating and absorbing large quantities of adsorbate molecules.

6.6.4 Nanoporous materials

A regular organic or inorganic framework supports a regular, porous structure in nanoporous materials. Pores are typically 100 nm or smaller in size. The majority of nanoporous materials fall into one of two categories: bulk materials or membranes. Bulk nanoporous materials, such as activated carbon and zeolites, can be thought of as nanoporous membranes, while cell membranes can be thought of as nanoporous membranes. A porous medium, often known as a porous material, is one that contains pores (voids). The "matrix" or "frame" refers to the skeletal portion of the material. Fluid is usually injected into the pores (liquid or gas). Natural nanoporous materials abound, but man-made nanoporous materials can also be created. Combining polymers with different melting points, for example, allows one polymer to deteriorate when heated. Only certain substances can pass through a nanoporous material with regularly sized pores, while others are blocked.

Porous silica, zeolites, clays, and porous metal oxides are examples of inorganic nanoporous materials. The creation of pores in a material can give it unique characteristics that aren't present in nonporous materials. Nanoporous materials have a diverse range of surface compositions and properties. The surface-to-volume ratios of nanoporous materials are quite high. These materials are useful in the sectors of adsorption, catalysis, purification, energy conversion, and sensing because of their exceptional properties and nanoporous framework architectures (Zhang and Jaroniec, 2020; Chen et al., 2020a,b; Szcześniak et al., 2020).

6.7 Quantum dots

Quantum dots (QDs) are single nanoparticles (nanocrystals) that are approximately 2–10 nm (nm) in diameter. Commonly, QDs are small particles or nanocrystals of semiconductors with a diameter ranging from 2 to 10 nm. QDs produce a distinctive fluorescence utilized for subcellular labeling and imaging. They are used in photonic, fluorescent, and electrochemical applications due to the tunability to generate different light outputs. They are also extremely energy efficient.

The following methods can be employed for the preparation of semiconductor quantum dots (O'Brien and Pickett, 2005):

- Precipitation methods: These methods are commonly used with aqueous media and at low temperatures.

- ➢ Reactive methods: These methods are commonly used with high boiling point solvents and usually involved metal-organic and organometallic compounds.
- ➢ Hydrothermal and solvothermal methods: These methods allow easy control of the synthesis parameters by which the size and shape of QDs can be controlled.
- ➢ Vapor-phase reactions: Such reactions allow easy control of the size and shape of QDs.
- ➢ Reactions in confined solids or constrained on surfaces or involving micelles or confined reaction spaces.
- ➢ Microwave irradiation methods: These methods are fast and easy.
- ➢ Ultrasound irradiation: This method is easy and requires less effort in synthesis.

6.8 Silicene

Silicene is a two-dimensional allotrope of silicon with a hexagonal honeycomb structure roughly similar to that of graphene nanostructures. It consists of one atomic layer thick with a special low-buckled structure. It is of interest theoretically because of its superior properties, which it shares with graphene. Because silicene is a conductor in its basic form due to the zero bandgap development, numerous approaches for producing the bandgap in silicene have been developed. In addition, silicene has been utilized to create a variety of electrical devices, including transistors and photodetectors (Sone et al., 2014; Guzmán-Verri and Lew Yan Voon, 2007).

6.9 MXenes

MXenes are a two-dimensional class of inorganic chemicals. These materials are made up of thin layers of transition metal carbides, nitrides, or carbonitrides that are only a few atoms thick (Naguib et al., 2014; Pang et al., 2019). The inclusion of initial transition metal carbides, metal nitrides, and metal carbonitrides from MAX-phases, defined as MXenes, has greatly increased the range of 2D materials. MXenes have a lateral dimension similar to graphene, ranging from nanoscale to micrometer, with a theoretic thickness of about 0.98 nm (single layer) and excellent chemical and physical properties.

6.10 2D pnictogens (phosphorene, arsenene, antimonene, and bismuthene)

2D pnictogens are known as phosphorene, arsenene, antimonene, and bismuthene. They are one-atom-thick and contain monolayers with a wide range of bandgaps, such as phosphorene from black phosphorus, and share similar fascinating features with graphene, such as vibrational and electrical properties.

6.11 Metal-organic framework

Metal-organic frameworks (MOFs) are organic–inorganic hybrid crystalline porous materials made up of a regular array of positively charged metal ions or clusters surrounded by (coupled with) organic 'linker' molecules to form one-, two-, or three-dimensional structures. The metal ions form nodes, which connect the arms of the linkers to form a cagelike structure. MOFs have a very large internal surface area due to their hollow structure (Batten et al., 2013). They are a type of coordination polymer that has the unique property of being porous. The organic ligands used are sometimes referred to as "struts" or "linkers," with 1,4-benzenedicarboxylic acid (BDC) being one example.

6.12 Core–shell nanoparticles

Nanoparticles made up of two or more materials are known as core–shell and composite nanoparticles, i.e., those that have an interior material that forms a core and an exterior material that forms a shell around the core structure. The core–shell particles are made in a variety of configurations, including inorganic/inorganic, organic/organic, organic/inorganic, and inorganic/organic (Schärtl, 2010; León Félix et al., 2017).

6.13 Carbon-based materials

Carbon-based class includes nanomaterials that have been intensively investigated for a variety of applications, including sorption, due to their unique properties (Fig. 8.7). The amazing capabilities of tunable carbon-based nanomaterials have sparked a lot of attention for their potential applications in new technologies and solving modern problems. CNTs, carbon-based quantum dots, fullerenes, carbon nanohorns, graphene, and a variety of other nanomaterials belong to the carbon family (Chen et al., 2020a,b; Fan et al., 2020). There are several types of each class (Table 8.1).

For example, the graphene family can be prepared either by top-down or by bottom-up approaches. The bottom-up method can produce defect-free graphene single sheets with good physical properties. Examples of this method are techniques such as epitaxial growth and CVD, chemical methods, electrochemical methods, or hydrothermal methods. Since the top-down methodology usually refers to the mechanical exfoliation of graphite to obtain few or single-sheet graphene. Graphene can also be obtained in the oxidation of graphite by strong oxidizing agents followed by an exfoliation to give the graphene oxide (GO). This can be reductively processed by thermal, chemical, microwave, photothermal, photochemical, or microbial/bacterial to give the

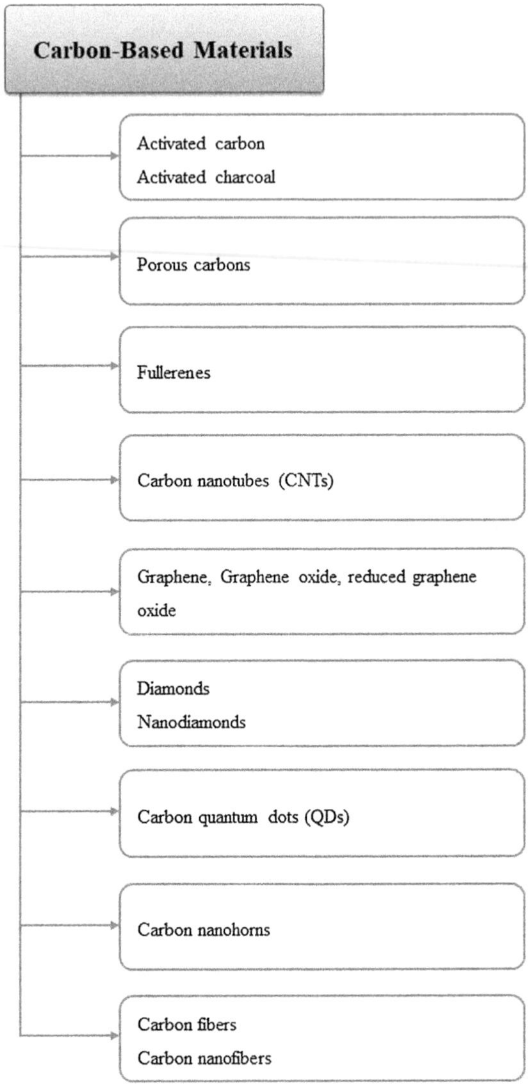
Carbon-Based Materials
Activated carbon
Activated charcoal
Porous carbons
Fullerenes
Carbon nanotubes (CNTs)
Graphene, Graphene oxide, reduced graphene oxide
Diamonds
Nanodiamonds
Carbon quantum dots (QDs)
Carbon nanohorns
Carbon fibers
Carbon nanofibers

FIGURE 8.7 Types of carbon-based materials.

reduced graphene oxide (rGO) (Bianco et al., 2013). Numerous top-down methods are described in the literature for the synthesis of graphene, ranging from liquid-phase exfoliation (Hernandez et al., 2008) to electrochemical exfoliation (Sun et al., 2011).

Depending on the preparation method, graphene derivatives can be formed (Fig. 8.8). The major types of "graphenes" include:

TABLE 8.1 Examples of some carbon-based nanostructures promising for large-scale production.

Name	Definition
Fullerenes	Fullerenes are an allotrope of carbon and symmetrical cages of sp^2-hybridized carbon atoms. They are of various sizes as per the number of carbon atoms, for instance, C100, C84, C76, C72, C70, and C60. The most abundant fullerene is C60 fullerene, consisting of a hollow structure with 12 pentagons and 20 hexagons formed of 60 carbon atoms which are linked to each other via covalent chemical bonding of sp^2 hybridized, with icosahedral symmetry.
Carbon nanotubes (CNTs)	CNT is rolled nanosheets of single-layered sp^2-hybridized carbon atoms. Allotropes of carbon have a cylindrical nanostructure with diameters ranging from <1 to 50 nm. The surfaces of the nanotubes comprise sp^2-hybridized carbon atoms which are arranged in hexagons. They are categorized as either single-walled nanotubes (SWNTs), double-walled nanotubes, or multiwalled nanotubes (MWNTs). The nanotubes retain a high aspect ratio. CNTs are used to improve the mechanical characteristics of composite materials because of their high tensile strength (Iijima, 1991).
Covalent functionalized CNTs	In covalent functionalization, the functional groups are covalently bonded to the sidewalls of CNTs. This can be performed via; **(i)** Activation of CNTs by creating reactive species like carboxylic, hydroxyl, and amine groups; **(Ii)** The direct covalent linking of the anticipated functionality via cycloaddition, radical addition, and electrophilic and nucleophilic addition reactions. The covalent functionalization of CNTs is divided into two subcategories: Sidewall functionalization and ends/defects functionalization.
Noncovalent functionalized CNTs	Noncovalent modification of CNTs is a simple way to modify the CNTs surface to improve their dispersibility and efficiency for desired uses. This modification is attained via $\pi-\pi$ interactions, CH$-\pi$ interactions, van der Waals attraction, and electrostatic interactions. The noncovalent

Continued

TABLE 8.1 Examples of some carbon-based nanostructures promising for large-scale production.—cont'd

Name	Definition
	approach is helpful since there is no damage to CNTs sp^2 carbon structure thus maintaining their characteristics. Noncovalent functionalization is accomplished using a variety of materials, including polymers, aromatic compounds, and head–tail surfactants.
Graphene	Graphene is an allotrope of carbon consisting of a single layer of atoms arranged in a two-dimensional honeycomb lattice nanostructure, with sp^2-hybridized carbon atoms planar sheets that are tightly packed to honeycomb-like lattices. Graphene nanosheets are thin nanomaterials, with exceptional mechanical strength.
Graphene	A single layer (or layers) of 2D carbon atoms. Consists of carbon atoms.
Graphene oxide (GO)	Heavily oxidized graphene, made of C, O, and H. GO is a monolayer material, chemically modified graphene synthetized by oxidation of the graphite and exfoliation that is accompanied by wide oxidative modification of the basal plane. The GO has a layered structure with hydrophilic polar groups at the sheet's borders, such as carbonyl (–C=O) and carboxyl (–COOH), hydroxyl (–OH) and epoxy groups (C–O–C), which are preferentially positioned in the basal plane.
Reduced graphene oxide (rGO)	A reduced form of graphene oxide; consists of C and H. Depending upon the synthesis process, it may contain less oxygen.
Nanodiamonds	Nanodiamonds are made up of sp^3-hybridized carbon with remarkable optical and mechanical capabilities, as well as large specific surface areas and complex surface structures. Carbide chlorination, chemical vapor deposition, high-energy ball milling, ion irradiation of graphite, and laser ablation are all processes for making nanodiamonds. Nanodiamonds have a core–shell structure, have a lot of surface chemistry, and have a lot of functional groups on their surface. A variety of functional groups are found on nanodiamond

TABLE 8.1 Examples of some carbon-based nanostructures promising for large-scale production.—cont'd

Name	Definition
	surfaces, including amide, alkene, aldehyde, carboxylic acid, ketone, hydroperoxide, carbonate ester, nitroso, and alcohol groups, which aid in further functionalization for desired applications.
Carbon quantum dots (QDs)	QDs are zero-dimensional discrete nanoparticles with less than 10 nm. They are quasi-spherical nanoparticles of carbon. They have numerous distinctive properties which make them extraordinary nanostructures for some uses such as tunable photoluminescence properties and good multiphoton excitation. QDs are of distinctive features relating to quantum confinement effects. They are soluble in water due to the presence of oxygen-containing functions.
Carbon nanohorns	Carbon nanohorns or carbon nanocones are conical carbon nanostructures consisting of sp^2 carbon sheets. For instance, single-walled carbon nanohorns comprise a tubular structure with a graphene sheet that has a conical end. They are made up of tubular structures with a conical end and graphene sheets. They are considered a subset of fullerenes because of their closed cage structure, and their elongated shape gives them a structural counterpart to short single-walled CNTs.

- single-layer graphene (single sheet),
- bilayer graphene (double layers),
- few-layer graphene, multiple sheets (layers),
- graphene oxide (GO), and
- reduced graphene oxide (rGO).

The last GO and rGO and their derivatives are the most used for environmental purposes.

7. Adsorption system

Adsorption is defined as a separation process in which fluid, liquid, or gas molecules bond to the outer and inner surfaces of a solid object termed the adsorbent. The separation is based on the selective adsorption (thermodynamic

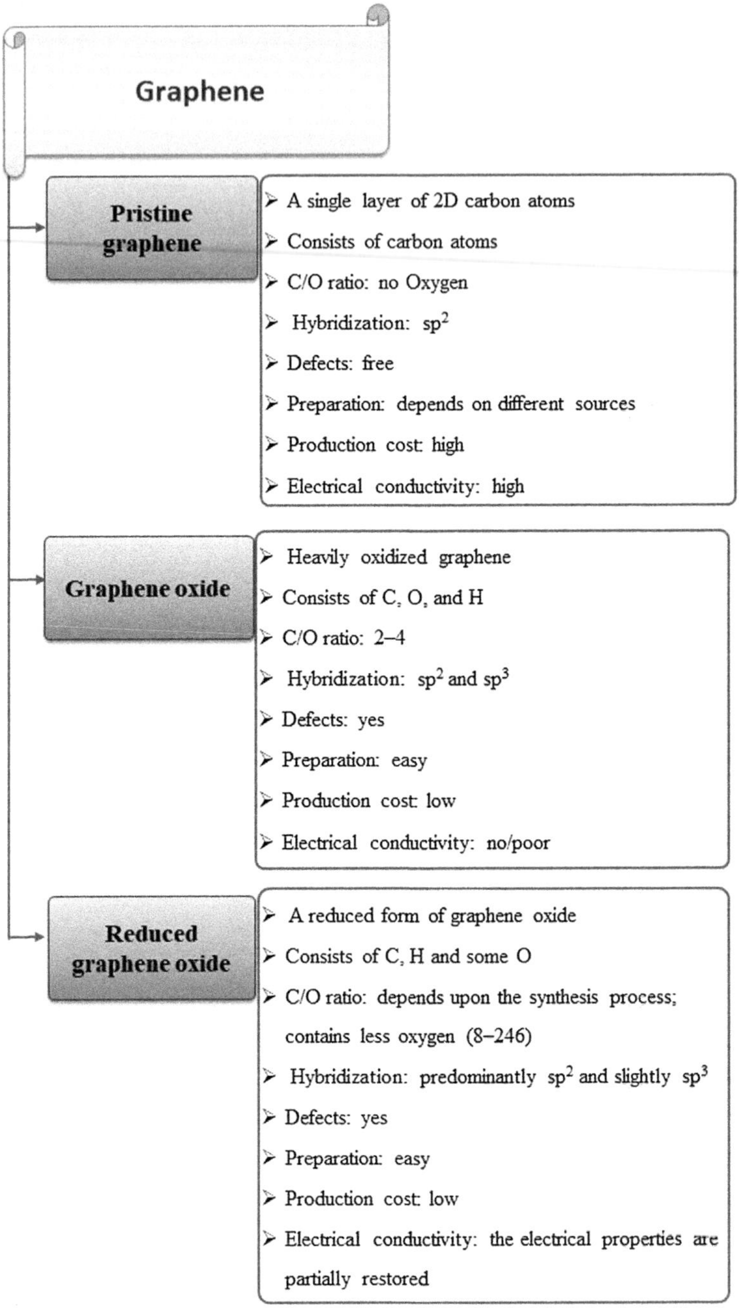
Graphene
Pristine graphene
A single layer of 2D carbon atoms
Consists of carbon atoms
C/O ratio: no Oxygen
Hybridization: sp2
Defects: free
Preparation: depends on different sources
Production cost: high
Electrical conductivity: high
Graphene oxide
Heavily oxidized graphene
Consists of C, O, and H
C/O ratio: 2–4
Hybridization: sp2 and sp3
Defects: yes
Preparation: easy
Production cost: low
Electrical conductivity: no/poor
Reduced graphene oxide
A reduced form of graphene oxide
Consists of C, H and some O
C/O ratio: depends upon the synthesis process; contains less oxygen (8–246)
Hybridization: predominantly sp2 and slightly sp3
Defects: yes
Preparation: easy
Production cost: low
Electrical conductivity: the electrical properties are partially restored

FIGURE 8.8 Types of graphene and their properties.

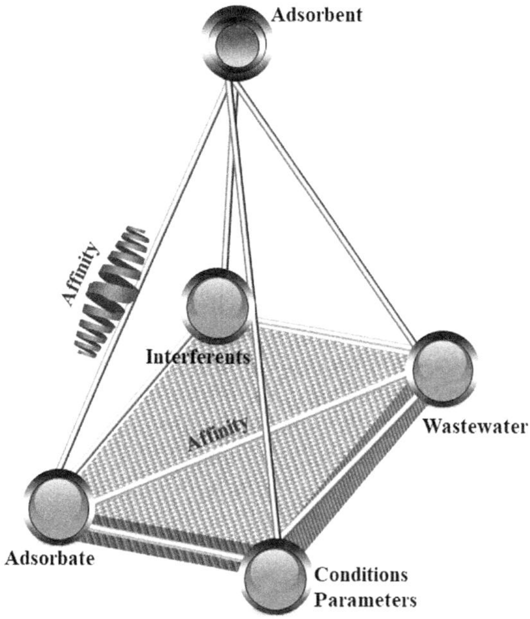
Adsorbent
Affinity
Interferents
Affinity
Wastewater
Adsorbate
Conditions
Parameters

FIGURE 8.9 Illustration of the relationships between the adsorption components.

as well as kinetic selectivity) of adsorbate by an adsorbent as a result of mass transfer and specific interactions between the adsorbent's surface and the contaminants (Dubinin, 1966). The intricate interactions among the components involved, namely the adsorbent, the adsorbate, and the wastewater, are manifested in this surface phenomena (e.g., effluent, synthetic solution, or water). The system is affected by conditions, parameters and interferents.

The system components and their interactions are depicted in Fig. 8.9 in the form of a schematic adsorption model. In this ternary system, the fundamental interaction force influencing adsorption is the affinity between the adsorbent and the adsorbate (Crini, 2005; Crini and Badot, 2010). Adsorption can also be influenced by the affinities between (i) the adsorbate and the solution, (ii) the adsorbent and the solution, and (iii) the pollutant molecules. Hydrophobic chemicals have limited solubility in an aqueous solution and are forced to the adsorbent surface. It is realistic to assume that the contact forces between the adsorption components will influence adsorption capacity.

8. Adsorption mechanisms

The main challenge in adsorption is to choose the best promising form based on low cost, high capacity (typically indicated by the capacity value (q_{max}), high adsorption rate, high selectivity, and fast kinetics. The following step is to recognize the adsorption mechanisms, specifically the interactions that occur at the adsorbent/adsorbate contact (Veglio' and Beolchini, 1997; Crini, 2005). The adsorption mechanisms involved in absorption might orientate the design of the desorption strategy. For instance, the recovery of specific contaminants such as valuable metal ions is also an essential parameter for the process' economics.

Depending on the material composition, the contaminant structure and characteristics, and the solution conditions, more than one of these interactions may take place simultaneously in an oriented-adsorption process utilizing a particular adsorbent (pH, ionic strength, temperature). Fig. 8.10 shows the classification of adsorption mechanisms (Crini, 2010). Physisorption, chemisorption, ion-exchange, and precipitation are the four basic mechanisms postulated. Some writers view ion-exchange process as a chemisorption mechanism.

The term ion-exchange does not specifically specify the binding process; rather, it is used to characterize the experimental findings. Microprecipitation refers to precipitation that occurs locally at the surface of a biosorbent as a result of specific conditions.

Generally, adsorption is a physicochemical process that involves a number of different mechanisms. Some precise binding mechanisms are:

- physical adsorption (physisorption), (such as hydrogen bond, electrostatic interactions, and van der Waals interactions),
- chemical binding (including ionic and covalent),
- surface adsorption,
- hydrophobic interactions ($\pi-\pi$ interactions, Yoshida's interactions),
- proton displacement,
- precipitation: this can be surface precipitation and microprecipitation,
- mineral nucleation,
- surface adsorption,
- electrostatic interactions (attraction interactions),
- acid–base interactions,
- chelation,
- binding or surface complexation,
- complexation (coordination),
- inclusion complex formation,
- diffusion into the network of the material,
- absorption, and
- ion-exchange.

9. Key features in nanoadsorbents

Ideal nanoadsorbents, such as mesoporous nanoparticles, hydrogel, polymeric nanoparticles, aerogel, or carbon nanotube–type materials, should have a

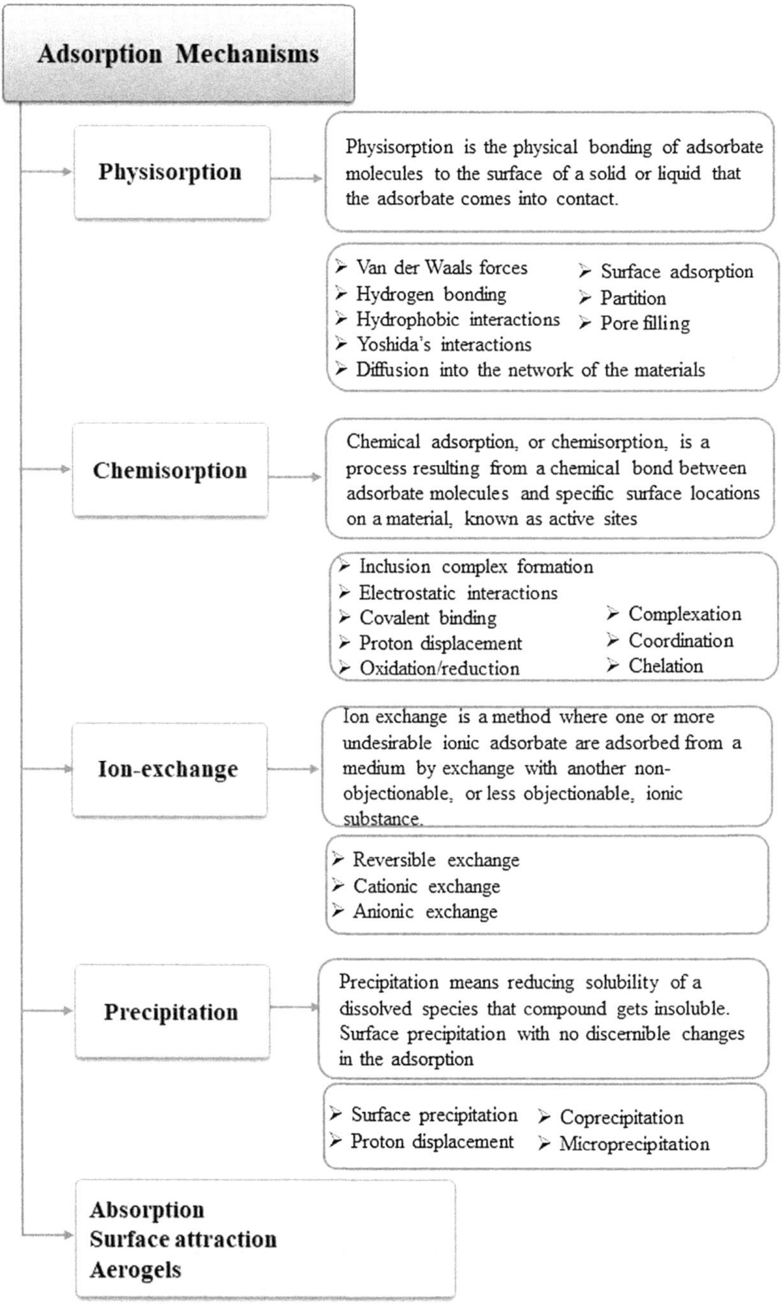
Adsorption Mechanisms
Physisorption
Physisorption is the physical bonding of adsorbate molecules to the surface of a solid or liquid that the adsorbate comes into contact.
➢ Van der Waals forces
➢ Hydrogen bonding
➢ Hydrophobic interactions
➢ Yoshida's interactions
➢ Diffusion into the network of the materials
➢ Surface adsorption
➢ Partition
➢ Pore filling
Chemisorption
Chemical adsorption, or chemisorption, is a process resulting from a chemical bond between adsorbate molecules and specific surface locations on a material, known as active sites
➢ Inclusion complex formation
➢ Electrostatic interactions
➢ Covalent binding
➢ Proton displacement
➢ Oxidation/reduction
➢ Complexation
➢ Coordination
➢ Chelation
Ion-exchange
Ion exchange is a method where one or more undesirable ionic adsorbate are adsorbed from a medium by exchange with another non-objectionable, or less objectionable, ionic substance.
➢ Reversible exchange
➢ Cationic exchange
➢ Anionic exchange
Precipitation
Precipitation means reducing solubility of a dissolved species that compound gets insoluble. Surface precipitation with no discernible changes in the adsorption
➢ Surface precipitation
➢ Proton displacement
➢ Coprecipitation
➢ Microprecipitation
Absorption
Surface attraction
Aerogels

FIGURE 8.10 Classification of mechanisms of adsorption.

number of key features to intend for scavenging different target pollutants from contaminated waters and wastewaters under variable chemical, physical, biological, and microbiological conditions (Perera and Ayres, 2020; Shen et al., 2019; Xiong et al., 2020; Mudhoo and Sillanpää, 2021). These features include:

- Chemical and thermal stability
- High adsorption and removal capacities
- High recovery rate of spent adsorbents, regeneration, and recyclability
- Adequate selectivity
- Adequate tunability of porosity
- Specific types of functionalization have the potential to change the surface chemistry
- High mechanical strength, structural integrity, and shape recovery potential
- Self-healing and self-cleaning properties
- Amenability for being produced in bulk through green synthetic routes
- Ability for being integrated in large-scale water/wastewater treatment processes
- Low-cost bulk production and regeneration

An ideal nanoadsorbent would competitively solve a reasonable part of the core technical, economic, and secondary pollution issues related to existing conventional water purification and wastewater treatment methods and conventional adsorbents.

10. Conclusions

To summarize, numerous ultrathin nanomaterials have been reported, including metal nanoparticles, metal oxides, silicene, borophene, carbon nanostructures, antimonene, MXenes, and MOFs. Nanoscale materials' unique properties have made them valued in many uses, considerably boosting the performance of a wide range of technologies and materials across a variety of disciplines. The size, shape, synthesis conditions, and appropriate functionalization of nanomaterials may all be accurately controlled using various synthesis methods to get the desired qualities. Furthermore, nanomaterial surfaces can be functionalized to fine-tune their characteristics for specific purposes. Nanomaterials are well-known members of the materials, and their huge surface areas, rapid charge transfer capabilities, and great mechanical strength have led to substantial research and development for a variety of applications. Due to the abundant availability of some nanomaterials like carbon nanostructures and metal oxides sources, these are promising for large-scale production to be used as adsorbents.

Nanotechnology can play an active role in several fields and industrial applications. Many industries and academics are interested in the characteristics of various nanomaterials and their applications in many domains. With the use of nanomaterials, more progress is anticipated, in some areas like in water treatment, environmental remediation, gas separation, oil refinery, and clean green energy.

References

Baer, D.R., Gaspar, D.J., Nachimuthu, P., Techane, S.D., Castner, D.G., 2010. Application of surface chemical analysis tools for characterization of nanoparticles. Anal. Bioanal. Chem. 396 (3), 983–1002.

Batten, S.R., Champness, N.R., Chen, X.M., Garcia-Martinez, J., Kitagawa, S., Öhrström, L., O'Keeffe, M., Suh, M.P., Reedijk, J., 2013. Terminology of metal–organic frameworks and coordination polymers (IUPAC Recommendations 2013). Pure Appl. Chem. 85 (8), 1715–1724.

Bianco, A., Cheng, H.-M., Enoki, T., et al., 2013. All in the graphene family—a recommended nomenclature for two-dimensional carbon materials. Carbon 65, 1–6.

Britto, P.J., Santhanam, K.S.V., Rubio, A., Alonso, J.A., Ajayan, P.M., 1999. Improved transfer at carbon nanotube electrodes. Adv. Mater. 11, 154.

Butler, S.Z., et al., 2013. Progress, challenges, and opportunities in two-dimensional materials beyond graphene. ACS Nano 7 (4), 2898–2926.

Chakraborty, I., Pradeep, T., 2017. Atomically precise clusters of noble metals: emerging link between atoms and nanoparticles. Chem. Rev. 117 (12), 8208–8271.

Chen, D., Wei, L., Li, J., Wu, Q., 2020a. Nanoporous materials derived from metal-organic framework for supercapacitor application. J. Energy Storage 30, 101525.

Chen, L., Zhao, S., Hasi, Q., Luo, X., Zhang, C., Li, H., Li, A., 2020b. Porous carbon nanofoam derived from pitch as solar receiver for efficient solar steam generation. Glob. Challenges 4, 1900098.

Crini, G., 2005. Recent developments in polysaccharide-based materials used as adsorbents in wastewater treatment. Prog. Polym. Sci. 30, 38–70.

Crini, G., 2010. Chapter 2: wastewater treatment by sorption. In: Sorption Processes and Pollution. PUFC, Besançon, pp. 39–78.

Crini, G., Badot, P.M. (Eds.), 2010. Sorption Processes and Pollution. PUFC, Besançon, 489 pp.

Datsyuk, V., Kalyva, M., Papagelis, K., Parthenios, J., Tasis, D., Siokou, A., Kallitsis, I., Galiotis, C., 2008. Chemical oxidation of multiwalled carbon nanotubes. Carbon 46, 833.

Drexler, E., 1986. Engines of Creation: The Coming Era of Nanotechnology. Anchor Books, Doubleday. ISBN: 0385199732.

Dubinin, M.M., 1966. Porous structure and adsorption properties of activated carbons. In: Walker, P.L. (Ed.), Chemistry and Physics of Carbon, vol. 2. Marcel Dekker, New-York, pp. 51–120.

Fan, X., Soin, N., Li, H., Li, H., Xia, X., Geng, J., 2020. Fullerene (C60) nanowires: the preparation, characterization, and potential applications. Energy Environ. Mater. 3, 469–491.

Ghabban, A.A., Zubaidi, A.B.A., Jafar, M., Fakhri, Z., 2018. Effect of nano SiO_2 and nano $CaCO_3$ on the mechanical properties, durability and flowability of concrete. IOP Conf. Ser. Mater. Sci. Eng. 454 (012016), 1–10.

Guzmán-Verri, G., Lew Yan Voon, L., 2007. Electronic structure of silicon-based nanostructures. Phys. Rev. B 76 (7), 075131.

Hernandez, Y., Nicolosi, V., Lotya, M., et al., 2008. High-yield production of graphene by liquid-phase exfoliation of graphite. Nat. Nanotechnol. 3, 563–568.

Iijima, S., 1991. Helical microtubules of graphitic carbon. Nature 354, 56–58.

Jiménez-Soto, J.M., Lucena, R., Cárdenas, S., Valcárcel, M., 2010. In: Marulanda, J.M. (Ed.), Solid Phase (Micro)extraction Tools Based on Carbon Nanotubes and Related Nanostructures, Carbon Nanotubes; Carbon Nanotubes. InTech, New York, ISBN 978-953-307-054-4.

Jose, V.R., Sakho, E.H.M., Parani, S., Thomas, S., Oluwafemi, O.S., Wu, J., 2019. Nanomaterials for Solar Cell Applications. Elsevier, pp. 75–95.

Kotb, E., Ahmed, A.A., Saleh, T.A., Ajeebi, A.M., Al-Gharsan, M.S., Aldahmash, N.F., 2020. Pseudobactins bounded iron nanoparticles for control of an antibiotic-resistant *Pseudomonas aeruginosa* ryn32. Biotechnol. Prog. 36 (1), e2907.

Krishnan, S.K., Singh, E., Singh, P., Meyyappan, M., Nalwa, H.S., 2019. A review on graphene-based nanocomposites for electrochemical and fluorescent biosensors. RSC Adv. 9, 8778–8881.

León Félix, L., Coaquira, J.A.H., Martínez, M.A.R., Goya, G.F., Mantilla, J., Sousa, M.H., Valladares, L.D.L.S., Barnes, C.H.W., Morais, P.C., 2017. Sci. Rep. 7, 41732.

Lipatov, A., et al., 2015. Few-layered titanium trisulfide (TiS3) field-effect transistors. Nanoscale 7 (29), 12291–12296.

Macwan, D.P., Pragnesh, N.D., Chaturvedi, S., 2011. A review on nano-TiO_2 sol-gel type syntheses and its applications. J. Mater. Sci. 46, 3669.

Mishra, R., Militky, J., Venkataraman, M., 2018. Nanotechnology in Textiles: Theory and Application, pp. 311–353.

Mudhoo, A., Sillanpää, M., 2021. Magnetic nanoadsorbents for micropollutant removal in real water treatment: a review. Environ. Chem. Lett. 19, 4393–4413.

Naguib, M., Mochalin, V.N., Barsoum, M.W., Gogotsi, Y., 2014. MXenes: a new family of two-dimensional materials. Adv. Mater. 26, 992–1005.

O'Brien, P., Pickett, N., 2005. In: Rao, C., Müller, A., Cheetham, A. (Eds.), The Chemistry of Nanomaterials: Synthesis, Properties and Applications. Wiley-VCH Verlag GmbH & Co. KGaA, Weinheim.

Pal, N., 2020. Nanoporous metal oxide composite materials: a journey from the past, present to future. Adv. Colloid Interface Sci. 280, 102156.

Pang, J., Mendes, R.G., Bachmatiuk, A., Zhao, L., Ta, H.Q., Gemming, T., Liu, H., Liu, Z., Rummeli, M.H., 2019. Applications of 2D MXenes in energy conversion and storage systems. Chem. Soc. Rev. 48, 72–133.

Peng, Y., Liu, H.W., 2006. Effects of oxidation by hydrogen peroxide on the structures of multiwalled carbon nanotubes. Ind. Eng. Chem. Res. 45, 6483.

Perera, M.M., Ayres, N., 2020. Dynamic covalent bonds in self-healing, shape memory, and controllable stiffness hydrogels. Polym. Chem. 11, 1410–1423.

Saba, N., Paridah, M.T., Abdan, K., Ibrahim, N.A., 2016. Effect of oil palm nano filler on mechanical and morphological properties of kenaf reinforced epoxy composites. Construct. Build. Mater. 123, 15–26.

Saleh, T.A., 2011. The influence of treatment temperature on the acidity of MWCNT oxidized by HNO_3 or a mixture of HNO_3/H_2SO_4. Appl. Surf. Sci. 257 (17), 7746–7751.

Saleh, T.A., 2015. Isotherm, kinetic, and thermodynamic studies on Hg (II) adsorption from aqueous solution by silica-multiwall carbon nanotubes. Environ. Sci. Pollut. Control Ser. 22 (21), 16721–16731. https://doi.org/10.1007/s11356-015-4866-z.

Saleh, T.A., 2017. Insights into carbon nanotube-metal oxide composite: embedding in membranes. Int. J. Sci. Res. Environ. Sci. Toxicol. 2 (1), 1–4.

Saleh, T.A., 2018. Simultaneous adsorptive desulfurization of diesel fuel over bimetallic nanoparticles loaded on activated carbon. J. Clean. Prod. 172, 2123–2132.

Saleh, T.A., 2020a. Nanomaterials: classification, properties, and environmental toxicities. Environ. Technol. Innovat. 20, 101067.

Saleh, T.A., 2020b. Characterization, determination and elimination technologies for sulfur from petroleum: toward cleaner fuel and a safe environment. Trends Environ. Anal. Chem. 25, e00080.

Saleh, T.A., 2021. Protocols for synthesis of nanomaterials, polymers, and green materials as adsorbents for water treatment technologies. Environ. Technol. Innovat. 24, 101821.

Saleh, T.A., 2022. Experimental and Analytical methods for testing inhibitors and fluids in water-based drilling environments. Trac. Trends Anal. Chem. 116543. https://doi.org/10.1016/j.trac.2022.116543.

Schärtl, W., 2010. Current directions in core−shell nanoparticle design. Nanoscale 2, 829.

Shen, Y., Zhu, C., Song, S., et al., 2019. Defect-abundant covalent triazine frameworks as sunlight-driven self-cleaning adsorbents for volatile aromatic pollutants in water. Environ. Sci. Technol. 53, 9091−9101.

Sone, J., Yamagami, T., Nakatsuji, K., Hirayama, H., 2014. Epitaxial growth of silicene on ultra-thin Ag(111) films. New J. Phys. 16 (9), 095004.

Sun, Y., Wu, Q., Shi, G., 2011. Graphene based new energy materials. Energy Environ. Sci. 4, 1113−1132.

Szcześniak, B., Borysiuk, S., Choma, J., Jaroniec, M., 2020. Mechanochemical synthesis of highly porous materials. Mater. Horiz. 7, 1457 —1473.

Tomar, R.S., Jyoti, A., Kaushik, S., 2020a. Nanobiotechnology: Concepts and Applications in Health, Agriculture, and Environment.

Tomar, R., Abdala, A.A., Chaudhary, R.G., Singh, N.B., 2020b. Photocatalytic degradation of dyes by nanomaterials. Mater. Today Proc. 29, 967−973.

Valcarcel, M., Simonet, B.M., 2011. Nanomaterials for improved analytical processes. Anal. Bioanal. Chem. 399 (1), 1−2.

Valenzuela, E.F., Menezes, H.C., Cardeal, Z.L., 2020. Passive and grab sampling methods to assess pesticide residues in water. Rev. Environ. Chem. Lett. 18, 1019−1048.

Veglio', F., Beolchini, F., 1997. Removal of metals by biosorption: a review. Hydrometallurgy 44, 301−316.

Webb, J.D., 2015. Selective functionalization of the mesopores of SBA-15. Microporous Mesoporous Mater. 203, 123−131.

Wigginton, N.S., Haus, K.L., Hochella, M.F., 2007. Aquatic environmental nanoparticles. J. Environ. Monit. 9, 1306.

Xiong, Y., Xu, L., Jin, C., Sun, Q., 2020. Cellulose hydrogel functionalized titanate microspheres with self-cleaning for efficient purification of heavy metals in oily wastewater. Cellulose 27, 7751−7763. Feynman's classic 1959 talk "There's Plenty of Room at the Bottom.

Xu, N., Xu, Y., Zhu, J., 2017. Topological insulators for thermoelectrics. Npj Quantum Mater 2 (1), 51.

Zaleska-Medynska, A., et al., 2016. Noble metal-based bimetallic nanoparticles: the effect of the structure on the optical, catalytic and photocatalytic properties. Adv. Colloid Interface Sci. 229, 80−107.

Zhang, L., Jaroniec, M., 2020. Strategies for development of nanoporous materials with 2D building units. Chem. Soc. Rev. 49, 6039−−6055.

Zhang, N.Y., Me, J., Varadan, V.K., 2002. Functionalization of carbon nanotubes by potassium permanganate assisted with phase transfer catalyst. Smart Mater. Struct. 11, 962.

Zhu, W., Guo, Y., Ma, B., Yang, X., Li, Y., Li, P., Zhou, Y., Shuai, M., 2020. Fabrication of highly dispersed Pd nanoparticles supported on reduced graphene oxide for solid phase catalytic hydrogenation of 1,4-bis(phenylethynyl) benzene. Int. J. Hydrogen Energy 45, 8385−−8395.

Chapter 9

Reactors and procedures used for environmental remediation

1. Introduction

Remediation is the process of cleaning, sanitizing, and restoring materials. Environmental remediation refers to reducing pollutants, for example, from contaminated soil, industrial water, groundwater, or surface water. This can isolate, immobilize, or remove the actual source of pollution, for example, by decontaminating areas, surfaces, and environmental media. Site remediation removes polluted or contaminated soil, sediment, surface water, or groundwater to reduce the impact on people and on the environment. Environmental remediation can be in situ and ex situ. There are several classes of environmental remediation, including soil remediation, groundwater and surface water remediation, sediment remediation, and gas or air remediation.

Water treatment improves the quality of the water. The procedures are determined by the quality of the water supply. In all cases, pollutants have to be separated from water, and water has to be disinfected to deactivate any existing microorganisms present in water.

2. Potential uses of nanomaterials

The ability to form nanomaterials with the desired properties in a specific synthesis route allows nanomaterials to be used across many fields, including medicine, energy, electronics, water, and environmental applications (Saleh, 2020a,b, 2021, 2022). The examples of the potential uses of nanomaterials are shown in Fig. 9.1. The healthcare field, for example, utilizes some nanomaterials such as silica and iron nanoparticles for drug delivery. Carbon nanotubes (CNTs) are used in antibody processes to form bacteria sensors. They are also used in the morphing of aircraft wings and to manufacture light bats with high performance. In nanocomposite form, they are used in response to the applications of electric voltage and nanowires. Other examples are

Surface Science of Adsorbents and Nanoadsorbents. https://doi.org/10.1016/B978-0-12-849876-7.00002-6

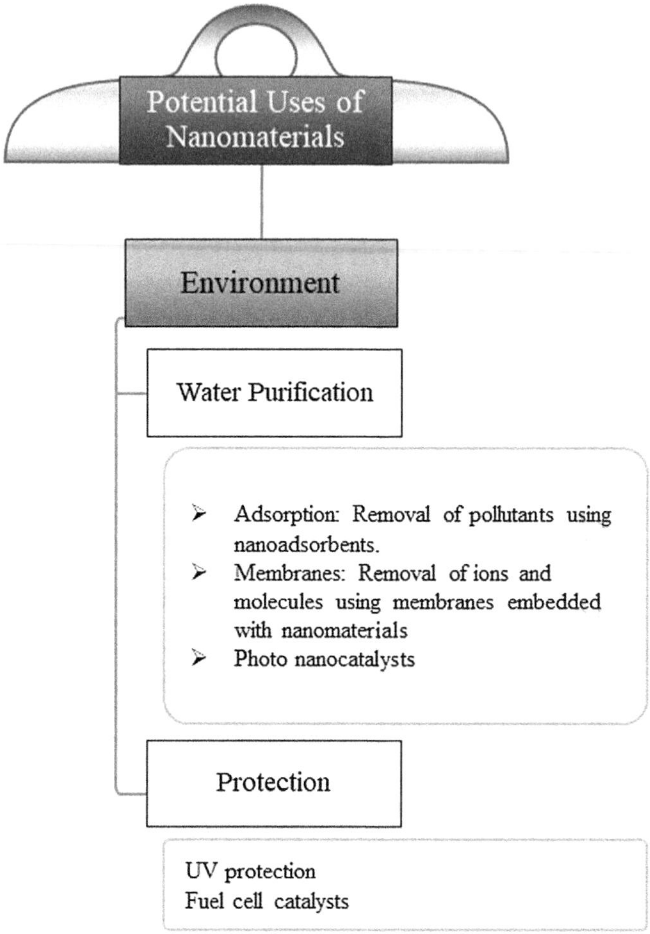
Potential Uses of Nanomaterials
Environment
Water Purification
Adsorption: Removal of pollutants using nanoadsorbents.
Membranes: Removal of ions and molecules using membranes embedded with nanomaterials
Photo nanocatalysts
Protection
UV protection
Fuel cell catalysts

FIGURE 9.1 Examples of potential uses of nanomaterials, composites, and polymeric hybrid materials in environmental remediation.

nanowires made of zinc oxide for solar cells. Due to high stability, titania nanoparticles are used in sunscreen for improved UV protection. They are also used as antimicrobial and antibacterial materials and in coating self-cleaning walls and surfaces, such as plastic garden chairs (Leaper et al., 2021; Qu et al., 2021).

3. Nanomaterial applications in water treatment

Wastewater treatment is essential for recovering freshwater for human activities and farming. Water is an essential compound worldwide and is established as a social and economic asset. Freshwater is required for life, clean industrial processes, agriculture, and environmental preservation. Scarcity and depletion of freshwater have an impact on climate change, interannual climate variability, and energy generation. In general, wastewater is discharged with dissolved substances or pollutants from mining, petrochemical, textile, industrial, agricultural (such as pesticides and fertilizers), and other activities with enormous effluent quantities.

The volume of hazardous waste has increased dramatically because of inappropriate garbage disposal that is not subject to environmental standards. Pesticide and fertilizer overuse, landfill leakage, industrial spills, and urban run-off penetration can all impact still water. Additionally, spills from industries, chemical spills, and oil spills contribute to the introduction of nonaqueous phase liquids containing substantial amounts of organic matrices as well as other constituents such as polyaromatic hydrocarbons with other contaminants. Several types of pollutants are present in wastewater such as heavy metals, pesticides, herbicides, oil contents, dyes, and pharmaceuticals. These compounds pose a major threat to the environment. Such pollutants accumulate, causing hazardous effects. In some places, accumulative pollution and the possible risk of polluted waters are aggravating. To overcome the related water scarcity problems, water regeneration is required from present wastewater using highly efficient treatment processes.

Wastewater remediation is a promising solution to purify water from industrial wastewater and to protect the environment. The selection and implementation of this remedial method essentially depend on several factors such as the level of contamination, the type of pollutants, and the source of wastewater. The direct wastewater-to-landfill disposal method is not permitted due to limited land and the large amounts of wastewater bodies, which may create both environmental and health concerns. Thus, water remediation is required, which is defined as the process of treating polluted waters and converting them into harmless products. Quantitative measurements of pollutants are necessary to select a suitable treatment technique. Several methods and technologies have been developed for water remediation, including chemical, physical, and biological approaches (Fig. 9.2). For effective wastewater remediation, most of the water treatment methods (decomposition, ion exchange, filtration, and membrane filtering) are combined. For example, when treating industrial effluents, the initial step is to remove solid wastes, after which further processes such as adsorption, ion exchange, bioremediation, sedimentation, and so on are performed.

Understanding the water treatment techniques, the materials, and the types of pollutants released in different wastewater is important. This will help select the technological solutions available for the treatment of these pollutants so that pollution can be controlled by adopting an appropriate treatment method and selecting suitable materials.

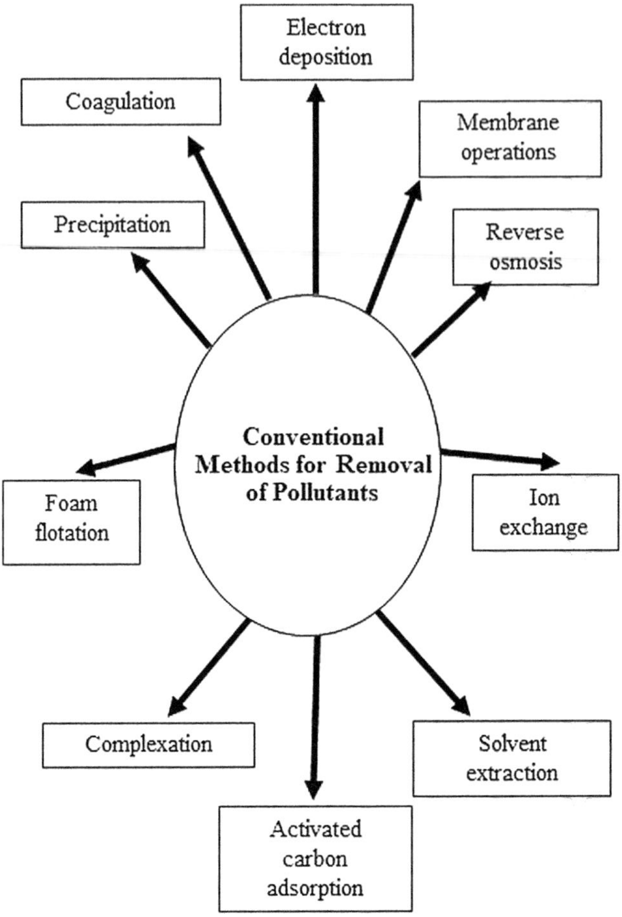
Electron deposition
Coagulation
Membrane operations
Precipitation
Reverse osmosis
Conventional Methods for Removal of Pollutants
Foam flotation
Ion exchange
Complexation
Solvent extraction
Activated carbon adsorption

FIGURE 9.2 Examples of the conventional methods used for the removal of pollutants from water.

4. Treatment technologies and methods

4.1 Layout of water treatment plant

Receiving water from the source and then settling it to eliminate solid particles and fine sands is the first step in the water treatment process. A screen separates the floating and suspended contaminants in the treatment unit. Aerators remove the gases from the water before exposing the raw water to the air. Chemical coagulation, flocculation, and clarifying operations are then undertaken. The coagulants are then added to the water in a coagulant tank. Using a flash mixer, the additional coagulants are thoroughly mixed (Table 9.1).

TABLE 9.1 Nanotechnology in water treatment.

Nanomaterial	Properties	Applications
Nanoadsorbents	High specific surface, higher adsorption rates—high production costs	Point-of-use, removal of organics, heavy metals, bacteria
Nanometals and nanometal oxides	Short intraparticle diffusion distance, abrasion-resistant—less reusable	Removal of heavy metals (arsenic) and radionuclides, media filters, slurry reactors, powders
Membranes and membrane processes	Reliable, largely automated process—relative high energy demand	All fields of water and wastewater treatment processes

To acquire clean water, water treatment methods to eliminate pollutants and microorganisms are required. Different water treatment technologies are used depending on the continent and the supply of water. The water treatment technologies, which range from simple to complicated, are aimed at providing low-cost sanitation and environmental protection while also allowing water reuse to provide other benefits. Physical, chemical, and biological approaches can all be used to treat water. Fig. 9.3 shows the layout of a water treatment plant with the functions of each unit.

In the following sections, the discussion will be focused on some reactors used in water treatment.

4.2 Types of reactors used in adsorption

There are several types of reactors used for water/wastewater treatment by the adsorption process. Examples are

- ➢ Batch
- ➢ Complete mix
- ➢ Complete mix with the recycling
- ➢ Plug flow
- ➢ Plug flow with the recycle
- ➢ Complete mix with reactors in series
- ➢ Packed bed
- ➢ Fluidized bed
- ➢ Column reactors in series

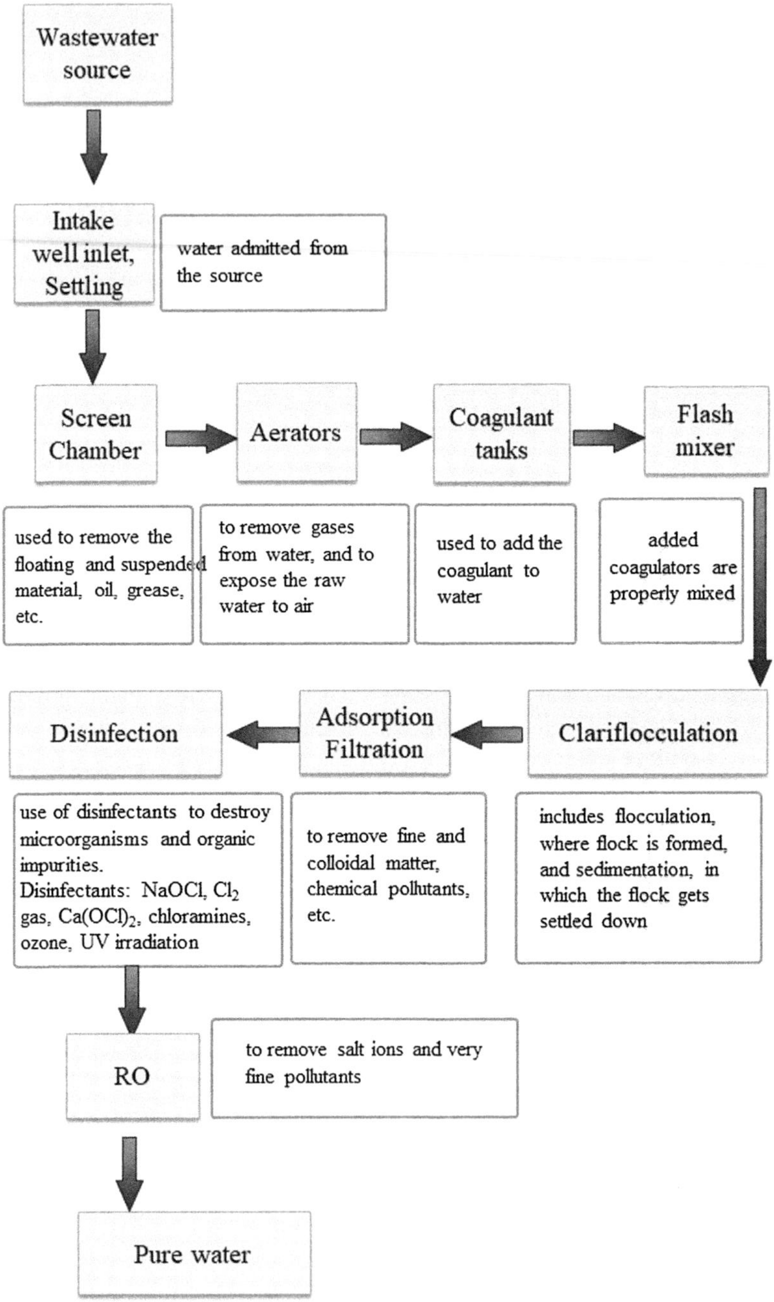
Wastewater source
Intake well inlet, Settling
water admitted from the source
Screen Chamber
Aerators
Coagulant tanks
Flash mixer
used to remove the floating and suspended material, oil, grease, etc.
to remove gases from water, and to expose the raw water to air
used to add the coagulant to water
added coagulators are properly mixed
Disinfection
Adsorption Filtration
Clariflocculation
use of disinfectants to destroy microorganisms and organic impurities.
Disinfectants: NaOCl, Cl_2 gas, $Ca(OCl)_2$, chloramines, ozone, UV irradiation
to remove fine and colloidal matter, chemical pollutants, etc.
includes flocculation, where flock is formed, and sedimentation, in which the flock gets settled down
RO
to remove salt ions and very fine pollutants
Pure water

FIGURE 9.3 The general layout of a water treatment plant with the functions of each unit.

5. Batch adsorption reactors for evaluating nanomaterials

Adsorption is one of the separation and purification techniques commonly used in the water treatment process, which involves a physicochemical mass transfer phenomenon in which a substance that begins in the liquid phase will accumulate or be adsorbed on a solid surface. Hence, dissolved pollutant substances will accumulate in the pores of the adsorbent until it reaches the equilibrium process. On the basis of the adsorption theory, the adsorption process is a surface phenomenon. Therefore, adsorbent materials generally have a porous structure with internal surface areas ranging from 100 to 1000 m^2. The adsorption process and capacity are closely related to several parameters such as the surface and pore area of the adsorbent, pH of the solution, temperature, concentration of adsorbed substances, and design of the reactor used. This method is widely used in water treatment because of the easy, flexible, and high-efficiency operating system for removing pollutants in the water.

5.1 Procedure and steps of testing

Batch adsorption tests should be carried out to analyze and test the effectiveness of any prepared adsorbent (Fig. 9.4). Fig. 9.5 depicts the protocol, which begins with the preparation of synthetic water laced with the contaminants of interest (metal ions, salts, organic pollutants, oil contents, and/or industrial components). Some factors, including adsorbent amount or dose, initial pollutant concentration, contact time, pH, volume, and temperature, should be optimized. The isotherm graph (a curve depicting the maximum adsorbed adsorbate [or pollutant] concentration per gram of adsorbent in wastewater at equilibrium) is the first stage. Kinetics, isotherm, and thermodynamic investigations are critical to understanding the adsorption mechanism. This enables the collection of data on adsorption behavior, processes, and capacity, which is useful for constructing columns or pilot size setups (Ali and Gupta, 2006).

5.2 Testing powdered and granular adsorbents

Generally, there are many physical forms of adsorbents and nanoadsorbents. These forms include granular and powdered forms. There are some parameters used to characterize the powder and granular properties; these include:

- ≻ Total surface area
- ≻ Bulk density
- ≻ Particle density, wetted in water
- ≻ Particle size range
- ≻ Effective size
- ≻ Uniformity coefficient

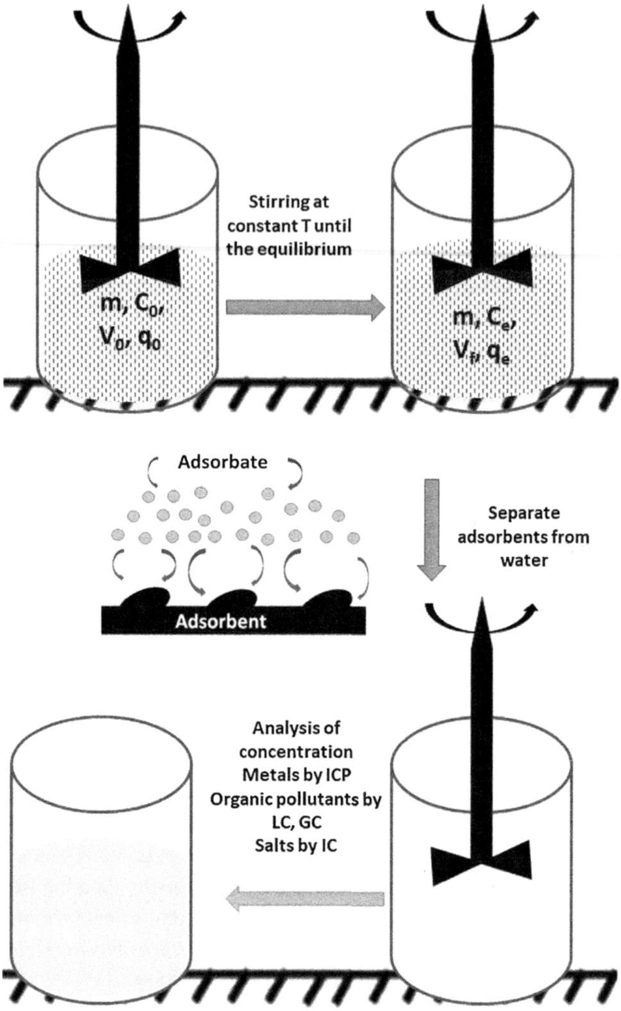
Stirring at
constant T until
the equilibrium
m, C0,
V0, q0
m, Ce,
Vf, qe
Adsorbate
Separate
adsorbents from
water
Adsorbent
Analysis of
concentration
Metals by ICP
Organic pollutants by
LC, GC
Salts by IC

FIGURE 9.4 Schematic representation of a discontinuous batch adsorption operation.

- ➢ Mean pore radius
- ➢ Iodine number
- ➢ Abrasion number
- ➢ Ash
- ➢ Moisture as packed

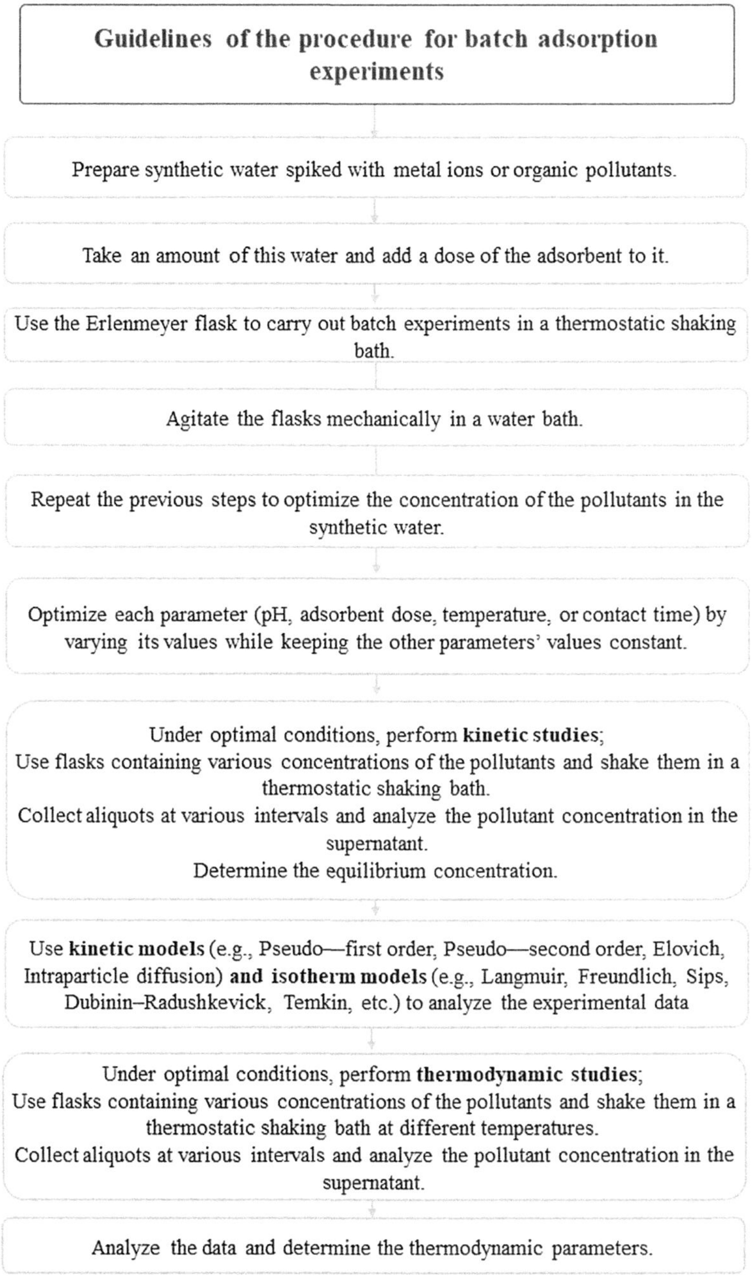
Guidelines of the procedure for batch adsorption experiments
Prepare synthetic water spiked with metal ions or organic pollutants.
Take an amount of this water and add a dose of the adsorbent to it.
Use the Erlenmeyer flask to carry out batch experiments in a thermostatic shaking bath.
Agitate the flasks mechanically in a water bath.
Repeat the previous steps to optimize the concentration of the pollutants in the synthetic water.
Optimize each parameter (pH, adsorbent dose, temperature, or contact time) by varying its values while keeping the other parameters' values constant.
Under optimal conditions, perform kinetic studies;
Use flasks containing various concentrations of the pollutants and shake them in a thermostatic shaking bath.
Collect aliquots at various intervals and analyze the pollutant concentration in the supernatant.
Determine the equilibrium concentration.
Use kinetic models (e.g., Pseudo—first order, Pseudo—second order, Elovich, Intraparticle diffusion) and isotherm models (e.g., Langmuir, Freundlich, Sips, Dubinin–Radushkevick, Temkin, etc.) to analyze the experimental data
Under optimal conditions, perform thermodynamic studies;
Use flasks containing various concentrations of the pollutants and shake them in a thermostatic shaking bath at different temperatures.
Collect aliquots at various intervals and analyze the pollutant concentration in the supernatant.
Analyze the data and determine the thermodynamic parameters.

FIGURE 9.5 Guidlines of the procedures for conducting water treatment batch adsorption experiments.

Usually, powdered adsorbents are in

- ➢ Fine powder, $d < 0.05$ mm or in nanosize
- ➢ High surface area
- ➢ Pore sizes (radii) down to around 1 nm

While granular adsorbents are in

- ➢ Particles of diameter: 0.5–4 mm
- ➢ A surface area equal to or a bit less than powdered adsorbents are mixed with raw polluted water. The system is allowed to reach equilibrium. After that, the spent adsorbent is removed by sedimentation or filtration. See Fig. 9.6.

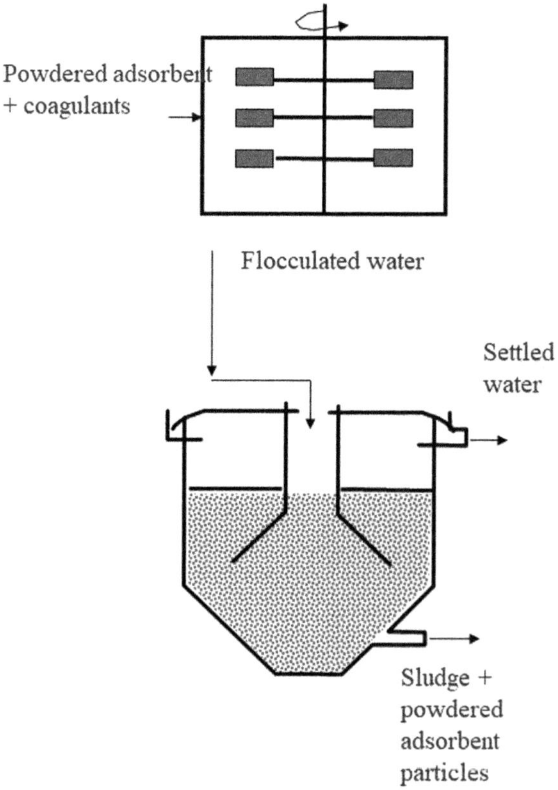
Powdered adsorbent
+ coagulants
Flocculated water
Settled
water
Sludge +
powdered
adsorbent
particles

FIGURE 9.6 The reactor design used for testing the efficiency of powdered adsorbent.

6. Column adsorption reactors for evaluating nanomaterials

Following the investigation of batch conditions, adsorbent characterization, adsorptive capacity, and adsorption mechanisms, certain pilot plant scales will be investigated to determine their practicality at the industrial level. A continuous stirred tank reactor (CSTR) is a model reactor in the form of a stirred tank, and the concentration of each component in the reactor is as uniform as the concentration of the flow coming out of the reactor. Batch experiments can be used to discover the best adsorption conditions for column tests. Fig. 9.7 depicts the technique for completing water treatment column testing.

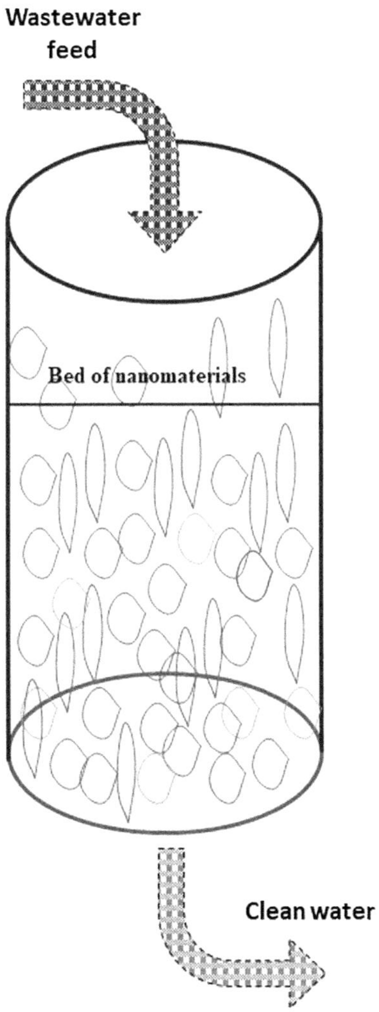
Wastewater feed
Bed of nanomaterials
Clean water

FIGURE 9.7 Illustration of simple column used for the adsorption of pollutants by the nanomaterial bed inside.

Fig. 9.8 lists the guidelines of the strategies for conducting water treatment column adsorption experiments. The adsorbents are packed in a bed column in the laboratory before the experiments begin. The dirty water is linked to one side of the column, and the eluted water is collected from the other side and tested for pollutants. The comparison of pollutant concentrations in water before and after elution offers information on the adsorbents' effectiveness and, as a result, the breakthrough capability. The pump can be linked to the

Guidelines of the procedure for column tests for water treatment

- Prepare synthetic water spiked with metal ions or organic pollutants
- Design a column of steel or pyrex glass on a laboratory scale.
- Pack the column with the adsorbent. Use mesh wire gauges at the pressure points. Use glass wool at the outlet to hold the adsorbent.
- Connect the vertical column to a pump to control the flow rates of the water.
- Start conditioning the column.
- Attach the synthetic water tank into the pump and to the column.
- Control the flow of the synthetic water tank with the pump.
- Collect aliquots at various intervals and analyze the pollutant concentration.
- Determine the breakthrough curves to find the column capacity at various flow rates, initial concentrations, etc.
- Regenerate the adsorbent by using acid, base, buffer, or organic. solvent. It depends on the type of pollutants.
- Based on the results, decide to use the adsorbents for pilot scale system.

FIGURE 9.8 Guidelines of the strategies for conducting water treatment column adsorption experiments.

column. Again, several parameters, such as the column's dimensions, packing zones, particle size, initial concentration, and flow rate, should be optimized. Mass transfer zones emerge in the column bed after the column system is operational and dirty water passes through. The qualities of the adsorbent, the kind of pollutant, and hydraulic parameters all influence the zone for each pollutant. After some time, the zone transfers down the column due to mass transfer, when the pollutants' concentrations in the effluent equal the influent concentration, and the column's breakthrough point occurs. A study of the time and shape of the breakthrough is possible.

Fig. 9.9 displays the fixed bed system with the illustration of breakthrough development (Gao, 2016; Mattson and Mark, 1971). Fig. 9.10 shows a fluidized bed system indicating a mass transfer zone. The time to breakthrough is decreased by:

- The increased particle size of carbon
- The higher concentration in the influent
- The increased pH of the water
- The increased flow rate
- The lower bed depth

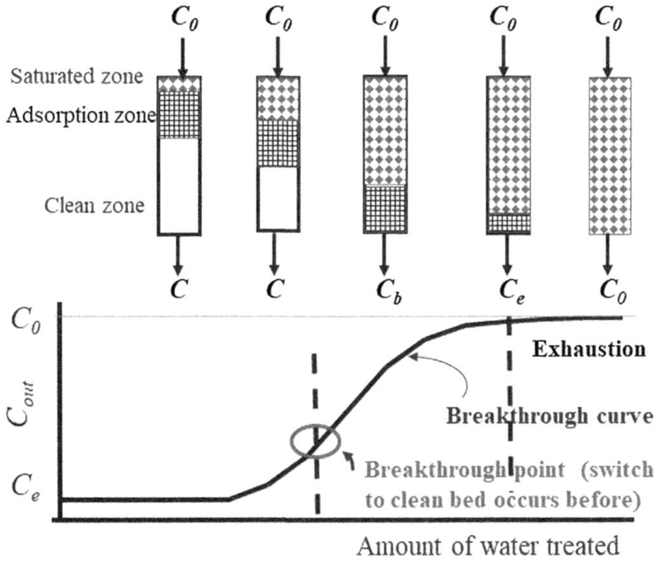

C_0
C_0
C_0
C_0
C_0
Saturated zone
Adsorption zone
Clean zone
C
C
C_b
C_e
C_0
C_0
C_{out}
C_e
Exhaustion
Breakthrough curve
Breakthrough point (switch to clean bed occurs before)
Amount of water treated

FIGURE 9.9 Fixed bed system with illustration of breakthrough development.

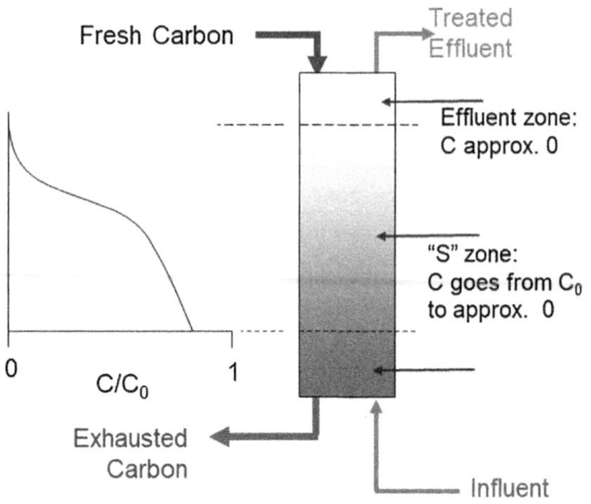
Fresh Carbon
Treated Effluent
Effluent zone:
C approx. 0
"S" zone:
C goes from C_0 to approx. 0
0
C/C_0
1
Exhausted Carbon
Influent

FIGURE 9.10 Fluidized bed system indicating mass transfer zone.

7. Reactors to evaluate nanomaterial-based membranes

The membrane filtration method has been widely used for water treatment processes, especially in separating components or suspended particles in liquid media. The treatment is generally based on a physical separation process due to differences in molecular size and membrane pores. Therefore, this method has been widely applied in water treatment ranging from the removal of relatively large particulate matter to dissolved particles. Fig. 9.11 shows a simple illustration of the water flow in a membrane system.

Nanoparticle membranes are utilized to improve chemical, physical, thermal, and mechanical properties. To assess the performance of produced membranes, they must be placed in a cell connected to a pump (Fan et al., 2021). Fig. 9.12 depicts one example. Depending on the raw wastewater, several factors shall be optimized. Some of the important factors are the flow rate, pH, initial concentrations of salts, and any other pollutants. Permeability, salt rejection, and oil rejection are the performance factors of modified membranes, which shall be studied using suitable experimental setup such as custom-made setups, cross-flow cells, Sterlitech, and so on. The membranes must be compacted with distilled water (DI) until it reaches a stable condition before obtaining wastewater measurements. Several metrics, including permeate flux, rejection rate, and conductivity, should be evaluated at various intervals (Al-Gamal et al., 2021).

There are numerous types of membranes with various chemical and mechanical structures. Membrane structures can vary, such as dense, microporous, asymmetric, composite, and membranes embedded with nanomaterials. They can comprise polymers, metal membranes, ceramic membranes (metal

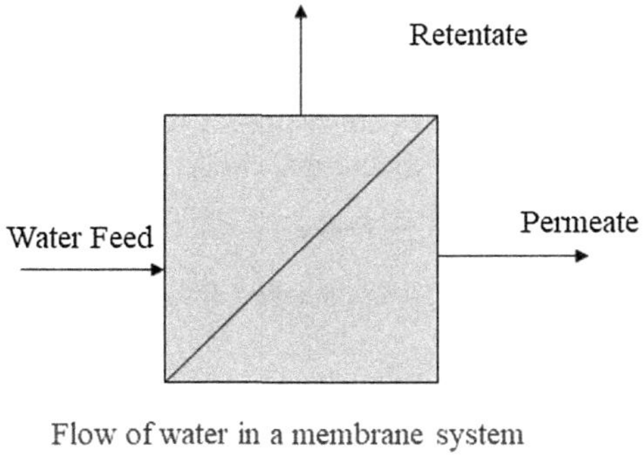
Retentate
Water Feed
Permeate
Flow of water in a membrane system

FIGURE 9.11 Illustration of the water flow in a membrane system.

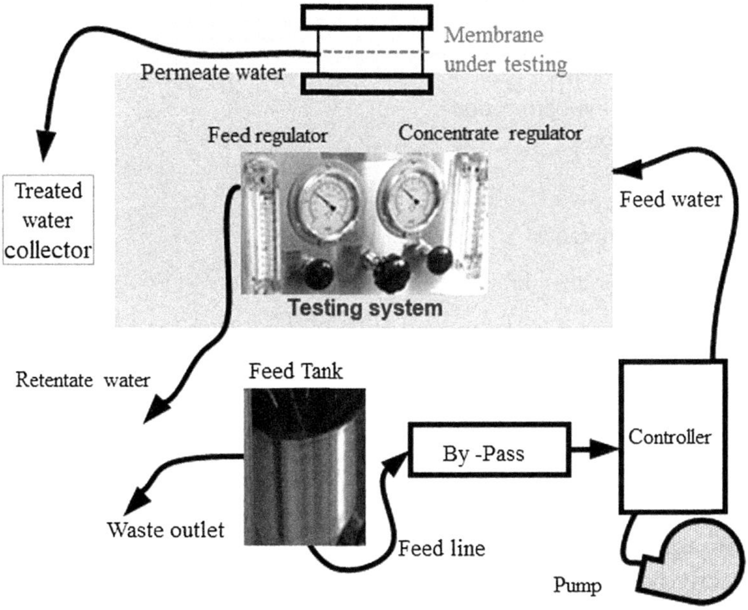
Membrane
under testing
Permeate water
Feed regulator
Concentrate regulator
Treated
water
collector
Feed water
Testing system
Retentate water
Feed Tank
By -Pass
Controller
Waste outlet
Feed line
Pump

FIGURE 9.12 The setup for a membrane testing is depicted in this diagram.

oxide, carbon, glass), and liquid membranes. The types of membranes are as follows:

- ➢ Microporous, with pore size from 0.01 to 10 microns in diameter. They can be used for the separation of solutes as a function of molecular size and pore size distribution
- ➢ Dense nonporous membranes
- ➢ Electrically charged microporous

Anisotropic (asymmetric) can be as follows:

- ➢ Thin active surface layer supported on a thicker porous layer
- ➢ Composite-based membranes consisting of different polymers in layers
- ➢ Others, including ceramic, metal, and liquid

Membrane modules can be as flows:

- ➢ Plate and frame, with flat sheets stacked into an element
- ➢ Tubular (tubes)
- ➢ Spiral wound designs using flat sheets
- ➢ Hollow fiber: They are down to 40 microns in diameter and might be several meters long, with an active layer on the exterior and a bundle of thousands of closely packed fibers enclosed in a cylinder

Fig. 9.13 illustrates the size spectrum of particles filtration using different membranes. Therefore, membrane processes can be as follows:

- ➢ Microfiltration
- ➢ Ultrafiltration
- ➢ Reverse osmosis
- ➢ Gas separation/permeation
- ➢ Pervaporation
- ➢ Dialysis
- ➢ Electrodialysis
- ➢ Liquid membranes

Nanofiltration membranes (Baker, 2004)

- ➢ Has a pressure-driven process
- ➢ Has a charge-based repulsion mechanism
- ➢ Reduces hardness, color, odor, and heavy metal ions
- ➢ Filters particles between 0.5−1 nm
- ➢ Can be used in desalination processes

8. Reactors to evaluate nanocatalyst photodegradation activity

The photocatalytic breakdown of the different contaminants in the water is facilitated by semiconductors. Waste-derived materials can be loaded, ornamented, doped, or coated with semiconductors to facilitate photocatalyst hybridization in composite materials (Kumar and Pandey, 2017; Alansi et al., 2018). Various operational parameters that affect photodegradation influence the oxidation rate and photocatalytic effectiveness. Several variables influence the photocatalysis reactions, and they should all be tuned to provide the best degradation conditions and hence a high capacity (Velempini et al., 2021). To determine the best operating parameters for catalyst dosage, pH, exposure

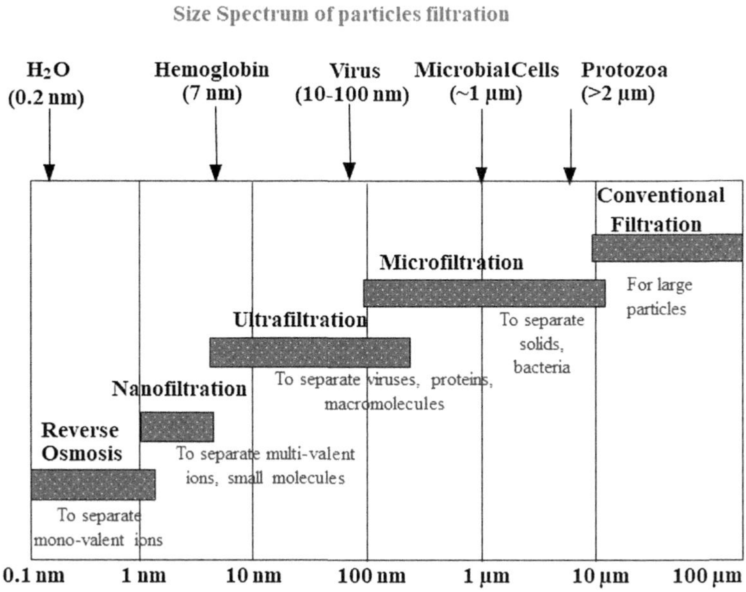
Size Spectrum of particles filtration
H_2O
(0.2 nm)
Hemoglobin
(7 nm)
Virus
(10-100 nm)
MicrobialCells
(~1 μm)
Protozoa
(>2 μm)
Conventional Filtration
For large particles
Microfiltration
To separate solids, bacteria
Ultrafiltration
To separate viruses, proteins, macromolecules
Nanofiltration
To separate multi-valent ions, small molecules
Reverse Osmosis
To separate mono-valent ions
0.1 nm
1 nm
10 nm
100 nm
1 μm
10 μm
100 μm

FIGURE 9.13 The size spectrum of particles filtration using different membranes.

(irradiation) period, and initial concentration, a series of studies are required. The proposed procedure, depicted in Fig. 9.14, can be used to assess the photocatalytic effectiveness of such composites. Cleaning the reactor and adding water and catalyst are the first steps. The lamp is turned on after attaining equilibrium in the dark to begin photodegradation. To track the efficacy of degradation and compute the capacity, aliquots are collected and examined.

Photocatalysis

- ≻ The oxidation process is advanced.
- ≻ The titania is widely utilized as a photocatalyst.
- ≻ The titania is irradiated by ultraviolet light.
- ≻ Electrons move into the conduction band.
- ≻ Electron hole pairs are created.
- ≻ It leads to a complex chain of oxidative–reductive reactions.
- ≻ Persistent compounds such as antibiotics or other micropollutants can be photocatalytically eliminated.
- ≻ Ultraviolet A radiation is only about 5% that of sunlight. Therefore, the photon efficiency is quite low, limiting its use on an industrial scale.

Guidelines for performing photocatalytic degradation tests

- Cleaning the reactor components.
- Filling the reactor with the sample (e.g., pollutants in water).
- Closing the reactor.
- Homogenization.
- Preliminary sampling to ascertain the initial conditions.
- Adding the catalyst and keeping the system in the dark to reach equilibrium.
- Degradation by switching the lamps on.
- Starting the timer to monitor the rate of photodegradation of the pollutants at different intervals.
- Sampling at different intervals and analysis.

FIGURE 9.14 Procedures for performing water treatment photocatalytic degradation experiments.

9. Hybrid technologies for water treatment

Hybrid technologies for water treatment are those technologies that combine more than one method or technique for water treatment. Some of the hybrid technologies for water treatment are discussed in the following sections as examples.

9.1 Photocatalytic membrane reactors

A new hybrid method for water and wastewater treatment is the photocatalytic membrane reactor (PMR), which combines photocatalysis with a membrane separation mechanism. Nontoxic and continuous running are just a few of the positive features of the system.

A PMR is a device that combines a photocatalyst and a membrane to produce chemical transformation. Two ways in which membrane is used in PMR are as follows:

- In slurry PMR, in which nanocatalysts are suspended in the reaction solution and the membrane only serves as an effective barrier (Fig. 9.15).
- In immobilized PMR, where the immobilization of the catalyst takes place on the membrane surface (Fig. 9.16).

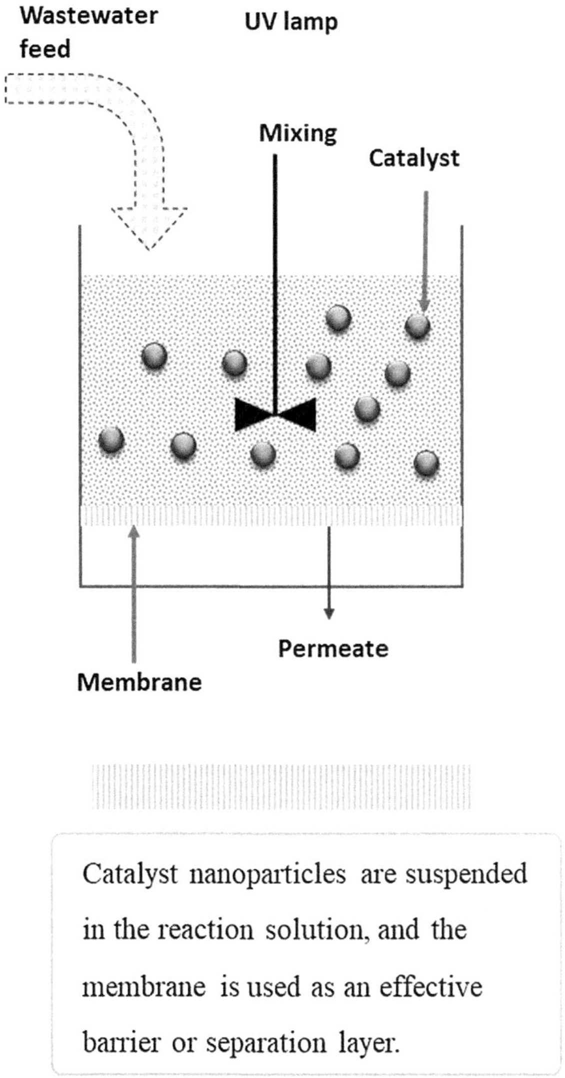
Wastewater feed
UV lamp
Mixing
Catalyst
Permeate
Membrane
Catalyst nanoparticles are suspended in the reaction solution, and the membrane is used as an effective barrier or separation layer.

FIGURE 9.15 Slurry photocatalytic membrane reactor.

The reactor should be capable of treating water containing strongly absorbing pollutants, as this is characteristic of several typical effluent streams arising from industrial activity.

Photocatalysis is defined as the beginning of a chemical reaction in the presence of a photocatalyst that absorbs light and is involved in the chemical

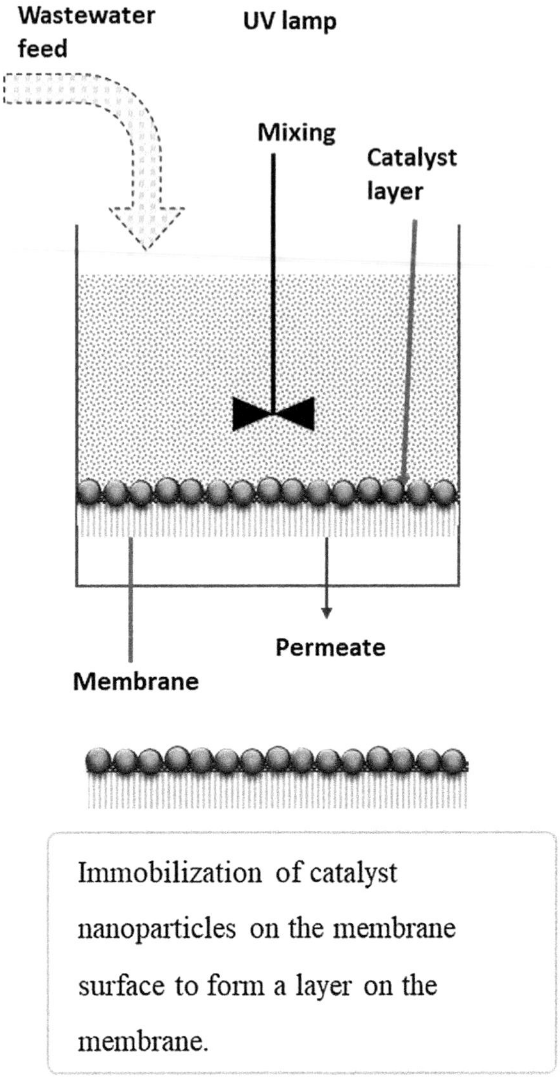
Wastewater feed
UV lamp
Mixing
Catalyst layer
Permeate
Membrane
Immobilization of catalyst nanoparticles on the membrane surface to form a layer on the membrane.

FIGURE 9.16 Immobilized photocatalytic membrane reactor.

transformation of the reactive pollutants under the action of ultraviolet, visible, or infrared radiation.

A photocatalyst is a material that may use ultraviolet, visible, or infrared energy to produce chemical transformations in reactive pollutants (Mozia, 2015; Pearce et al., 2007; Shi et al., 2014; Trägårdh, 1989).

Regarding the role of membranes in PMRs, the membrane can assume many roles, such as

- ➢ Separation
- ➢ Retention of the catalyst
- ➢ Catalyst support
- ➢ Recovery of catalyst

9.1.1 Photocatalytic membrane fouling

Fouling is the coverage of the membrane surface (external and internal) by deposits that adsorb or simply accumulate during operation. Membrane fouling can be due to several reasons:

- ➢ Photocatalytic interaction of organic pollutants and photocatalysts with the membrane surface
- ➢ Adsorption occurring without ultraviolet (UV) irradiation
- ➢ Pollutant degradation under UV irradiation

9.1.2 Foulant types

Substances that deposit on the membrane surface and cause its fouling are called foulants. Fouling is a membrane separation phenomenon resulting from several mechanisms: precipitation of sparingly soluble salts, adsorption, cake or gel formation, and pore blockage. Fouling is often divided into external and internal fouling, depending on where the foulant is deposited (Fig. 9.17).

The type of feedwater determines the severity of fouling. Generally speaking, there are several types of fouling including particulate/colloidal, organic fouling, inorganic fouling/scaling, and biofouling (or microbial/biological fouling) (Fig. 9.18).

Organic and inorganic components, as well as microbes, can cause fouling at the same time, and these components may interact in terms of mechanism (Amjad 1992). Because microorganisms can reproduce over time, biofouling is the Achilles heel of the membrane process. Even if 99.9% of them are removed, there are still enough cells left to grow at the expense of biodegradable compounds in the feedwater (Flemming et al., 1997). Biofouling is a type of organic fouling caused by organic matter obtained from microbial cellular debris, whereas abiotic fouling is caused by organic matter derived from microbial cellular debris (Amy, 2008). Biofouling has been recognized as a serious problem in nanofiltration (NF) and reverse osmosis (RO) membrane filtration (Vrouwenvelder and van der Kooij, 2002) and has been known to contribute to more than 45% of all membrane fouling (Komlenic, 2010). As a result, the focus of this review is on biofouling in NF and RO systems.

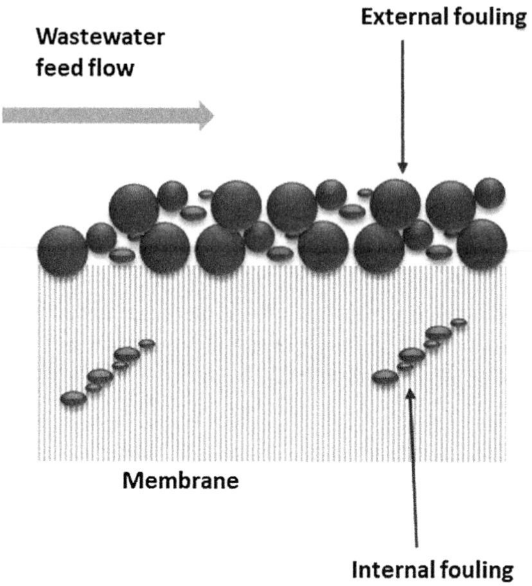
External fouling
Wastewater
feed flow
Membrane
Internal fouling

FIGURE 9.17 Illustration of external and internal fouling.

Fouling can also be categorized into irreversible fouling and reversible fouling, pore blocking, and cake layer, and these can be in different forms of fouling.

- ➢ Adsorption
- ➢ Deposition
- ➢ Gel/cake formation
- ➢ Pore blocking

The mechanisms of photocatalytic membrane fouling in PMRs can be mainly divided into two stages: (a) pore blocking and (b) cake layer.

Membrane fouling consequences

- ➢ Membrane pores are blocked
- ➢ Permeate flux decreases
- ➢ Production efficiency decreases
- ➢ Operation time increases
- ➢ Dense cake layer of catalyst forms

Membrane fouling prevention

- ➢ Pretreatment of feed solution
- ➢ Membrane modification with nanomaterials
- ➢ Self-cleaning process

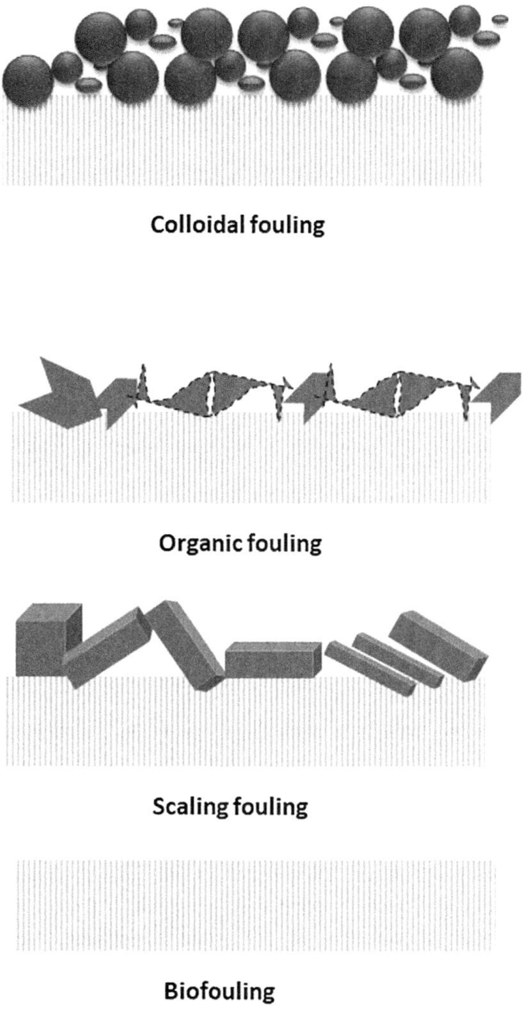
Colloidal fouling
Organic fouling
Scaling fouling
Biofouling

FIGURE 9.18 Examples of some types of membrane fouling.

- ➢ Optimization of operating parameters
- ➢ Reduction by physical and chemical methods or by the modification of the membrane or pretreatment of the feed solution

9.2 Photoelectrochemical reactor

The photocatalytic degradation process is combined with an electrochemical process in this system. Electrochemical reactors are devices that use electric

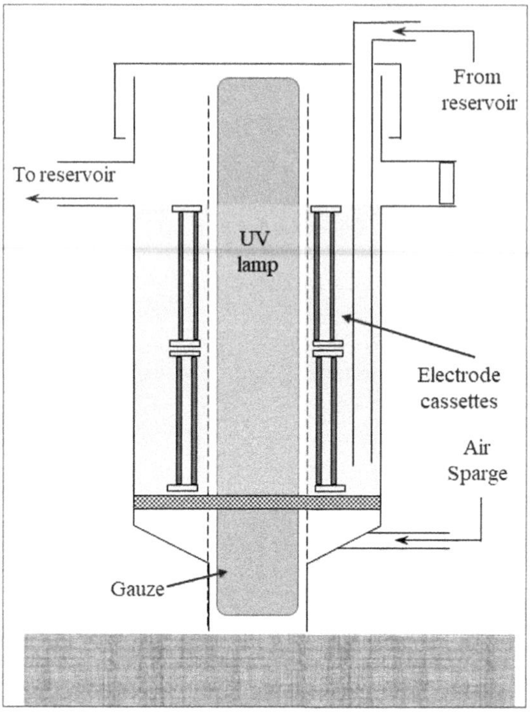
From
reservoir
To reservoir
UV
lamp
Electrode
cassettes
Air
Sparge
Gauze

FIGURE 9.19 Example of the design of the photoelectrochemical reactor.

currents to force material changes. The anode undergoes oxidation, while the cathode undergoes reduction (Wehinger et al., 2018; Alkire et al., 2015). Reactors of this type have a variety of designs. An example is shown in Fig. 9.19. The basic concepts of the electrochemical methods for water treatment are explained first, followed by examples of applications. A description of the reactors follows. This overview article does not provide technical operational data or design details such as current densities, voltages, or electrode spacings because it would be beyond its scope. Due to the diverse (waste)water compositions and the intended cleaning purpose, reactor designs are particularly unique in their use.

9.3 Other types of hybrid systems for water treatment using membrane

There are many other hybrid systems used for water treatment such as gas separation or the cleaning air system, which include the following mechanisms:

- ➢ Adsorption and membrane system
- ➢ Adsorption, membrane, and biological processes
- ➢ Adsorption and photodegradation reactors (Zhang et al., 2012)
- ➢ Electrochemical and adsorption

10. Conclusions

For any new materials and nanomaterials to be used prepared for environmental remediation, especially for wastewater treatment, the testing should be performed to evaluate its performance and efficiency in adsorption, separation in the removal of unwanted components, contaminants, or pollutants. There are several types of the experimental setups, reactors, designs, and procedures for testing. These include the testing using batch experimental setups, columns reactors, photoreactors, and membrane systems. Interestingly, the hybrid systems of more than one setup such as photoelectrochemical reactor, adsorption combined with photoreactor, adsorption membrane, and photoreactor with membrane can be used for testing. Such type of reactors provides more information on the performance in several steps rather than using a single setup.

References

Al-Gamal, A.Q., Falath, W.S., Saleh, T.A., 2021. Enhanced efficiency of polyamide membranes by incorporating TiO_2-Graphene oxide for water purification. J. Mol. Liq. 323, 114922.

Alansi, A.M., Al-Qunaibit, M., Alade, I.O., Qahtan, T.F., Saleh, T.A., 2018. Visible-light responsive BiOBr nanoparticles loaded on reduced graphene oxide for photocatalytic degradation of dye. J. Mol. Liq. 253, 297–304.

Ali, I., Gupta, V., 2006. Advances in water treatment by adsorption technology. Nat. Protoc. 1, 2661–2667.

Alkire, R.C., Bartlett, P.N., Lipkowski, J., 2015. Electrochemistry of Carbon Electrodes, first ed., vol 16. Wiley-VCH, Weinheim.

Amjad, Z., 1992. Reverse Osmosis, Membrane Technology, Water Chemistry and Industrial Application. Van Nostrand Reinhold, New York, NY, USA.

Amy, G., 2008. Fundamental understanding of organic matter fouling of membranes. Desalination 231, 44–51. https://doi.org/10.1016/j.desal.2007.11.037.

Baker, R.W., 2004. Membrane Technology and Applications, second ed. John Wiley.

Fan, Y., Ma, J., Zhu, Q., Qin, J., 2021. Fluorescent and mechanical properties of UiO-66/PA composite membrane. Colloids Surf. A Physicochem. Eng. Asp. 127083.

Flemming, H.-C., Griebe, T., Schaule, G., Schmitt, J., Tamachkiarowa, A., 1997. Biofouling—the Achille's heel of membrane processes. Desalination 113, 215–225.

Gao, T.Y., 2016. Water and Wastewater Treatment, fourth ed. Higher Education Press, Beijing.

Komlenic, R., 2010. Rethinking the causes of membrane biofouling. Filtr. Sep. 47, 26–28.

Kumar, A., Pandey, G., 2017. A review on the factors affecting the photocatalytic degradation of hazardous materials. Mater. Sci. Eng. Int. J. 1 (3), 106–114. https://doi.org/10.15406/mseij.2017.01.00018.

Leaper, S., Abdel-Karim, A., Gorgojo, P., 2021. The use of carbon nanomaterials in membrane distillation membranes: a review. Front. Chem. Sci. Eng. 15, 755–774. https://doi.org/10.1007/s11705-020-1993-y.

Mattson, J.S., Mark, H.B., 1971. Activated Carbon. Surface Chemistry and Adsorption from Dekker, New York.

Mozia, S., 2015. On photocatalytic membrane reactors in water and wastewater treatment and organic synthesis. Copernican Lett. 6, 17–23.

Pearce, G., 2007. Introduction to membranes: fouling control. Filtrat. Separ. 44 (6), 30–32.

Qu, M., Pang, Y., Li, J., Wang, R., He, D., Luo, Z., Shi, F., Peng, L., He, J., 2021. Eco-friendly superwettable functionalized-fabric with pH-bidirectional responsiveness for controllable oil-water and multi-organic components separation. Colloids Surf. A Physicochem. Eng. Asp. 624, 126817.

Saleh, T.A., 2020a. Characterization, determination and elimination technologies for sulfur from petroleum: toward cleaner fuel and a safe environment. Trends Environ. Anal. Chem. 25, e00080.

Saleh, T.A., 2020b. Nanomaterials: classification, properties, and environmental toxicities. Environ. Technol. Innovat. 20, 101067.

Saleh, T.A., 2021. Protocols for synthesis of nanomaterials, polymers, and green materials as adsorbents for water treatment technologies. Environ. Technol. Innovat. 24, 101821.

Saleh, T.A., 2022. Experimental and Analytical methods for testing inhibitors and fluids in water-based drilling environments. Trac. Trends Anal. Chem. 116543. https://doi.org/10.1016/j.trac.2022.116543.

Shi, X., Tal, G., et al., 2014. Fouling and cleaning of ultrafiltration membranes: a review. J. Water Proc. Eng. 1, 121–138.

Trägårdh, G., 1989. Membrane cleaning. Desalination 71 (3), 325–335.

Velempini, E., Prabakaran, T., Pillay, K., 2021. Recent developments in the use of metal oxides for photocatalytic degradation of pharmaceutical pollutants in water—a review. Mater. Today Chem. 19, 100380.

Vrouwenvelder, J.S., van der Kooij, D., 2002. Diagnosis of fouling problems of NF and RO membrane installations by a quick scan. Desalination 153, 121–124.

Wehinger, G., Kunz, U., Turek, T., 2018. In: Reschetilowski, W. (Ed.), Handbuch Chemische Reaktoren. Springer Spektrum, Berlin. https://doi.org/10.1007/978-3-662-56444-8_37-1.

Zhang, G., Yao, L., et al., 2012. Photocatalytic membrane reactor used for water and wastewater treatment. Recent Pat. Eng. 6 (2), 127–136, 21.

Chapter 10

Applications of nanomaterials to environmental remediation

1. Introduction on the adsorption process

Adsorption is a process where an adsorbate binds to a surface either via a physical bond (physisorption) or a chemical bond (chemisorption) (Table 10.1). Adsorbents are widely used in a variety of applications, including wastewater treatment, separation, metal extraction, catalysis, and membrane technology, because they are a convenient and low-cost way to remove pollutants. There are several material classes for pollutants adsorption (Fig. 10.1). Metal oxides, supported metal oxides, metal hydroxides, and zeolites are just a few of the material types as possible pollution adsorbents. Adsorption kinetics, surface transformation processes, and reaction equilibria are all closely linked, and knowing the linkages that connect them is crucial to creating an effective sorbent (Fig. 10.2).

This chapter focuses more on analyzing various adsorbents and nanoadsorbents used for environmental remediation especially water treatment.

2. Adsorption

2.1 Adsorption from solution phase

Solids can adsorb solutes from solutions. When a polluted water sample is shaken with adsorbent, pollutants are adsorbed by adsorbent and the amounts of pollutants decrease in the water. The following observations have been made in the case of adsorption from the liquid phase:

- ➢ The extent of adsorption mostly decreases with a controlled increase in temperature.
- ➢ The extent of adsorption enhances with an increase in the surface area of the adsorbent.
- ➢ The extent of adsorption depends on the concentration of the solute (pollutants) in the solution.

Surface Science of Adsorbents and Nanoadsorbents. https://doi.org/10.1016/B978-0-12-849876-7.00005-1

TABLE 10.1 A description of numerous terms associated with adsorption and nanoadsorption.

Terminology	Definition
Adsorption technology	Adsorption technology is the branch of science and engineering devoted to the study of the properties of adsorbent and adsorbate and their relationships and interactions.
Adsorption	Adsorption is the adhesion of atoms, ions, or molecules (adsorbate) from a gas, liquid, or dissolved solid to a surface (adsorbent). This process creates a film of the adsorbate on the surface of the adsorbent. Adsorption can also be defined as a physical binding of adsorbate molecules or ions onto the adsorbent.
Surface complex	The surface complex is a phenomenon where adsorption takes place and results in the formation of a stable molecular phase at the interface. Surface complex exists as inner- and outer-sphere surface complexes.
Sorption	It is a term used for absorption or adsorption. It is used for describing a system where a sorbate (for instance, an ion or a molecule) interacts with a sorbent, i.e., a solid surface, resulting in an accumulation at the sorbate–sorbent interface.
Biosorption	Biosorption or bio-adsorption is defined as the removal of substances from solution by biological materials. Biosorption refers to the passive or physicochemical attachment of a sorbate to a biosorbent, basically the binding of chemical species to biopolymers.
Bioaccumulation	Bioaccumulation is the process of precipitation or crystallization of metals that can take place within and around cell walls and the production by biomass of metal-binding polysaccharides.

- ➢ The extent of adsorption relies on the nature of both adsorbent and adsorbate.
- ➢ Kinetics and isotherms models can be used to describe the process. For example, Freundlich's equation approximately describes the behavior of adsorption from water with a difference that instead of pressure, pollutant concentration is considered. This can be tested experimentally by taking solutions of various concentrations of pollutants. Then, equal volumes of solutions are added to equal amounts of adsorbent in various flasks. The

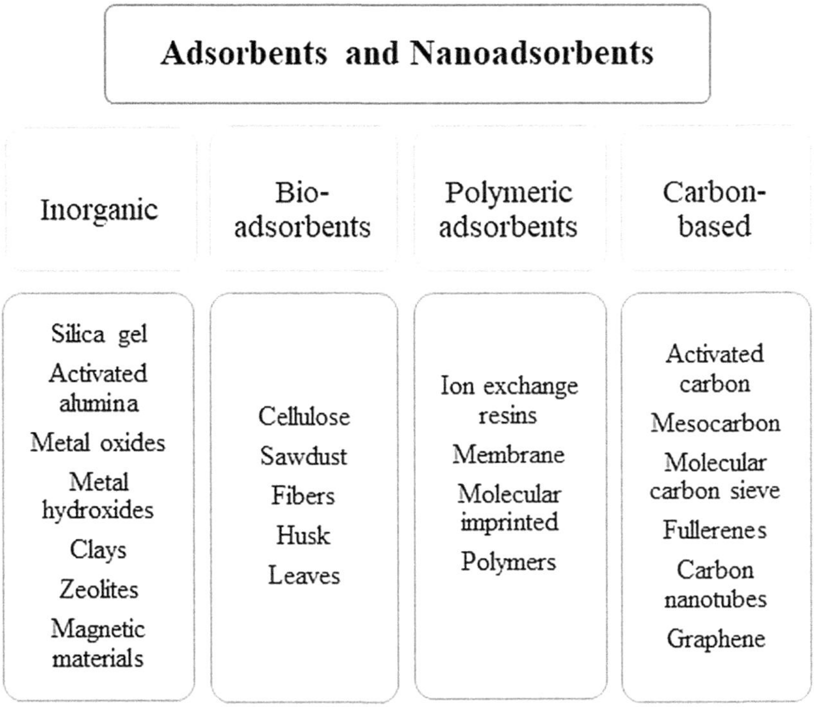
Adsorbents and Nanoadsorbents
Inorganic
Bio-adsorbents
Polymeric adsorbents
Carbon-based
Silica gel
Activated alumina
Metal oxides
Metal hydroxides
Clays
Zeolites
Magnetic materials
Cellulose
Sawdust
Fibers
Husk
Leaves
Ion exchange resins
Membrane
Molecular imprinted
Polymers
Activated carbon
Mesocarbon
Molecular carbon sieve
Fullerenes
Carbon nanotubes
Graphene

FIGURE 10.1 Examples of adsorbents and nanoadsorbents used in environmental remediation.

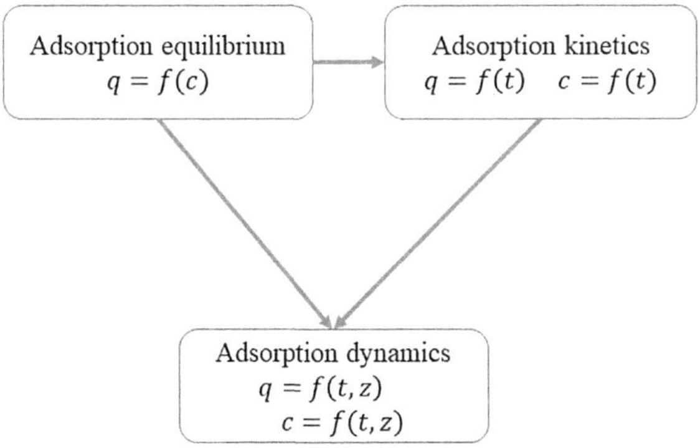
Adsorption equilibrium
q = f(c)
Adsorption kinetics
q = f(t) c = f(t)
Adsorption dynamics
q = f(t,z)
c = f(t,z)

FIGURE 10.2 Main elements of the adsorption theory.

final concentration is measured in each flask after adsorption. The difference between initial and final concentrations provides the amounts of adsorbed pollutants. Using the model equation, the validity of the Freundlich isotherm is recognized.

2.2 Applications of adsorption

There are several applications of the adsorption process. These include:

- ➢ Water purification: By the addition of adsorbent (for example, alum stone) to the water, impurities (pollutants or contaminants) get adsorbed on the adsorbent, and water gets purified.
- ➢ Removal of coloring pollutants from solutions: For example, carbon nanostructures adsorb colored impurities from water, removing the color.
- ➢ Heterogeneous catalysis: The rate of reaction is increased by the adsorption of reactants on the solid surface of catalysts. Solid catalysts are used in a variety of gaseous processes that are important in the industry.
- ➢ Separation of inert gases: A mixture of noble gases can be separated by adsorption on coconut charcoal at different temperatures due to the difference in the degree of adsorption of gases by charcoal.
- ➢ Adsorption indicators: Certain precipitates, such as silver halides, have the ability to adsorb dyes such as eosin, fluorescein, and others, resulting in a distinct color at the endpoint.
- ➢ Chromatographic analysis: Adsorption chromatography based on the phenomenon of adsorption finds a number of applications in analytical and industrial fields.
- ➢ Ion exchange method: The method aims at the removal of the hardness of water; calcium and magnesium ions get adsorbed on the surface of ion exchange resin.
- ➢ Gas and air masks: Gas masks (devices consisting of silica gel, activated charcoal, or a mixture of sorbents) are commonly used for breathing in coal mines to adsorb poisonous gases or small particles that get sorbed on the surface of such materials.
- ➢ Control of humidity: Aluminum and silica gels are used as sorbents to remove moisture and control humidity.

2.3 Factors affecting adsorption

Adsorption takes place on the surface of solids. Nevertheless, the extent of adsorption of gases or pollutants on the surface of solids depends mostly on some factors including:

- ➢ The nature of the adsorbent
- ➢ The nature of the adsorbed gas
- ➢ The pressure of the gas of concentration of adsorbate in solution
- ➢ The temperature of the solution
- ➢ Other environmental conditions

3. Adsorption in removing pollutants from water

When filtering impurities and contaminants out of wastewater, mechanical filtrations usually are used. Nonetheless, if there are particles in water that are dissolved or the particles are too small and cannot be removed by mechanical filtration, then other treatment processes are required. For soluble pollutants in water, different techniques are used, among them adsorption is used. Adsorption is used to remove several types of pollutants such as toxic metal ions, pesticides, oil, industrial solvents, disinfection by-products, and inorganic and organic chemicals.

3.1 Current purification methods

Water purification can be achieved by several wastewater treatment methods (Fig. 10.3). The methods used for water purifications can be classified as shown in Fig. 10.4, including physical, chemical, and biological methods. There are several reactors, processes, and techniques used for water treatment as discussed in the previous chapter.

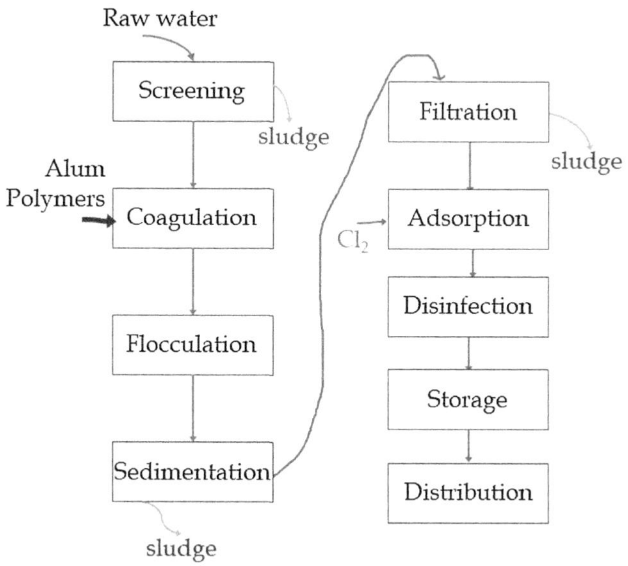
Raw water
Screening
sludge
Alum
Polymers
Coagulation
Flocculation
Sedimentation
sludge
Filtration
sludge
Cl_2
Adsorption
Disinfection
Storage
Distribution

FIGURE 10.3 General water treatment scheme.

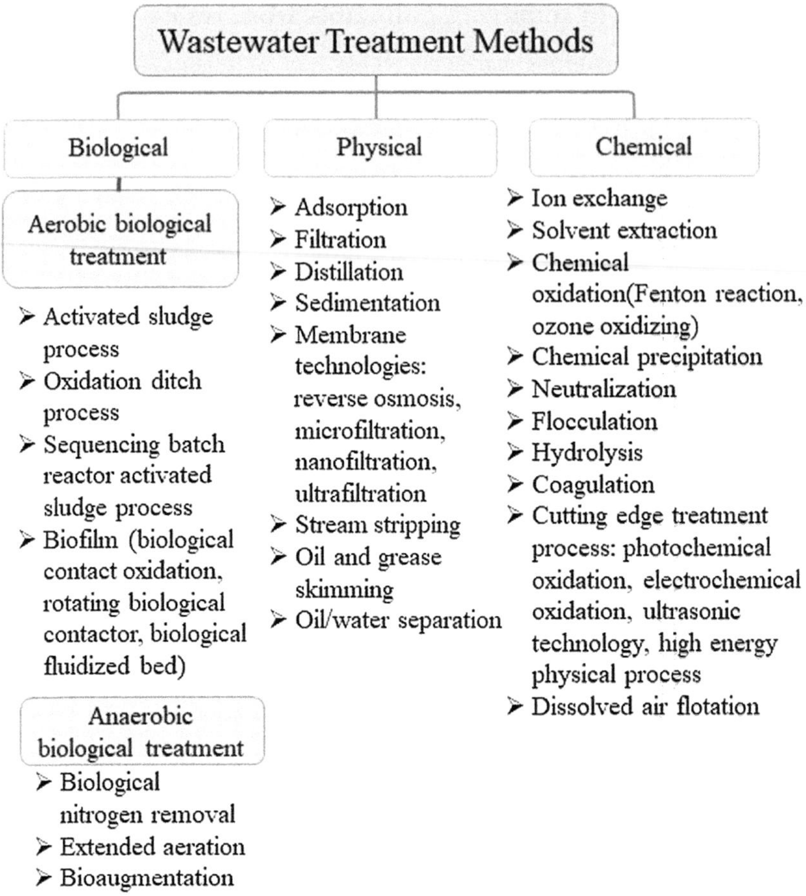
Wastewater Treatment Methods
Biological
Physical
Chemical
Aerobic biological treatment
➢ Activated sludge process
➢ Oxidation ditch process
➢ Sequencing batch reactor activated sludge process
➢ Biofilm (biological contact oxidation, rotating biological contactor, biological fluidized bed)
Anaerobic biological treatment
➢ Biological nitrogen removal
➢ Extended aeration
➢ Bioaugmentation
➢ Adsorption
➢ Filtration
➢ Distillation
➢ Sedimentation
➢ Membrane technologies: reverse osmosis, microfiltration, nanofiltration, ultrafiltration
➢ Stream stripping
➢ Oil and grease skimming
➢ Oil/water separation
➢ Ion exchange
➢ Solvent extraction
➢ Chemical oxidation(Fenton reaction, ozone oxidizing)
➢ Chemical precipitation
➢ Neutralization
➢ Flocculation
➢ Hydrolysis
➢ Coagulation
➢ Cutting edge treatment process: photochemical oxidation, electrochemical oxidation, ultrasonic technology, high energy physical process
➢ Dissolved air flotation

FIGURE 10.4 Types of methods commonly used in water and wastewater treatment.

3.2 Drawbacks in the present water treatment process

There are some drawbacks in present water treatment processes, including

- ➢ Most of the methods available have some limitations or have some impact either on our health or on the environment
- ➢ Time-consuming
- ➢ Fails to remove the heavy metal ions and chemical pollutants
- ➢ Some methods are not able to remove microorganisms
- ➢ Application of UV lightning requires longer radiation time
- ➢ Some methods remove the useful minerals and nutrients

- ➢ Additionally, the surface area or active sites, the lack of selectivity, and the adsorption kinetics are all factors that restrict the efficiency of current conventional adsorbents.

3.3 Nanotechnology in water treatment

Nanotechnology plays a key role in the development of several fields, including water purification. In several water treatment techniques and methods, nanomaterials are used for water purification. These include the use of nanomaterials in:

- ➢ Adsorption
- ➢ Membrane nanotechnology
- ➢ Photocatalysis
- ➢ Disinfection and microbial control
- ➢ Sensing and monitoring

Recently, nanoadsorbents offer important development with their:

- ➢ Extremely high specific surface area
- ➢ Associated sorption sites
- ➢ Tunable pore size
- ➢ Enhanced surface chemistry

Examples:

- ➢ Nanocrystalline zeolites can be used to remediate water-containing cationic species such as ammonium and heavy metals.
- ➢ Magnetic nanoparticles bind with contaminants, such as oil and arsenic, and are removed using a magnet.
- ➢ Polymeric-based nanoadsorbents and metal-based nanoadsorbent

3.4 Adsorption mechanisms

It is very important to understand the interactions between the adsorbents and adsorbates in order to optimize the parameters and conditions and thus obtain high efficient water treatment process. Adsorption mechanisms are the possibility to be established by:

- ➢ The use of various analytical techniques such as FTIR, SEM, nitrogen adsorption–desorption isotherms, Raman spectroscopy, TGA/DTA, DSC, ^{29}Si, and ^{13}C solid-state NMR, XRD, XPS, pH_{PZC}, pH_{IEP}, CHN element analysis, Boehm titration, and solution calorimetry.
- ➢ Understating the chemical nature of the adsorbent and adsorbate, the adsorbent's surface, and physical or chemical interactions between the adsorbate and adsorbent. The use of characterization instruments in

addition to the adsorptive thermodynamic data (such as changes in enthalpy and entropy) and activation and adsorption energies are required to confirm whether the adsorption of contaminants in an aqueous solution is a physical process or chemical process (Wang et al., 2020).

➢ The adsorption mechanisms depend on the adsorption forces (interaction between adsorbate and adsorbent) which could include interactions such as:
  - Dispersion forces
  - London or the Van der Waals active forces
  - The electrostatic attraction forces (Coulomb)
  - Coulombic-unlike charges
  - Point charge and a dipole
  - Dipole–dipole interactions
  - Point charge-neutral species
  - Hydrogen bonding
  - $\pi-\pi$ interaction
  - Ligand exchange
  - Electron donor–acceptor mechanisms
  - Metal complexation
  - Hydrophobic interactions
  - Hydrophilic interactions
  - Covalent bonding with reaction
  - Other interactions between the ionic moieties.

To be somehow more specific, the type of interaction depends on the nature of the adsorbent and the nature of adsorbate or pollutants. For metal ions adsorption, the possible interaction mechanisms are:

➢ Electrostatic attraction: cationic attraction ($pH > pH_{pzc}$) and anionic attraction ($pH < pH_{pzc}$)
➢ Ion exchange
➢ Complexation
➢ Coprecipitation

For organic contaminants, the possible adsorption interaction mechanisms are:

➢ Pore-filling
➢ Hydrophobic effect
➢ Electrostatic attraction: cationic attraction and anionic attraction
➢ Hydrogen bonds
➢ Partition onto unoccupied areas

4. Nanoadsorbents

There are several types of nanomaterials that can be used as adsorbents. Nanomaterials include metal nanoparticles (silver, gold, iron, copper, platinum, palladium, nickel), metal oxides (alumina, silica, titania, etc.), nanostructured mixed oxides (nanostructured binary iron–titanium mixed oxide nanoparticles), and magnetic nanoparticles (iron di- and trioxide). Carbonaceous nanomaterials are another class of nanomaterials that include carbon nanotubes (CNTs), carbon nanoparticles, carbon nanosheets, graphene, carbon quantum dots, fullerenes, and carbon nanofibers. Silicon nanomaterials include silicon nanotubes, silicon nanoparticles, and silicon nanosheets. Other classes of nanomaterials include nanofibers, nanoclays, polymer-based nanomaterials, MXenes, xerogels, aerogels, nanocomposites, and hybrid materials. Most of the nanomaterials can be used as adsorbents for several applications. In the following sections, some examples are discussed.

4.1 Carbon nanostructures

Carbon nanostructures such as carbon nanotubes (CNTs), graphene, carbon nanospheres, and their composites are promising for their use in environmental remediation. Carbon nanostructures are promising solutions for the removal of pollutants via adsorption (Bergmam and Machado, 2015). Their adsorption efficacy depends fundamentally on their textural properties such as pore size (micropore and mesopore), associated with the specific surface area and the pore volume (Machado et al., 2016).

The structural properties of carbon nanostructures allow a strong interaction with pollutants through noncovalent forces. Such interactions include hydrophobic interactions, hydrogen bonding, $\pi-\pi$ stacking, electrostatic interaction, and van der Waals forces. Moreover, CNTs permit the incorporation of walls of one or more functional groups. This increases the selectivity of the resulting system (Saleh et al., 2022).

Single- and multiwalled CNTs were evaluated as sorbents for the sorption of Alizarin Red dye from water (Machado et al., 2016). The adsorption data were best suited by the general-order kinetic model. A Liu model was used to produce a decent match of adsorption isotherms. The authors established that the adsorption process was exothermic and spontaneous at all temperatures using thermodynamic simulations. Single walls showed higher capacity than MWCNTs. The adsorption of the dye by CNTs may be due to electrostatic interaction, according to ab initio simulations. These findings corroborate the findings of the experiments. Furthermore, the first principle simulation revealed that when the carbon nanotube diameter increases, the binding energies between dye and SWCNTs rise due to the increased $\pi-\pi$ interactions.

The functional groups anchored on the carbon nanostructures such as CNTs and graphene oxides (GOs) surface, provide the high negative charge density, and thus can have reactive sites for the adsorption of a diversity of adsorbates, such as heavy metal ions, cationic species, cationic dyes, and synthetic dyes (Xiao et al., 2016). This effect can be accentuated in acidic or high pH solutions since in this condition the carbon nanostructures surface contains more negative charges, which enhances its electrostatic interaction with positively charged of the cationic species. Furthermore, because of its aromatic structure, the carbon nanostructures can straightforwardly adsorb pollutants on its surface owing to interactions of the $\pi-\pi$ stacking. The carbon nanostructures surface can be functionalized (or decorated) with nanoparticles such as metal and metal oxides to increase a possible affinity for adsorbates. For example, SiO_2−CNT showed good removal of Hg(II) removal and the experimental data fit a pseudo-second-order model with good regeneration (Fig. 10.5) (Saleh, 2015).

For the removal of methylene blue (MB) dye, GO shows exponentially enhanced adsorption capacity with the increase of oxidation degree. The adsorption behavior of GO changes from a Freundlich adsorption model to a Langmuir adsorption model as the oxidation degree increases (Yan et al., 2014). This can be attributed to more active adsorption sites on GO. When the oxidation degree was low, the dye interacted with the GO primarily through parallel−stacking interactions, forming multilayer adsorption. However, when the oxidation degree was higher, the dye interacted with the GO predominantly through electrostatic interactions, forming monolayer adsorption.

Reduced GO (rGO) has properties that make it an excellent adsorbent for wastewaters including oil and organic solvents (Jauris et al., 2016). The

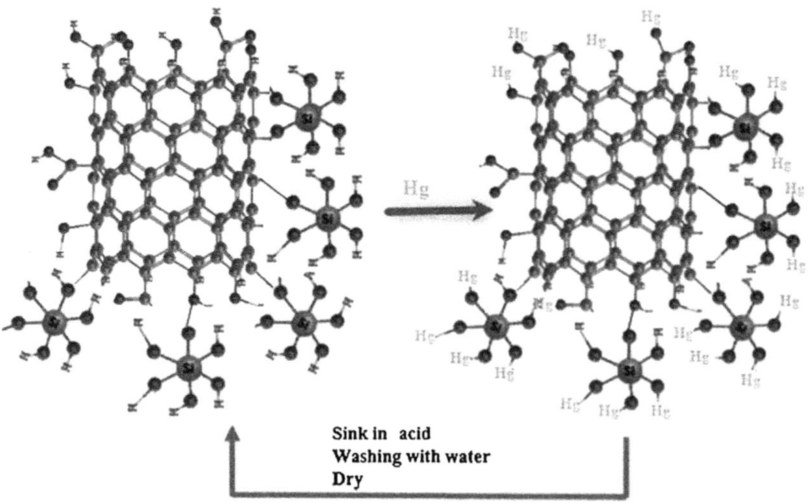
Hg
Sink in acid
Washing with water
Dry

FIGURE 10.5 SiO_2−CNT for the Hg(II) adsorption.

electrostatic interactions between charged pollutants and the surface oxygen-containing groups of rGO, as well as the π−π stacking between sp^2 regions of rGO and aromatic structures of pollutants, may further aid in the adsorption of these contaminants. The rGO, unlike the GO, does not have a strongly negative surface charge, which makes it an interesting adsorbent for anionic synthetic dyes. The various interactions of MB and Acid Red dyes with the rGO active sites cause diverse adsorption behaviors, resulting in a significant difference in adsorption capacities and kinetics as well. The equilibrium data for MB adsorption are fitted to the Langmuir isotherm model, but the equilibrium isotherm of Acid Red I is best described by the Freundlich model. The pseudo-second-order kinetic model is shown to be followed by both dyes' adsorption rates (Kim et al., 2015). Due to high specific interactions, the rGO is more advantageous for the sorption of cationic dyes than anionic dyes.

4.2 Clays and modified clays

Because of their abundance, low cost, high adsorption, and ion exchange capabilities, clay materials have gained popularity as adsorbents. Clay materials are applied for the adsorption of inorganic and organic pollutants. Examples of clay materials are smectites (saponite, montmorillonite), kaolinite, mica (illite), serpentine, pyrophyllite (talc), vermiculite, and sepiolite. Clay materials have strong adsorption capacities due to the net negative charge that comes from the chemical structure of minerals. Because of this negativity, clay materials can absorb positively charged species. Furthermore, because of their large surface area and porosity, they have a high sorption behavior.

The performance of the clays as adsorbents can be improved by decorating with metal or metal oxides nanoparticles or by grafting polymers on their surfaces (Fig. 10.6). With this purpose, to improve the adsorption efficiency of clays as a low-cost adsorbent can be modified by various polymers such as poly(acrylic acid), melamine-formaldehyde, polyaniline, chitosan-poly(vinyl alcohol), poly(acrylic acid-co-acrylamide), and polyamide. For example, poly(ethylene diamine-trimesoyl chloride)-modified diatomite shows efficient removal of rhodamine dye from wastewaters (Saleh et al., 2021). The experimental data are well designated by the pseudo-second-order kinetic model and the thermodynamic results indicate that the dye adsorption had an exothermic character.

4.3 Metal oxide−based nanomaterials

Inorganic nanoparticles such as metal or metal oxide−based nanomaterials are extensively utilized to remove heavy metal ions and dyes. Metals or metal oxides nanoparticles, such as titania, zinc oxides, manganese oxides, and iron oxides, provide high surface area and specific affinity. Metal oxides have low

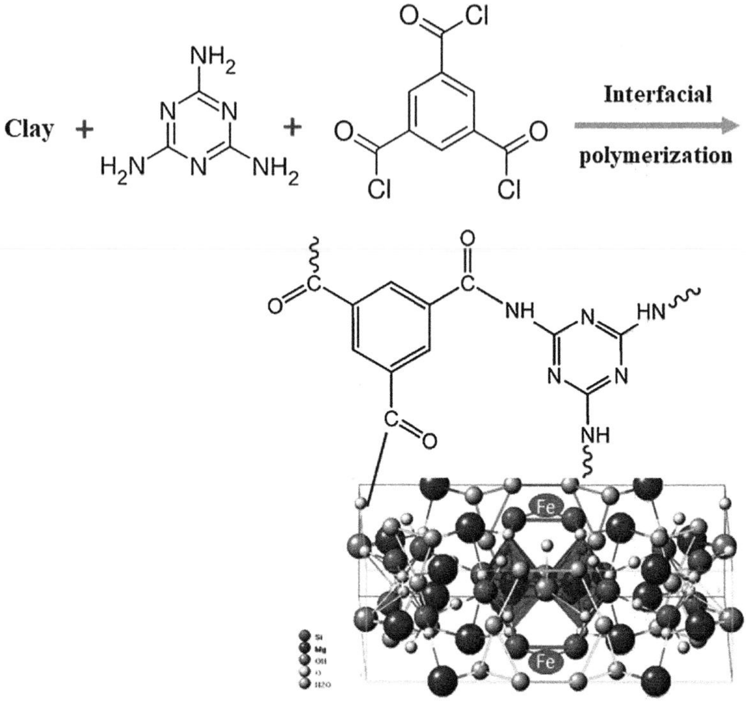
Clay +
NH2
N
N
H2N
N
NH2
+
O
Cl
O
Cl
O
Cl
Interfacial
polymerization
O
C
O
C
NH
N
HN
N
N
NH
C
O
Fe
Fe

FIGURE 10.6 Structure of adsorbent formed by polymerization of trimesoyl chloride with melamine and a clay.

solubility, have a limited environmental impact, and are not involved in the development of secondary pollutants.

4.4 Zeolites, alumina, and silica

Other inorganic adsorbents include zeolites, alumina, and silica. Zeolites are microporous minerals that can be found naturally as silicate minerals or synthesized as magnetically modified zeolite or bio-zeolite. Zeolites are categorized based on the ratio of silicon to alumina (s/a ratio), in primary building units (PBU) into:

(i) high silica ($s/a > 5$),
(ii) intermediate silica ($2 < s/a \leq 5$), and
(iii) low silica ($2 \geq s/a$).

Zeolites of low and intermediate silica have good electrostatic fields in the cavities which support their uses in the adsorption of polar molecules. High silica zeolites are characterized by their hydrophobicity, which supports their uses in micropollutants removal from industrial wastewater effluents, including personal care products and pharmaceuticals. The ability of the PBU to produce diverse configurations of the secondary building units (SBU) results in the development of many frameworks of natural and synthetic zeolite and zeotypes.

Clinoptilolite is the most abundant zeolite in nature (Ali et al., 2020). Due to their high ion exchange capability, they are used as an adsorbent. Magnetically modified zeolite prepared exhibited high adsorption capabilities for the lead ions and good chemical inertness in the pH range 5−11 (Nah et al., 2006). Bio-zeolite comprised of mixed bacteria, and reformed zeolites can be used for the removal of organic pollutants such as pyridine and quinoline. The organic pollutant was decomposed by bacteria, and the modified zeolite eliminated the ammonium ion produced by pyridine and quinoline breakdown (Bai et al., 2010).

The alumina nanoparticles prepared by immobilization of 2,4-dinitrophenylhydrazine on nanoalumina coated with sodium dodecyl sulfate were used for the adsorption and extraction of lead and copper ions (Afkhami et al., 2011). Factors affecting the efficiency are pH, flow rates of samples and eluent, type of eluent, breakthrough volume, and potentially interfering ions. Nanoalumina on single-walled CNTs prepared by sol−gel method were used for the determination of cadmium ions and its adsorption capacity (Kalfa et al., 2009). The adsorption capacity of the composite to cadmium ion was higher than that of CNTs and the physical condition of the material is more appropriate to use in a column technique.

Mesoporous silica has been reported to be a highly efficient adsorbent for antibiotic water pollutants such as oxytetracycline. Moreover, the presence of hexagonal channels within the silica-based adsorbent facilitates rapid mass transport during adsorption. The development of efficient adsorbents with high reusability and cost-competitiveness is crucial. To improve their efficiency and capability to adsorb anionic ions and organics, silica and zeolites can be modified by decorating with nanoparticles, functionalization with amino groups, or by grafting polymers on their surfaces (Truong et al., 2021; Melnyk et al., 2021). For example, nanosilica surface modified with a strong positive charge via a novel coating of poly(3-methacryloylamino propyltrimethylammonium chloride) (Truong et al., 2021) showed approximately 90% removal of beta-lactam cefixime. Nanosilica modified with polyamide as a nanocomposite showed high removal of MB and heavy metal ions (Fig. 10.7).

Such nanomaterials can be used also for membrane separation with thin-film membranes to enable the removal of wastewater pollutants and for desalination applications. The desirable features include high permeability, high fouling resistance, and good mechanical strength. The modification of the active polyamide layers of thin-film composite forward osmosis membranes was modified with various nanosilica (Nguyen et al., 2021). Because of their enhanced hydrophilicity and antifouling ability, together with minimal particle

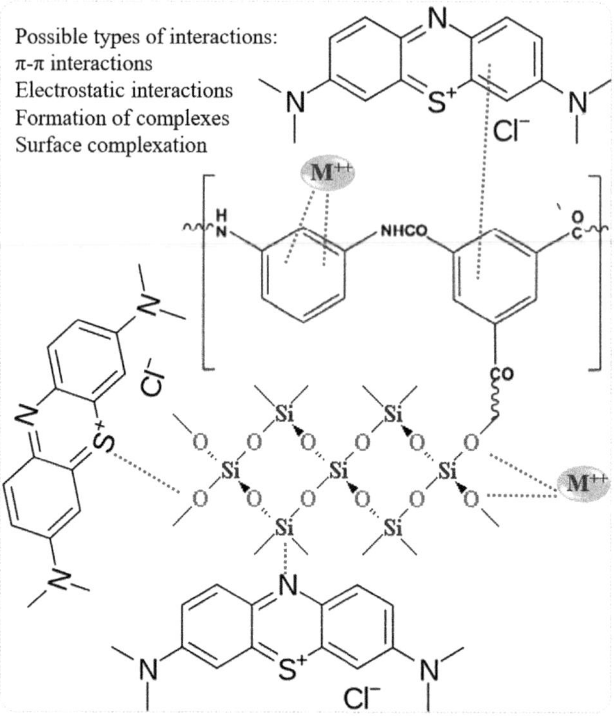
Possible types of interactions:
π-π interactions
Electrostatic interactions
Formation of complexes
Surface complexation
M++
NHCO
Cl⁻
Si
O
N
S⁺

FIGURE 10.7 Schematic illustration of the possible interactions modes of MB dye and metal ions adsorption on silica/polymer nanocomposite.

aggregation, the modified membranes had a high permeate flux and a significant reverse solute flux. Poly(vinylidene fluoride) (PVDF) with nanosilica membrane exhibited about 98.5% efficiency when applied to hexadecane-in-water emulsion coalescence separation (Yang et al., 2021). Notably, new trends in the area include the design of nanomembranes with multifunctional capabilities, such as simultaneous self-cleaning and photocatalytic activity.

They can also be used for ion-exchange applications including the design of ion-exchange resins that can facilitate the separation and recovery of metal ions from wastewater such as noble metals from electroplating wastewater streams. An example is the mesoporous silica ion-exchange resin prepared by in situ polymerization, which displayed excellent adsorption capacity of lead with 5 times reusability with ion exchange between H^+ (from $-COOH$ from the incorporated acrylic acid monomer) and lead ions (Zhang et al., 2021). Various silica-based composite adsorbents, such as amino-functionalized mesoporous SiO_2 integrated into sodium alginate, SiO_2 decorated with nano-ferrous oxalate, and silicate-modified oil tea camellia shell-derived biochar, can be used for metal ions removal. Photodegradation relies also on the adsorption of organic compounds on the catalyst surface followed by degradation. Silica and zeolites can be used as supports for titania or zinc oxides catalysts, in the UV and visible light ranges. Several strategies are used to improve the slow kinetics owing to the fast recombination of electrons and holes. Carbon quantum dots (CQDs) with highly tunable photoluminescence properties are among the emerging nanomaterials for photocatalytic degradation and can be further functionalized to enable pollutants photodegradation.

4.5 Magnetic nanoadsorbents

Magnetic nanoadsorbents are proving to be highly effective functional materials with superior micropollutant sequestration and rapid adsorption kinetics. They are commonly characterized with high specific surface areas (e.g., 1188 m^2/g for magnetic coal-based activated carbon (Liu et al., 2021), high pore volumes (Masunga et al., 2019), robust structures, and extensively interconnected porous networks (Fan et al., 2021). Such properties promote high adsorption capacity for micropollutants (Icten and Ozer, 2021).

Interestingly, magnetic nanoadsorbents can be easily separated in situ from waters in the form of a magnetic adsorbent—adsorbate sludge with the use of a magnetic field (Mashile et al., 2020; Balbino et al., 2020). Such property allows to

- perform an integrated one-step capture and purification of specific species,
- process of high throughputs, and
- process of the removal with low energy requirement and associated cost entailed by mostly a continuous process ran at low pressure (Schwaminger et al., 2019).

There are several magnetic nanomaterials and modified magnetic nanoparticles reported as adsorbents for several applications. Magnetic nanomaterials are good in reducing detergents and chemical oxygen demand. They can also be used to remove nitrogen, phosphates, and toxic metal ions (Castelo-Grande et al., 2021). For example, Ag-magnetic nanoparticles had noticeably high effectiveness to disinfect effluent. They also show high performance in advanced treatment with enhanced removal of chemical oxygen demand. Compared to magnetic nanomaterials; Ag-loaded magnetic materials show a 0.06 increase in total coliforms, fecal coliforms, and heterotrophic bacteria log reductions, with around a 6% increase in the removal of chemical oxygen demand (Najafpoor et al., 2020). Magnesium—zinc ferrites were reported as an effective adsorbent for removing nickel ions from water.

4.6 Nanocomposites

There are several types of nanocomposites used as adsorbents. Nanocomposites are formed by a combination of inorganic with organic components in order to improve some of the chemical, physical, thermal, and mechanical properties.

Polydopamine modified nanoparticles as a nanocomposite was reported with an adsorption capacity of 609 mg g^{-1} for methylene blue (MB) for pH ranging 3–10 and at 45°C (Chen et al., 2020). The high capacity is ascribed to the high electrostatic interactions between the negatively charged adsorbent and the cationic MB molecules. This adsorbent showed poor adsorption of Rhodamine B due to mostly steric hindrance generated by the longer lateral alkyl chain connected to the nitrogen center, which weakened $\pi-\pi$ stacking interactions and electrostatic attractions between adsorbent and the dye molecules. Sulfur-functionalized polyamidoamine dendrimer/magnetic materials exhibited adsorption selectivity of mercury ions in the presence of nickel, zinc, and manganese ions. The maximum adsorption capacity for mercury ions and silver ions was 0.80 and 1.29 mmol g^{-1} (Luan et al., 2021). Another magnetic molecular imprint polymer composite prepared from vinyl-functionalized magnetic nanomaterials showed high binding capacity toward erythromycin and ciprofloxacin at 70 and 32 mg g^{-1} (Kuhn et al., 2020). The networks with high binding capacity, selectivity, and recyclability can be employed for monitoring as well as the removal of antibiotic pollutants from water and food.

Co-multiwalled CNT nanocomposite showed adsorption of MB that is endothermic and followed pseudo-second-order kinetic model with 324 mg g^{-1} capacity (Çalımlı, 2021). Hydroxypropyl-β-cyclodextrin-polyurethane/GO magnetic nanoconjugates reported with 987 and 1399 mg g^{-1} adsorption capacities of chromium and lead ions with data fit pseudo-second-order kinetics model (Nasiri and Alizadeh, 2021). Magnetic tubular carbon nanofibers were reported with porous morphology, and high surface area showed fast removal of copper ions, with 376 mg g^{-1} adsorption capacity (Ahmad et al., 2020). With spontaneous and exothermic adsorption, magnetic sodium alginate gel beads displayed a high adsorption capacity of 1252 mg g^{-1} (Li and Lin, 2021).

Poly(acrylamide acrylic acid) grafted on steel slag showed an efficient magnetic adsorbent for cationic and anionic dyes (Basaleh et al., 2021a). Poly(acrylamide acrylic acid)/baghouse dust magnetic composite hydrogel is an efficient adsorbent for metals and MB (Basaleh et al., 2021b).

The polymer-modified carbon nanostructures as nanocomposites have a great potential in various applications. They are considered as promising sorbents of contaminants owing to their unique physical and chemical behaviors. For example, a composite of oxidized carbon nanotube (CNT) grafted with polyethylene glycol (PEG) showed high efficiency in the removal of phenol and metal ions such as Cu, Hg, Cr, Fe, Co, Ni, Al, and Pb from industrial wastewater at a contact time of 60 min. Although adsorption is a complex process involving multiple mechanisms, it is possible to speculate that the mechanisms of adsorption are hydrogen bond interactions, $\pi-\pi$ interactions, and electrostatic interactions. Based on this, the phenol removal is

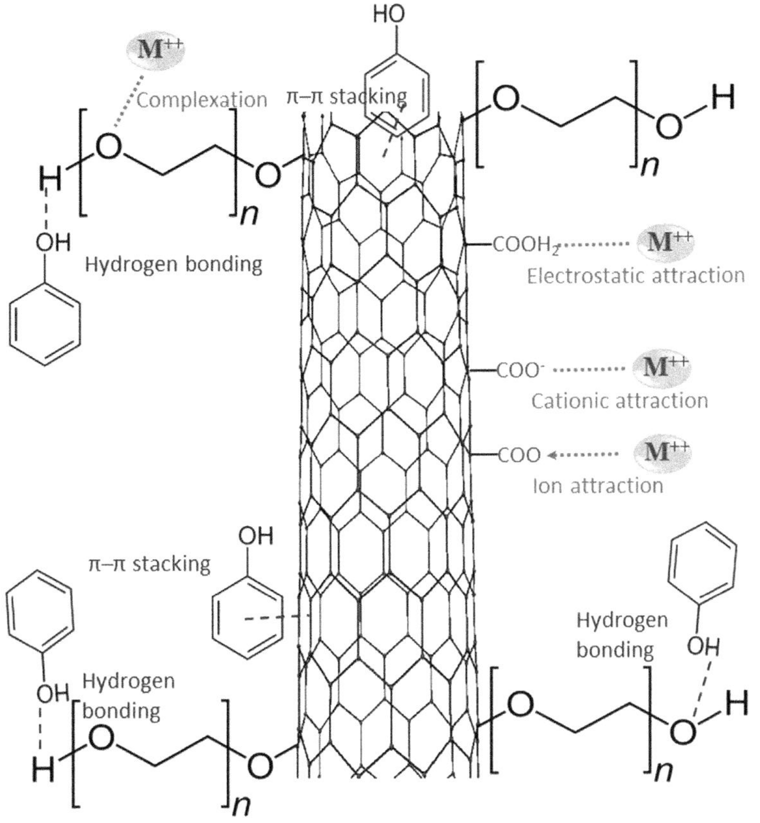
HO
M++
Complexation
π–π stacking
O
H
O
n
H
OH
Hydrogen bonding
COOH2
M++
Electrostatic attraction
COO-
M++
Cationic attraction
COO
M++
Ion attraction
OH
π–π stacking
Hydrogen bonding
OH
Hydrogen bonding
OH
H
n

FIGURE 10.8 Possible interaction mechanisms between phenol and metal ions with the adsorbent.

suggested to comprise the steps illustrated in Fig. 10.8. Adsorption on the surface of the CNT/PEG adsorbent containing oxygen-containing groups can form hydrogen bonds between the oxygen and hydrogen from phenol and adsorbent. This enhances the adsorption process through phenol interaction with oxygen-containing groups on the adsorbent and the adsorption thus occurs on the entire surface and brings about a uniform distribution of adsorbed phenol on the surface of the CNT/PEG adsorbent. Furthermore, attractions such as π–π interaction are formed between carbonyl groups of the modified CNT and aromatic rings of the adsorbed phenol molecules.

4.7 Regeneration and reuse

After the adsorption of pollutants, the adsorbent can be treated in very different ways depending on the targets. Following are the possible ways:

- If the adsorbent was used to remove metals from water, then the adsorbent can be treated with a solvent like nitric acid to dissolve the metals from the adsorbent, then the metal can be extracted and separated from the filtrate to end up with metal extraction process. While the adsorbent can be reused further in water purification.
- If the adsorbent was used to remove metals from water, then the spent adsorbent loaded with metals can be used in some other applications as raw materials. For example, it can be used in road instructions, buildings, etc. Note that, in such applications, leaching tests and toxicity tests should be employed to satisfy the environmental and safety requirements.
- If the adsorbent was used to remove metals from water, then the spent adsorbent loaded with metals can be further chemically or thermally treated to improve to form composite materials of adsorbent/metal form. This material can be used in other applications such as in water purification of organic pollutants.
- If the adsorbent was used to remove organic pollutants from water, then the spent adsorbent loaded with organic molecules can be treated with a solvent suitable to dissolve the organic adsorbate. Then, the organic adsorbate molecules can be extracted and separated for other possible uses. An example is the separation of oil from water in produced water treatment. Another example is the separation of dyes from water.

There are various methods for the regeneration of the spent nanoadsorbent. Examples of these methods include:

- chemical methods,
- supercritical extraction,
- microbial regeneration (Momina Shahadat and Isamil, 2018),
- solvent extraction (Dutta et al., 2019),
- microwave and ultraviolet irradiation (Sun et al., 2017), and
- thermal (Aguedal et al., 2019).

For example, spent carbon nanostructures can be thermally regenerated at 500°C under low oxygen conditions in the presence of steam. There could be mostly some loss in the activity of adsorbents. For instance, carbon-based adsorbents lose about 5%–15% for each regeneration. It should be mentioned that adsorbed organics are volatilized and oxidized during the regeneration process unless the conditions are controlled otherwise.

Factors affecting the regeneration process:

- Surface charge;
- Molecular structure of adsorbate;
- Functional groups present;
- pH: The materials should be withstanding extreme pH;

- ➢ Temperature: The materials should be withstanding high temperatures;
- ➢ The materials should be withstanding long processing times;
- ➢ Materials should not experience pore blockage or filling.

It is recommended to optimize the regeneration process to ensure maximum selectivity, potential stability, and improved adsorption efficacy of the regenerated nanoadsorbent during its next set of multiple adsorptive interactions with the pollutants (Mudhoo and Sillanpää, 2021).

5. Pilot scale

Although there have been a number of laboratory-scale research on the performance of nanomaterials as adsorbents, there have been comparatively few pilot scale investigations on nanomaterials employed in water treatment.

The optimization of pilot scale should be based on the evaluation of capture efficacies of various separator configurations, and consider the following factors:

- ➢ capture element sizing,
- ➢ particle radius,
- ➢ particle mass density,
- ➢ particle permeability,
- ➢ channel diameter,
- ➢ water mass density and water dynamic viscosity, and
- ➢ average flow velocity.

For example, for water treatment under turbulent wastewater flow systems, an open-gradient magnetic separator was constructed. This system consists of identical electromagnets working as capture elements. The working parameters of the system were optimized and experimentally investigated (Belounis et al., 2015). A pilot scale nanomaterials separator that mimicked large-scale wet magnetic separator regimes was recently developed to show that magnetic extraction of vivianite from sludge is possible (Prot et al., 2019). Magnetic separation was reported to be able to concentrate vivianite by a factor of 2–3 and reduce organic contents from 40 down to around 20%. When the accompanying investment and operation expenses are taken into consideration, the benefits (especially the expected reduction in waste sludge volume) appear to be in balance with installing a magnetic separator (Prot et al., 2019). A new design nanoadsorbents recovery apparatus was developed and evaluated experimentally and by computational fluid dynamics modeling for its performance for treating water in-line for continuous flow (Powell et al., 2020). The efficacy was dependent on the device configuration and hydraulic flow conditions, as well as the magnetic adsorbents uptake (Gehrke et al., 2015).

The application of nanomaterials as in real-scale and industrial-scale water treatment units will necessitate significant research and development efforts in primary interrelated components. These include:

- ➢ Increasing the amount of untreated wastewater captured and channeling it to large-scale water and wastewater treatment plants;
- ➢ Increasing the capture of untreated waters and channeling them to the large-scale wastewater treatment facilities;
- ➢ Selecting *intelligent* nanomaterials for industrial uses;
- ➢ System modeling, simulation, and process optimization of real water and wastewater remediation units by nanomaterials;
- ➢ Explaining and further understanding the lifecycle environmental impacts of the use of nanoadsorbents and separation systems in large-scale water purification and wastewater treatment systems;
- ➢ More collaboration of key industry partners and the research community will be equally crucial in research and development activities related to the design and pilot scale testing of effective magnetic separation systems in the existing water treatment facilities.

6. Limitations of nanomaterials for water applications and future trends

6.1 Limitations

Although nanomaterials have shown promising results in wastewater treatment, there are considerable obstacles standing in the way of these promises being realized. Nanomaterial toxicity, cost-effectiveness, the transformation of nanomaterials, and social acceptability are among the roadblocks. These limitations of nanomaterials are related to

- ➢ Their stability under scale-up process and possible downstream toxicity levels (You et al., 2021);
- ➢ The efficacy of regeneration methods, and recovery for reuse will limit the selection of nanomaterials for specific industrial-scale water purification and wastewater treatment process;
- ➢ Difficulties with the recovery of micropollutant-saturated nanoadsorbents at potentially high capture efficiencies;
- ➢ Difficulties with the adaption of the techniques to large-scale areas;
- ➢ Harmful effects on aquatic organisms;
- ➢ There is a need for online monitoring systems to provide reliable real-time measurement data on the quality and quantity of nanomaterials present only in trace amounts in water;
- ➢ For photocatalysis the ultraviolet radiation is only about 5% that of sunlight; the photon efficiency is quite low, limiting the use of nanomaterials as catalysts on an industrial scale;

- The estimation of the costs involved in the scaling-up of synthetic schemes for large production (Peralta et al., 2021; Neha et al., 2021). For example, some nanomaterials like CNTs and graphene are still of relatively high production costs.

6.2 Future research trends

Sustainable application of nanomaterials requires a controlled synthesis route to obtain continuous uniform pore structures with a narrow particle size distribution and favorable textural properties according to the application requirements. This will further enhance the ability to correlate the properties of nanomaterials with their performance and hence facilitate the prediction of the performance in various fields including environmental remediation, especially water purification. Detailed kinetics studies on flotation and aggregation will go a long way to establish the mechanism of separation of adsorbents from wastewater and provide the optimal sets of operating conditions. Moreover, it will be helpful to carry out parametric studies to uncover the combined effects of the various parameters involved in the synthesis of nanomaterials for water purification.

More robust studies following a standard design of experiments that show the synergy of the factors influencing the properties of adsorbents will be interesting. Further studies on kinetics and biochemical interactions of nanoparticles within organisms are imperative. There is limited existing research regarding ecological and environmental implications of natural and anthropogenic nanoparticle pollution.

7. Conclusion

Nanoscience offers a possible platform for improving water and wastewater treatment processes at the nanoscale level. It facilitates the production of multifunctional nanomaterials by allowing for the controlled integration of active sites in a single nanostructured material. As a result, high-performance modular water and wastewater treatment processes may be simply planned and developed. Although nanomaterials have good properties such as large surface area, their selectivity and competitive adsorption to various ions require further development. Thus, modification and functionalization of nanomaterials are needed to improve their selectivity and efficiency to adsorb pollutants such as organics and anions.

As nanoscience research is gaining momentum, large amounts of data are being generated from different research fields to tailor the properties of nanomaterials to achieve optimal performance. One of the expectations of nanomaterials research is the emergence of a reliable correlation between the performance and properties of the nanomaterials. The application of nanomaterials in wastewater treatment is partly because of their hydrophobic nature and mechanical strength to withstand harsh wastewater environments.

Nanomaterials have the ability to anchor relevant functional groups and possess a high surface area and excellent dispersibility. The relationship between the aforementioned properties and the performance of the final nanomaterials (nanoadsorbents) should be further investigated. In brief:

- There is a significant need to develop efficient nanomaterials for novel advanced water technologies.
- Nanoengineered materials offer the potential for novel water technologies that can be easily adapted to customer-specific applications.
- Nanomaterials enable higher process efficiency owing to their unique properties.
- In the realm of effluent monitoring systems for nanomaterials, there is still a lot of room for innovation.
- Nanoengineered water technologies are rarely adaptable to mass processes.
- In many circumstances, these are not cost-effective when compared to traditional therapy options.
- Nanomanufacturing, on the other hand, has a lot of promise for water improvements in future.

References

Afkhami, A., Saber-Tehrani, M., Bagheri, H., Madrakian, T., 2011. Lame atomic absorption spectrometric determination of trace amounts of Pb(II) and Cr(III) in biological, food and environmental samples after preconcentration by modified nano-alumina. Microchim. Acta (1–2), 172.

Aguedal, H., Iddou, A., Aziz, A., et al., 2019. Effect of thermal regeneration of diatomite adsorbent on its efficacy for removal of dye from water. Int. J. Environ. Sci. Technol. 16, 113–124.

Ahmad, M., Wang, J., Xu, J., et al., 2020. Magnetic tubular carbon nanofibers as efficient Cu(II) ion adsorbent from wastewater. J. Clean. Prod. 252, 119825.

Ali, M.E., Hoque, M.E., Safdar Hossain, S.K., et al., 2020. Nanoadsorbents for wastewater treatment: next generation biotechnological solution. Int. J. Environ. Sci. Technol. 17, 4095–4132.

Bai, Y., Sun, Q., Xing, R., Wen, D., Tang, X., 2010. Removal of pyridine and quinoline by bio-zeolite composed of mixed degrading bacteria and modified zeolite. J. Hazard Mater. 181 (1–3), 916–922.

Balbino, T.A.C., Bellato, C.R., da Silva, A.D., et al., 2020. Preparation and evaluation of iron oxide/hydrotalcite intercalated with dodecylsulfate/β-cyclodextrin magnetic organocomposite for phenolic compounds removal. Appl. Clay Sci. 193, 105659.

Basaleh, A.A., Al-Malack, M.H., Saleh, T.A., 2021a. Poly (acrylamide acrylic acid) grafted on steel slag as an efficient magnetic adsorbent for cationic and anionic dyes. J. Environ. Chem. Eng. 9 (2), 105126.

Basaleh, A.A., Al-Malack, M.H., Saleh, T.A., 2021b. Poly (acrylamide acrylic acid)/Baghouse dust magnetic composite hydrogel as an efficient adsorbent for metals and MB; synthesis, characterization, mechanism, and statistical analysis. Sustainable Chemistry and Pharmacy 23, 100503.

Belounis, A., Mehasni, R., Ouil, M., et al., 2015. Design with optimization of a magnetic separator for turbulent flowing liquid purifying applications. IEEE Trans. Magn. 51, 1–8.

Bergmann, C.P., Machado, F.M., 2015. Carbon Nanomaterials as Adsorbents for Environmental and Biological Applications. Springer International Publishing, New York.

Çalımlı, M.H., 2021. Magnetic nanocomposite cobalt-multiwalled carbon nanotube and adsorption kinetics of methylene blue using an ultrasonic batch. Int. J. Environ. Sci. Technol. 18, 723–740.

Castelo-Grande, T., Augusto, P.A., Rico, J., et al., 2021. Magnetic water treatment in a wastewater treatment plant: Part I—sorption and magnetic particles. J. Environ. Manag. 281, 111872.

Chen, B., Cao, Y., Zhao, H., et al., 2020. A novel Fe^{3+}-stabilized magnetic polydopamine composite for enhanced selective adsorption and separation of Methylene blue from complex wastewater. J. Hazard Mater. 392, 122263.

Dutta, T., Kim, T., Vellingiri, K., et al., 2019. Recycling and regeneration of carbonaceous and porous materials through thermal or solvent treatment. Chem. Eng. J. 364, 514–529.

Fan, S., Qu, Y., Yao, L., et al., 2021. MOF-derived cluster-shaped magnetic nanocomposite with hierarchical pores as an efficient and regenerative adsorbent for chlortetracycline removal. J. Colloid Interface Sci. 586, 433–444.

Gehrke, I., Geiser, A., Somborn-Schulz, A., 2015. Innovations in nanotechnology for water treatment. Nanotechnol. Sci. Appl. 8, 1–17.

Icten, O., Ozer, D., 2021. Magnetite doped metal–organic framework nanocomposites: an efficient adsorbent for removal of bisphenol-A pollutant. New J. Chem. 45, 2157–2166.

Jauris, I.M., Matos, C.F., Saucier, C., et al., 2016. Adsorption of sodium diclofenac on graphene: a combined experimental and theoretical study. Phys. Chem. Chem. Phys. 18, 1526–1536.

Kalfa, O.M., Yalcinkaya, O., Tuerker, A.R., 2009. Synthesis and characterization of nano-scale alumina on single walled carbon nanotube. Inorg. Mater. 45 (9).

Kim, H., Kang, S.-O., Park, S., Park, H.S., 2015. Adsorption isotherms and kinetics of cationic and anionic dyes on three-dimensional reduced graphene oxide macrostructure. J. Ind. Eng. Chem. 21, 1191–1196.

Kuhn, J., Aylaz, G., Sari, E., et al., 2020. Selective binding of antibiotics using magnetic molecular imprint polymer (MMIP) networks prepared from vinyl-functionalized magnetic nanoparticles. J. Hazard Mater. 387, 121709.

Li, B.G., Lin, W.J., 2021. Preparation of magnetic alginate-based biogel composite cross-linked by calcium ions and its super efficient adsorption for direct dyes. Mater. Sci. Forum 1035, 1022–1029.

Liu, Y., Zhu, Z., Cheng, Q., et al., 2021. One-step preparation of environment-oriented magnetic coal-based activated carbon with high adsorption and magnetic separation performance. J. Magn. Magn Mater. 521, 167517.

Luan, L., Tang, B., Liu, Y., et al., 2021. Selective capture of Hg(II) and Ag(I) from water by sulfur-functionalized polyamidoamine dendrimer/magnetic Fe_3O_4 hybrid materials. Separ. Purif. Technol. 257, 117902.

Machado, F.M., Carmalin, S.A., Lima, E.C., et al., 2016. Adsorption of Alizarin Red S dye by carbon nanotubes: an experimental and theoretical investigation. J. Phys. Chem. C 120, 18296–18306.

Mashile, G.P., Mpupa, A., Nqombolo, A., et al., 2020. Recyclable magnetic waste tyre activated carbon-chitosan composite as an effective adsorbent rapid and simultaneous removal of methylparaben and propylparaben from aqueous solution and wastewater. J. Water Proc. Eng. 33, 101011.

Masunga, N., Mmelesi, O.K., Kefeni, K.K., Mamba, B.B., 2019. Recent advances in copper ferrite nanoparticles and nanocomposites synthesis, magnetic properties and application in water treatment: Review. J. Environ. Chem. Eng. 7, 103179.

Melnyk, I.V., Tomina, V.V., Stolyarchuk, N.V., et al., 2021. Organic dyes (acid red, fluorescein, methylene blue) and copper(II) adsorption on amino silica spherical particles with tailored surface hydrophobicity and porosity. J. Mol. Liq. 336, 116301.

Momina Shahadat, M., Isamil, S., 2018. Regeneration performance of clay-based adsorbents for the removal of industrial dyes: a review. RSC Adv. 8, 24571−24587.

Mudhoo, A., Sillanpää, M., 2021. Magnetic nanoadsorbents for micropollutant removal in real water treatment: a review. Environ. Chem. Lett. 19, 4393−4413.

Nah, I.W., Hwang, K.Y., Jeon, C., Choi, H.B., 2006. Removal of Pb ion from water by magnetically modified zeolite. Miner. Eng. 19 (14), 1452−1455.

Najafpoor, A., Norouzian-Ostad, R., Alidadi, H., et al., 2020. Effect of magnetic nanoparticles and silver-loaded magnetic nanoparticles on advanced wastewater treatment and disinfection. J. Mol. Liq. 303, 112640.

Nasiri, S., Alizadeh, N., 2021. Hydroxypropyl-β-cyclodextrin-polyurethane/graphene oxide magnetic nanoconjugates as effective adsorbent for chromium and lead ions. Carbohydr. Polym. 259, 117731.

Neha, R., Adithya, S., Jayaraman, R.S., et al., 2021. Nano-adsorbents an effective candidate for removal of toxic pharmaceutical compounds from aqueous environment: a critical review on emerging trends. Chemosphere 272, 129852.

Nguyen, T.Q., Tung, K.L., Lin, Y.L., et al., 2021. Modifying thin-film composite forward osmosis membranes using various SiO_2 nanoparticles for aquaculture wastewater recovery. Chemosphere 281, 130796.

Peralta, M.E., Mártire, D.O., Moreno, M.S., et al., 2021. Versatile nanoadsorbents based on magnetic mesostructured silica nanoparticles with tailored surface properties for organic pollutants removal. J. Environ. Chem. Eng. 9, 104841.

Powell, C.D., Atkinson, A.J., Ma, Y., et al., 2020. Magnetic nanoparticle recovery device (MagNERD) enables application of iron oxide nanoparticles for water treatment. J. Nanoparticle Res. 22, 48.

Prot, T., Nguyen, V.H., Wilfert, P., et al., 2019. Magnetic separation and characterization of vivianite from digested sewage sludge. Separ. Purif. Technol. 224, 564−579.

Saleh, T.A., Sarı, A., Tuzen, M., 2022. Simultaneous removal of polyaromatic hydrocarbons from water using polymer modified carbon. Biomass Convers. Biorefinery 1−10.

Saleh, T.A., Tuzen, M., Sarı, A., 2021. Evaluation of poly (ethylene diamine-trimesoyl chloride)-modified diatomite as efficient adsorbent for removal of rhodamine B from wastewater samples. Environ. Sci. Pollut. Control Ser. 28 (39), 55655−55666.

Saleh, T.A., 2015. Isotherm, kinetic, and thermodynamic studies on Hg(II) adsorption from aqueous solution by silica- multiwall carbon nanotubes. Environ. Sci. Pollut. Res. 22, 16721−16731.

Schwaminger, S.P., Fraga-García, P., Eigenfeld, M., et al., 2019. Magnetic separation in bioprocessing beyond the analytical scale: from biotechnology to the food industry. Front. Bioeng. Biotechnol. 7, 1−12.

Sun, Y., Zhang, B., Zheng, T., Wang, P., 2017. Regeneration of activated carbon saturated with chloramphenicol by microwave and ultraviolet irradiation. Chem. Eng. J. 320, 264−270.

Truong, T.T.T., Vu, T.N., Dinh, T.D., et al., 2021. Adsorptive removal of cefixime using a novel adsorbent based on synthesized polycation coated nanosilica rice husk. Prog. Org. Coating 158, 106361.

Wang, L., et al., 2020. Rational design, synthesis, adsorption principles and applications of metal oxide adsorbents: a review. Nanoscale 12, 4790–4815.

Xiao, J., Lv, W., Xie, Z., Tan, Y., Song, Y., Zheng, Q., 2016. Environmentally friendly reduced graphene oxide as a broad-spectrum adsorbent for anionic and cationic dyes via $\pi-\pi$ interactions. J. Mater. Chem. 4, 12126–12135.

Yan, H., Tao, X., Yang, Z., et al., 2014. Effects of the oxidation degree of graphene oxide on the adsorption of methylene blue. J. Hazard Mater. 268, 191–198.

Yang, Y., Li, Y., Cao, L., et al., 2021. Electrospun PVDF-SiO_2 nanofibrous membranes with enhanced surface roughness for oil-water coalescence separation. Separ. Purif. Technol. 269.

Zhang, S., Ning, S., Liu, H., et al., 2021. Preparation of ion-exchange resin via in-situ polymerization for highly selective separation and continuous removal of palladium from electroplating wastewater. Separ. Purif. Technol. 258, 117670.

Index

'*Note:* Page numbers followed by "f" indicate figures and "t" indicate tables.'

V

W

X

Z

9780128498767